Modeling of Polymer Composites and Nanocomposites

Modeling of Polymer Composites and Nanocomposites

Guest Editor

Rafał Grzejda

Basel • Beijing • Wuhan • Barcelona • Belgrade • Novi Sad • Cluj • Manchester

Guest Editor
Rafał Grzejda
Faculty of Mechanical
Engineering and Mechatronics
West Pomeranian University
of Technology in Szczecin
Szczecin
Poland

Editorial Office
MDPI AG
Grosspeteranlage 5
4052 Basel, Switzerland

This is a reprint of the Special Issue, published open access by the journal *Polymers* (ISSN 2073-4360), freely accessible at: https://www.mdpi.com/journal/polymers/special_issues/Y39W46TDJU.

For citation purposes, cite each article independently as indicated on the article page online and as indicated below:

Lastname, A.A.; Lastname, B.B. Article Title. *Journal Name* **Year**, *Volume Number*, Page Range.

ISBN 978-3-7258-4853-9 (Hbk)
ISBN 978-3-7258-4854-6 (PDF)
https://doi.org/10.3390/books978-3-7258-4854-6

Contents

polymers MDPI

Editorial

Modeling of Polymer Composites and Nanocomposites

Rafał Grzejda

Faculty of Mechanical Engineering and Mechatronics, West Pomeranian University of Technology in Szczecin, 19 Piastow Ave., 70-310 Szczecin, Poland; rafal.grzejda@zut.edu.pl

1. Introduction

The importance of polymer composites (PCs) and nanocomposites (PNs) has increased significantly in recent years due to their enhanced material performance, sustainability and versatile applications [1–5]. These composites are constructed by combining a polymer matrix, reinforcement (e.g., fibers) and filler, most commonly either by introducing a monomer into the system and polymerizing it, or by directly introducing the polymer into the system [6,7]. Natural polymers are also used as a matrix for composites [8]. In addition, natural fibers are applied as reinforcements in PCs as an alternative to synthetic fibers, particularly in terms of improving the mechanical properties of the matrix [9–11]. Furthermore, natural fiber-reinforced composites (FRCs) show exceptional economic efficiency and are environmentally friendly, making them highly suitable for a wide range of applications [12–14].

Recent advancements in the field of PCs and PNs have focused on a number of aspects, including processing techniques, product innovations, surface and color treatments and the investigation of critical properties [15–17]. Moreover, research is being carried out on layered hybrid composites with bio-based raw materials in order to provide sustainable alternatives in structural applications with an emphasis on diversifying the sourcing of raw materials [18,19]. To optimize functional characteristics, specific additives are introduced into the composite formulation, which play a key role in enhancing interfacial adhesion, mechanical strength, dimensional stability and resistance to environmental degradation, thus increasing the overall durability and functionality of the material [20–24].

Composite products find a wide range of applications in the automotive, transport, sports, shipbuilding, packaging, medical and engineering industries, among others [25,26]. Advances in polymer composite technology continue to lead to improvements in mechanical performance, biodegradability and recyclability [27,28]. Nanotechnology and fiber modification techniques are being explored to further enhance the strength, thermal stability and fire resistance of composites. All this underlines how important PCs and PNs are today and that their modeling and further extensive research on them is needed. Some of this research is presented in the Special Issue of *Polymers* entitled "Modeling of Polymer Composites and Nanocomposites", the introduction to which is this Editorial.

2. Overview of Papers Published in the Special Issue

Umer et al. (Contribution No. 1) documented the feasibility of converting clean and renewable energy sources into electricity through the use of piezoelectric materials. They used a micromechanical study of the effect of adding silica nanoparticles to a polyimide matrix on the piezoelectric–elastic response of piezoelectric fiber-reinforced nanocomposites [29]. It was found that the addition of silica nanoparticles to the polymer improves

 https://doi.org/10.3390/polym17141944

the elastic and piezoelectric properties of the piezoelectric fiber nanocomposites. At the fiber volume fraction of 60%, the nanocomposite with polyimide filled with 3% silica exhibits a 39% improvement in the transverse Young's modulus, which is almost a 32% improvement in transverse shear modulus and 37% improvement in the piezoelectric coefficient compared to the composite without nanoparticles. It was also shown that the piezoelectric–elastic properties, described by the above-mentioned piezoelectric constants, can be enhanced with a reduction in the diameter of the nanoparticles.

The second paper by Umer et al. (Contribution No. 2) deals with multifunctional PCs containing micro/nano hybrid reinforcements, which are currently attracting intense attention in the field of materials science and engineering. In the paper, a multi-phase analytical model was developed to study the effective electrical conductivity [30] of PCs with silicon carbide microcarriers and carbon nanoparticles. Before the percolation threshold, it was observed that the addition of nanoparticles with uniform dispersion can improve the electrical conductivity of the PCs tested. It was also shown that the electrical conductivity increases more by reducing the size of the nanoparticles.

Issues related to carbon fiber-reinforced polymer concrete beams for safety design are addressed in the paper by Maidi et al. (Contribution No. 3). It presents the structural response of such beams integrated with rigid frames during seismic events. Multi-scale simulations and parametric analyses were performed to optimize the residual state and global performance. The main parameters considered were tensile and compressive reinforcement, concrete strength, height-to-width ratio, cross-sectional coverage and confinement level, all of which are important for understanding their impact on seismic performance [31]. The results of the parametric analysis highlight the increased ductility and higher load carrying capacity of the tested reinforced beams compared to reinforced concrete beams [32]. This indicates that it is possible to design carbon fiber-reinforced polymer concrete components that could enhance ductile frames with increased energy dissipation and be suitable for use in non-corrosive seismic-resistant buildings.

Marashdeh and Madkhali (Contribution No. 4) considered heavy metal doping as a means of enhancing the radiation protection efficiency of *Nigella sativa* eumelanin polymers. The herbal polymer was doped with iron (Fe), copper (Cu) and zinc (Zn) and it was examined how this doping affects the gamma ray shielding properties of the polymer. The doping of eumelanin polymers with the aforementioned components has been shown to be particularly effective in enhancing gamma ray protection at low energies, with Cu providing the most significant overall improvement, making these composites suitable for applications requiring enhanced radiation protection at lower gamma ray energies [33,34].

To improve the thermal and electrical properties of organic electronics, one potential research area is the study of composite layers that combine semiconducting organic polymers with inorganic nanoparticles, namely gold nanoparticles (AuNPs). One such promising composite system is based on two semiconducting polymers, PDPP4T and PCPDTBT, which are known for their small optical band gaps and high charge carrier mobility, making them ideal candidates for organic solar cells [35]. The inclusion of AuNPs in the blends of these polymers offers a unique opportunity to explore how nanoparticles affect the thermal transformations in these materials, ultimately affecting material properties and device performance. With this in mind, Jarka et al. (Contribution No. 5) used variable temperature ellipsometry [36] to examine thermal transitions in PDPP4T/PCPDTBT/AuNPs composite layers. They made their conclusions based on the developed phase diagram [37] for the PDPP4T/PCPDTBT blend using variable temperature spectroscopic ellipsometry with differential scanning calorimetry serving as a reference technique. The incorporation of AuNPs was evaluated to significantly affect the thermal stability and crystallization of the material, which is essential for organic field-effect transistor applications.

Understanding the structural dynamics of semiflexible polymers in an implicit solvent under different conditions provides valuable insights into their behavior in diverse environments. For this reason, Williams and Gray (Contribution No. 6) investigated the effect of the angular width of the bending potential on structural state behavior and conformational variation using microcanonical analysis [38]. A range of angular widths was considered, with the widest value corresponding directly to the classical semiflexible polymer model [39]. It was shown that as angular width decreases, structural variability within states decreases, structural state overlap decreases and conformations become more stable, leading to an expansion of the parameter space dominated by individual structures.

It is well known that adding carbon and glass fibers to polymer matrices improves their properties such as strength, stiffness and impact resistance. However, most of the existing research on the effects of these additives mostly concerns the study of single factors—such as fiber orientation, filling pattern or interlayer adhesion—without fully considering the interaction of these parameters in practical load-bearing applications. Furthermore, the intrinsic anisotropy of parts produced by Fused Deposition Modeling (FDM) [40], compounded by inconsistencies in layer bonding and fiber deposition, increases the complexity of predicting their mechanical properties. To address these challenges, Nemes et al. (Contribution No. 7) conducted a comprehensive, full-field deformation and damage analysis of carbon and glass fiber-reinforced 3D printing filaments based on Polylactic Acid (PLA) and Polyethylene Terephthalate Glycol (PETG) [41]. The effects of printing parameters—including infill pattern, density and structure orientation—were investigated under both bending and compressive loading. The study found that although carbon and glass composites provide a more stable structure, their mechanical properties are generally inferior to PLA and PETG. These materials are less sensitive to changes in fill pattern and fill factor, so their strength is more uniform across different printing settings. Tests showed that the carbon and glass fiber PLA filaments behaved similarly, but in terms of both tolerable force and achievable deformation, the carbon fiber PLA filament performed slightly better. If bending is the expected stress for the structure, it is better to use a basic PLA material, which showed better results than fiber-reinforced filaments. Carbon fiber PLA and glass fiber PLA filaments have also been shown to perform better with mesh infill patterns and lower infill densities. This can be of practical importance when a low infill density has to be chosen (to optimize weight) [42] and when bending stresses are expected. According to tests, the optimum infill value for these materials is between 20 and 30%, so it was suggested that for load-bearing components there is no reason to use a higher infill.

The growing demand for fully recyclable composites has spurred extensive research into thermoplastics, valued for their recyclability and excellent mechanical properties. Recently, thermoplastic polymer blends have gained attention for their enhanced recyclability and sustainability, as well as their ability to improve thermal stability, viscosity and manufacturability. However, limited data are available on the mechanical characterization of composites that incorporate these blends, particularly for thermoplastic recycling. To expand on this knowledge, Hao et al. (Contribution No. 8) investigated the stress–strain behavior of the following three polymer blends relevant to structural applications: PES/PEEK, PPS/PEEK and HDPE/PP. They then performed numerical analysis to predict the mechanical properties of unidirectional FRCs using each of these blends as a matrix. Representative Volume Element-based simulations [43,44] were used for this purpose. Finally, the suitability of the blends tested to produce fully matrix-recycled composites was critically evaluated. This has provided valuable preliminary insight into the mechanical viability and sustainability benefits of using recycled thermoplastic blends as matrices for unidirectional FRCs.

The effect of resin composition on the photopolymerization of zirconia ceramics produced by digital light processing [45] using additive manufacturing (AM) was studied by Kuang et al. (Contribution No. 9). An oligomer was considered as an additive to the photosensitive resin. The dependence of the content of this additive on the viscosity and curing properties of ceramic suspensions was investigated. The results demonstrated that the introduction of oligomers is conducive to enhancing the cross-linking density and reducing defects. Furthermore, zirconia ceramics fabricated by photopolymerization with oligomer photosensitive resin have been shown to exhibit excellent mechanical properties, greatly expanding the potential applications of high-performance zirconia ceramic components using AM.

Rachtanapun et al. (Contribution No. 10) successfully developed antistatic and anti-flammable biodegradable PCs by melt-blending polybutylene succinate with epoxy resin, polybutylene adipate-co-terephthalate and MgO particles. The addition of MgO improved the thermal decomposition behavior and water resistance of the blends, which could be attributed to the high thermal stability and hydrophobicity of the metal particles. The produced PCs showed an improvement in the V-1 rating [46] of flame retardancy, indicating an enhancement in the flame retardancy of biodegradable composite films. MgO served as a flame retardant, increasing the strength of the residual charcoal from the cross-linked matrix. The antistatic properties of the composites were improved using both plasma technology and the sparking process [47]. The antistatic effect of the plasma sputtering process was found to be better than that of the sparking process, as the sputtering method increased the uniform dispersion of metal nanoparticles on the composite surface. The resulting antistatic and anti-flammable biodegradable PCs with improved properties have potential for use in packaging, electronics and automotive applications.

3. Conclusions

The collective contributions to this Special Issue reflect a comprehensive and future-oriented exploration of sustainable practices and innovative technologies in the field of polymer composite and nanocomposite materials and products. The following conclusions can be drawn from the content presented in the papers covered in this Special Issue:

1. The introduction of suitable nanoparticles or dopants into polymer composites leads to improved piezoelectric properties and increased electrical conductivity efficiency, but also, for example, to effectively increase protection against gamma radiation at low energies.
2. The application of carbon fiber-reinforced polymer concrete offers hope for the possibility of designing ductile frames with enhanced energy dissipation, suitable for use in non-corrosive seismic-resistant buildings.
3. Adding carbon and glass fibers to polymer matrices used in 3D printing can improve their properties such as strength, stiffness and impact resistance.
4. Thermoplastic polymer blends are gaining attention for their increased recyclability and sustainability, as well as their ability to improve thermal stability or even flame retardancy and broad manufacturing capabilities.

Conflicts of Interest: The author declares no conflicts of interest.

Abbreviations

The following abbreviations are used in this paper:

AM	Additive manufacturing
AuNPs	Gold nanoparticles
FDM	Fused Deposition Modeling
FRCs	Fiber-reinforced composites
PCs	Polymer composites
PETG	Polyethylene Terephthalate Glycol
PLA	Polylactic Acid
PNs	Polymer nanocomposites

List of Contributions:

1. Umer, U.; Abidi, M.H.; Mian, S.H.; Alasim, F.; Aboudaif, M.K. Effects of silica nanoparticles on the piezoelectro-elastic response of PZT-7A–polyimide nanocomposites: Micromechanics modeling technique. *Polymers* **2024**, *16*, 2860. https://doi.org/10.3390/polym16202860.
2. Umer, U.; Abidi, M.H.; Almutairi, Z.; Aboudaif, M.K. A multi-phase analytical model for effective electrical conductivity of polymer matrix composites containing micro-SiC whiskers and nano-carbon black hybrids. *Polymers* **2025**, *17*, 128. https://doi.org/10.3390/polym17020128.
3. Maidi, M.; Sherzer, G.L.; Gal, E. Multiscale numerical study of enhanced ductility ratios and capacity in carbon fiber-reinforced polymer concrete beams for safety design. *Polymers* **2025**, *17*, 234. https://doi.org/10.3390/polym17020234.
4. Marashdeh, M.; Madkhali, N. Enhancing radiation shielding efficiency of *Nigella sativa* eumelanin polymer through heavy metals doping. *Polymers* **2025**, *17*, 609. https://doi.org/10.3390/polym17050609.
5. Jarka, P.; Hajduk, B.; Kumari, P.; Janeczek, H.; Godzierz, M.; Tsekpo, Y.M.; Tański, T. Investigations on thermal transitions in PDPP4T/PCPDTBT/AuNPs composite films using variable temperature ellipsometry. *Polymers* **2025**, *17*, 704. https://doi.org/10.3390/polym17050704.
6. Williams, M.J.; Gray, M.C. Microcanonical analysis of semiflexible homopolymers with variable-width bending potential. *Polymers* **2025**, *17*, 906. https://doi.org/10.3390/polym17070906.
7. Nemes, V.; Szalai, S.; Szívós, B.F.; Sysyn, M.; Kurhan, D.; Fischer, S. Deformation characterization of glass fiber and carbon fiber-reinforced 3D printing filaments using digital image correlation. *Polymers* **2025**, *17*, 934. https://doi.org/10.3390/polym17070934.
8. Hao, P.; Tang, N.; Tiscar, J.M.; Gilabert, F.A. Untapped potential of recycled thermoplastic blends in UD composites via finite element analysis. *Polymers* **2025**, *17*, 1168. https://doi.org/10.3390/polym17091168.
9. Kuang, N.; Qi, H.; Zhao, W.; Wu, J. Influence of resin composition on photopolymerization of zirconia ceramics fabricated by digital light processing additive manufacturing. *Polymers* **2025**, *17*, 1354. https://doi.org/10.3390/polym17101354.
10. Rachtanapun, P.; Suhr, J.; Oh, E.; Thajai, N.; Kanthiya, T.; Kiattipornpithak, K.; Kaewapai, K.; Photphroet, S.; Worajittiphon, P.; Tanadchangsaeng, N.; Wattanachai, P.; Jantanasakulwong, K.; Sawangrat, C. Flame retardance and antistatic polybutylene succinate/polybutylene adipate-co-terephthalate/magnesium composite. *Polymers* **2025**, *17*, 1675. https://doi.org/10.3390/polym17121675.

References

1. Gopinath, K.P.; Rajagopal, M.; Krishnan, A.; Sreerama, S.K. A review on recent trends in nanomaterials and nanocomposites for environmental applications. *Curr. Anal. Chem.* **2021**, *17*, 202–243. [CrossRef]
2. Toboła, W.; Papis, M.; Jastrzębski, D.; Perz, R. Experimental research of energy absorbing structures within helmet samples made with the additive manufacturing method—Preliminary study. *Acta Bioeng. Biomech.* **2023**, *25*, 127–136. [CrossRef]
3. Balakrishnan, T.S.; Sultan, M.T.H.; Shahar, F.S.; Basri, A.A.; Shah, A.U.M.; Sebaey, T.A.; Łukaszewicz, A.; Józwik, J.; Grzejda, R. Fatigue and impact properties of kenaf/glass-reinforced hybrid pultruded composites for structural applications. *Materials* **2024**, *17*, 302. [CrossRef] [PubMed]

4. Puchała, K.; Moneta, G.; Lichoń, D.; Grzejda, R.; Bednarz, A.; Mielniczek, W.; Łopatka, M.; Szymczyk, E.; Ignatovych, S.; Mykhailyshyn, R. Aerial medical platform for soldiers and civils evacuation—Concept, implementation plan and assessment of adaptation possibility of existing technologies. *Adv. Sci. Technol. Res. J.* **2025**, *19*, 28–50. [CrossRef]
5. Nowakowski, M.; Kosiuczenko, K.; Viliš, J. Unmanned vehicle mobility improvement against ballistic threats during special missions: A simulation study. *Transp. Probl.* **2025**, *20*, 139–151. [CrossRef]
6. Wysmulski, P.; Teter, A.; Debski, H. Effect of load eccentricity on the buckling of thin-walled laminated C-columns. *AIP Conf. Proc.* **2018**, *1922*, 080008.
7. Wysmulski, P.; Debski, H. The analysis of sensitivity to eccentric load of compressed thin-walled laminate columns. *AIP Conf. Proc.* **2019**, *2078*, 020006. [CrossRef]
8. Madkhali, N.; Algessair, S. Exploring the optical properties of metal-modified melanin following ultraviolet irradiation: An experimental and theoretical study using density functional theory. *Heliyon* **2024**, *10*, e29287. [CrossRef]
9. Magibalan, S.; Naveen, N.; Pradeep, N.; Vijayakumar, G.; Nithish Kumar, R. Experimental Investigations on Mechanical Properties of Sisal and Coir Fiber Reinforced Hybrid Bio Composites. *Mater. Today Proc.* **2023**, *in press*. [CrossRef]
10. Ibrahim, N.I.; Sultan, M.T.H.; Łukaszewicz, A.; Shah, A.U.M.; Shahar, F.S.; Józwik, J.; Najeeb, M.I.; Grzejda, R. Characterization and isolation method of *Gigantochloa scortechinii* (Buluh Semantan) cellulose nanocrystals. *Int. J. Biol. Macromol.* **2024**, *272*, 132847. [CrossRef]
11. Huang, Y.; Sultan, M.T.H.; Shahar, F.S.; Grzejda, R.; Łukaszewicz, A. Hybrid fiber-reinforced biocomposites for marine applications: A review. *J. Compos. Sci.* **2024**, *8*, 430. [CrossRef]
12. Bakhori, S.N.M.; Hassan, M.Z.; Bakhori, N.M.; Jamaludin, K.R.; Ramlie, F.; Daud, M.Y.M.; Aziz, S.A. Physical, mechanical and perforation resistance of natural-synthetic fiber interply laminate hybrid composites. *Polymers* **2022**, *14*, 1322. [CrossRef]
13. Shahar, F.S.; Sultan, M.T.H.; Grzejda, R.; Łukaszewicz, A.; Oksiuta, Z.; Krishnamoorthy, R.R. Harnessing the potential of natural composites in biomedical 3D printing. *Materials* **2024**, *17*, 6045. [CrossRef]
14. Huang, Y.; Sultan, M.T.H.; Shahar, F.S.; Łukaszewicz, A.; Oksiuta, Z.; Grzejda, R. Kenaf fiber-reinforced biocomposites for marine applications: A review. *Materials* **2025**, *18*, 999. [CrossRef] [PubMed]
15. Roy, N.; Sengupta, R.; Bhowmick, A.K. Modifications of carbon for polymer composites and nanocomposites. *Prog. Polym. Sci.* **2012**, *37*, 781–819. [CrossRef]
16. Sharma, S.; Sudhakara, P.; Omran, A.A.B.; Singh, J.; Ilyas, R.A. Recent trends and developments in conducting polymer nanocomposites for multifunctional applications. *Polymers* **2021**, *13*, 2898. [CrossRef]
17. Azlin, M.N.M.; Ilyas, R.A.; Zuhri, M.Y.M.; Sapuan, S.M.; Harussani, M.M.; Sharma, S.; Nordin, A.H.; Nurazzi, N.M.; Afiqah, A.N. 3D printing and shaping polymers, composites, and nanocomposites: A review. *Polymers* **2022**, *14*, 180. [CrossRef]
18. Gumowska, A.; Robles, E.; Bikoro, A.; Wronka, A.; Kowaluk, G. Selected properties of bio-based layered hybrid composites with biopolymer blends for structural applications. *Polymers* **2022**, *14*, 4393. [CrossRef]
19. Paszkiewicz, S.; Walkowiak, K.; Barczewski, M. Biobased polymer nanocomposites prepared by in situ polymerization: Comparison between carbon and mineral nanofillers. *J. Mater. Sci.* **2024**, *59*, 13805–13823. [CrossRef]
20. Mohan, N.; Senthil, P.; Vinodh, S.; Jayanth, N. A review on composite materials and process parameters optimisation for the fused deposition modelling process. *Virtual Phys. Prototyp.* **2017**, *12*, 47–59. [CrossRef]
21. de Campos, A.; Corrêa, A.C.; Claro, I.C.; de Morais Teixeira, E.; Marconcini, J.M. Processing, characterization and application of micro and nanocellulose based environmentally friendly polymer composites. In *Sustainable Polymer Composites and Nanocomposites*; Inamuddin, Thomas, S., Mishra, R.K., Asiri, A.M., Eds.; Springer: Cham, Switzerland, 2019; pp. 1–35.
22. Agwuncha, S.C.; Anusionwu, C.G.; Owonubi, S.J.; Sadiku, E.R.; Busuguma, W.A.; Ibrahim, I.D. Extraction of cellulose nanofibers and their eco/friendly polymer composites. In *Sustainable Polymer Composites and Nanocomposites*; Inamuddin, Thomas, S., Mishra, R.K., Asiri, A.M., Eds.; Springer: Cham, Switzerland, 2019; pp. 37–64.
23. Yuan, S.; Shen, F.; Chua, C.K.; Zhou, K. Polymeric composites for powder-based additive manufacturing: Materials and applications. *Prog. Polym. Sci.* **2019**, *91*, 141–168. [CrossRef]
24. Paszkiewicz, S.; Pypeć, K.; Irska, I.; Piesowicz, E. Functional polymer hybrid nanocomposites based on polyolefins: A review. *Processes* **2020**, *8*, 1475. [CrossRef]
25. Georgios, K.; Silva, A.; Furtado, S. Applications of Green Composite Materials. In *Biodegradable Green Composites*; Kalia, S., Ed.; John Wiley & Sons: Hoboken, NJ, USA, 2016; pp. 312–337.
26. Bhong, M.; Khan, T.K.H.; Devade, K.; Krishna, B.V.; Sura, S.; Eftikhaar, H.K.; Thethi, H.P.; Gupta, N. Review of Composite Materials and Applications. *Mater. Today Proc.* **2023**, *in press*. [CrossRef]
27. Diakun, J. Recycling Product Model and its application for quantitative assessment of product recycling properties. *Sustainability* **2024**, *16*, 2880. [CrossRef]
28. Diakun, J.; Grzejda, R. Product design analysis with regard to recycling and selected mechanical properties. *Appl. Sci.* **2025**, *15*, 512. [CrossRef]

29. Keramati, Y.; Ansari, R.; Hassanzadeh-Aghdam, M.K.; Umer, U. Micromechanical simulation of thermal expansion, elastic stiffness and piezoelectric constants of graphene/unidirectional $BaTiO_3$ fiber reinforced epoxy hybrid nanocomposites. *Acta Mech.* **2023**, *234*, 6251–6270. [CrossRef]
30. Abidi, M.H.; Al-Ahmari, A.M.; Umer, U.; Rasheed, M.S. Multi-objective optimization of micro-electrical discharge machining of nickel-titanium-based shape memory alloy using MOGA-II. *Measurement* **2018**, *125*, 336–349. [CrossRef]
31. Maidi, M.; Shufrin, I. Evaluation of existing reinforced concrete buildings for seismic retrofit through external stiffening: Limit displacement method. *Buildings* **2024**, *14*, 2781. [CrossRef]
32. Maidi, M.; Sherzer, G.L.; Gal, E. Enhancing ductility in carbon fiber reinforced polymer concrete sections: A multi-scale investigation. *Comput. Concr.* **2024**, *33*, 385–398.
33. Marashdeh, M.; Al-Hamarneh, I.F. Evaluation of gamma radiation properties of four types of surgical stainless steel in the energy range of 17.50–25.29 keV. *Materials* **2021**, *14*, 6873. [CrossRef]
34. Akhdar, H.; Mahmoud, K.A.; Madkhali, N.; Marashdeh, M.; Abu El-Soad, A.M.; Tharwat, M. Engineering multifunctional polypropylene nanocomposites: Tailoring structural, thermal, and gamma-ray shielding properties with $Ni_{0.9}Zn_{0.1}Fe_2O_4$ doping. *Prog. Nucl. Energy* **2024**, *177*, 105478. [CrossRef]
35. Hajduk, B.; Jarka, P.; Tański, T.; Bednarski, H.; Janeczek, H.; Gnida, P.; Fijalkowski, M. An investigation of the thermal transitions and physical properties of semiconducting PDPP4T:PDBPyBT blend films. *Materials* **2022**, *15*, 8392. [CrossRef] [PubMed]
36. Hajduk, B.; Bednarski, H.; Trzebicka, B. Temperature-dependent spectroscopic ellipsometry of thin polymer films. *J. Phys. Chem. B* **2020**, *124*, 3229–3251. [CrossRef]
37. Hajduk, B.; Bednarski, H.; Jarząbek, B.; Nitschke, P.; Janeczek, H. Phase diagram of P3HT:PC70BM thin films based on variable-temperature spectroscopic ellipsometry. *Polym. Test.* **2020**, *84*, 106383. [CrossRef]
38. Qi, K.; Liewehr, B.; Koci, T.; Pattanasiri, B.; Williams, M.J.; Bachmann, M. Influence of bonded interactions on structural phases of flexible polymers. *J. Chem. Phys.* **2019**, *150*, 054904. [CrossRef]
39. Williams, M.J.; Bachmann, M. System-size dependence of helix-bundle formation for generic semiflexible polymers. *Polymers* **2016**, *8*, 245. [CrossRef]
40. Joshua, R.J.N.; Raj, S.A.; Sultan, M.T.H.; Łukaszewicz, A.; Józwik, J.; Oksiuta, Z.; Dziedzic, K.; Tofil, A.; Shahar, F.S. Powder bed fusion 3D printing in precision manufacturing for biomedical applications: A comprehensive review. *Materials* **2024**, *17*, 769. [CrossRef]
41. Szalai, S.; Herold, B.; Kurhan, D.; Németh, A.; Sysyn, M.; Fischer, S. Optimization of 3D printed rapid prototype deep drawing tools for automotive and railway sheet material testing. *Infrastructures* **2023**, *8*, 43. [CrossRef]
42. Szalai, S.; Szívós, B.F.; Nemes, V.; Fischer, S. Sustainable 3D-printing filaments and their applications. *Chem. Eng. Trans.* **2024**, *114*, 967–972.
43. Gilabert, F.A. An efficient anisotropization technique to transform isotropic nonlinear materials into unidirectional and bidirectional composites. *Mater. Des.* **2021**, *206*, 109772. [CrossRef]
44. Laheri, V.; Hao, P.; Gilabert, F.A. Direct embedment of RVE-based microscale into lab size coupons to research fracture process in unidirectional and bidirectional composites. *Compos. Sci. Technol.* **2023**, *235*, 109949. [CrossRef]
45. Kuang, N.; Xiao, M.; Qi, H.; Zhao, W.; Wu, J. Optimization of resin composition for zirconia ceramic digital light processing additive manufacturing. *Polymers* **2025**, *17*, 797. [CrossRef] [PubMed]
46. Sawangrat, C. *Anti-Static and Anti-Flaming Improvement of Biodegradable Polymer Blends by Nanocoating via Plasma Technology and Sparking Process*; Technical Report; Chiang Mai University: Chiang Mai, Thailand, 2024.
47. Jantanasakulwong, K.; Thanakkasaranee, S.; Seesuriyachan, P.; Singjai, P.; Saenjaiban, A.; Photphroet, S.; Pratinthong, K.; Phimolsiripol, Y.; Leksawasdi, N.; Chaiyaso, T.; et al. Sparking nano-metals on a surface of polyethylene terephthalate and its application: Anti-coronavirus and anti-fogging properties. *Int. J. Mol. Sci.* **2022**, *23*, 10541. [CrossRef] [PubMed]

polymers

MDPI

Article

Effects of Silica Nanoparticles on the Piezoelectro-Elastic Response of PZT-7A–Polyimide Nanocomposites: Micromechanics Modeling Technique

Usama Umer [1,*], Mustufa Haider Abidi [1], Syed Hammad Mian [1], Fahad Alasim [1,2] and Mohammed K. Aboudaif [1]

[1] Advanced Manufacturing Institute, King Saud University, P.O. Box 800, Riyadh 11421, Saudi Arabia
[2] Industrial Engineering Department, College of Engineering, King Saud University, P.O. Box 800, Riyadh 11421, Saudi Arabia
* Correspondence: uumer@ksu.edu.sa

Abstract: By using piezoelectric materials, it is possible to convert clean and renewable energy sources into electrical energy. In this paper, the effect on the piezoelectro-elastic response of piezoelectric-fiber-reinforced nanocomposites by adding silica nanoparticles into the polyimide matrix is investigated by a micromechanical method. First, the Ji and Mori–Tanaka models are used to calculate the properties of the nanoscale silica-filled polymer. The nanoparticle agglomeration and silica–polymer interphase are considered in the micromechanical modeling. Then, considering the filled polymer as the matrix and the piezoelectric fiber as the reinforcement, the Mori–Tanaka model is used to estimate the elastic and piezoelectric constants of the piezoelectric fibrous nanocomposites. It was found that adding silica nanoparticles into the polymer improves the elastic and piezoelectric properties of the piezoelectric fibrous nanocomposites. When the fiber volume fraction is 60%, the nanocomposite with the 3% silica-filled polyimide exhibits 39%, 31.8%, and 37% improvements in the transverse Young's modulus E_T, transverse shear modulus G_{TL}, and piezoelectric coefficient e_{31} in comparison with the composite without nanoparticles. Furthermore, the piezoelectro-elastic properties such as E_T, G_{TL}, and e_{31} can be improved as the nanoparticle diameter decreases. However, the elastic and piezoelectric constants of the piezoelectric fibrous nanocomposites decrease once the nanoparticles are agglomerated in the polymer matrix. A thick interphase with a high stiffness enhances the nanocomposite's piezoelectro-elastic performance. Also, the influence of volume fractions of the silica nanoparticles and piezoelectric fibers on the nanocomposite properties is studied.

Keywords: PZT–polyimide composite; silica nanoparticle; piezoelectro-elastic properties; interphase; micromechanics modeling

1. Introduction

Piezoelectric-fiber-reinforced polymer composites are used for various engineering applications, such as in clean energy harvester devices from environmental sources and in vibration control, structural health monitoring, and structural morphing [1–4]. The improved mechanical flexibility, good stiffness/strength-to-weight ratio, reliability, and tailorable properties have made piezoelectric fibrous composites more favorable for the abovementioned applications as compared to neat piezoelectric materials [4–8]. Nevertheless, advances in electro-mechanical technologies have led to fast growth in the demand for piezoelectric fibrous composites with better functionalities. So, the design of new piezoelectric composites is an important issue for researchers and scientists.

The concept of using nano/micro-hybrid reinforcements in polymer composites has emerged for the sake of improving their multifunctional properties. In these composites, nano-sized reinforcements such as graphene nanoplatelets, carbon nanotubes (CNTs), and silica nanoparticles are introduced alongside traditional micron-sized fibers [8–13]. For example, Cui et al. [13] observed that carbon fiber–silica nanoparticle composites have better performance regarding interlaminar shear strength, impact strength, and flexural strength compared with that of carbon fiber composites. Hwayyin et al. [14] indicated an enhancement in the mechanical properties of carbon fiber–polyester composites at different weights of nano-silicon dioxide. The outcomes showed an increment in the tensile stress of 11.45% after an increase in the nanoparticle content from 0.16% wt. to 0.2% wt. [14]. Zheng et al. [15] studied the influence of silica nanoparticles on the mechanical properties of glass-fiber-reinforced epoxy composites. It was observed that the increase in the silica nanoparticle content yields an enhancement in the tensile modulus and compression strength. Gang et al. [16] reported that an increase from 0% to 5% of the nanoparticle volume fraction yielded an enhancement in the tensile modulus of a carbon fiber–silica nanoparticle–polyimide composite from 2475 MPa to 2780 MPa. Tang et al. [17] measured tensile properties in the transverse direction, interlaminar shear strength, and the mode I and mode II interlaminar fracture toughness of carbon fiber composites with 10 wt% and 20 wt% silica nanoparticles dispersed into epoxy. The transverse tensile properties and mode I interlaminar fracture toughness were improved by increasing the silica nanoparticle content in the epoxy [17].

Such a concept has been employed to study the advantages of nanofiller-containing matrices in improving the equivalent properties of piezoelectric fibrous composites. Keramati et al. [12] analyzed a nanocomposite in which $BaTiO_3$ fibers were placed inside a graphene nanosheet (GNS)-filled polymer. The addition of GNSs inside the polymer matrix led to an improvement in the elastic properties, transverse coefficient of thermal expansion, and piezoelectric coefficients e_{31} and e_{15}. The mechanical and piezoelectric characteristics of piezoelectric fiber–CNT-reinforced nanocomposites were investigated by Hasanzadeh et al. [18]. CNTs were randomly oriented into a polymer matrix. The mechanical properties and piezoelectric coefficient e_{31} of the piezoelectric nanocomposite containing CNTs were improved over those of the piezoelectric composite without CNTs [18]. Godara and Mahato [19] studied the elastic and piezoelectric coefficients of a nanocomposite in which piezoelectric fibers were embedded into a CNT-reinforced polymer [19]. The use of CNTs in piezoelectric-fiber-reinforced composites can enhance the structural/functional properties [19].

Generally, evaluating the engineering constants of piezoelectric fibrous nanocomposites containing silica nanoparticles is crucial in designing structures constructed with these materials. Many microstructural factors such as the amount, size, dispersion quality, variation in properties, and nanoparticle–matrix interfacial interaction affect the overall properties of silica-nanoparticle-containing composites [20–22]. Therefore, conducting studies in this area is of significance [9,18,23–26].

To the best of the authors' knowledge, the piezoelectro-elastic properties of piezoelectric fiber–nanoparticle–polymer nanocomposites with regard to the agglomeration and size of the silica nanoparticles and the silica–polymer interphase have not yet been sufficiently investigated. The novelty of this work comes from developing a hierarchical micromechanical method to comprehensively investigate the elastic and piezoelectric constants of PZT-7A–silica–polyimide nanocomposites with variables of important microstructures. So, the current research aims to evaluate the properties of a piezoelectric nanocomposite made of unidirectional piezoelectric fibers embedded in a polyimide matrix with silica nanoparticles. To achieve this, we developed a model using Ji's approach and the Mori–Tanaka method. After confirming the accuracy of the model, we studied how the volume, size, and clumping of nanoparticles, as well as the thickness and stiffness of the silica–polymer interface, affect the composite's elastic and shear moduli and piezoelectric coefficients (e_{31} and e_{33}). One potential use for these nanocomposites is in energy-harvesting devices.

2. Micromechanical Analysis of Piezoelectric Fibrous Nanocomposites

The nanocomposite consists of unidirectional piezoelectric microfibers, silica nanoparticles, and a polymer matrix. Figure 1 shows a representation of a lamina made of this piezoelectric fibrous nanocomposite. The configuration of such a nanocomposite is such that piezoelectric fibers are embedded inside a nanoparticle-filled polymer. Note that the piezoelectric fibers are aligned along the 3-direction. The dispersion of the silica nanoparticles into the polymer matrix can be uniform or non-uniform. The interphase region shown in this figure is considered due to the interaction between the nanoparticles and the polymer matrix. In general, the representative volume element (RVE) of the nanocomposite system may be treated as consisting of two phases, in which the reinforcement is the piezoelectric fiber and the matrix is the nanoparticle-filled polymer. The micromechanical modeling to predict the effective properties of piezoelectric fibrous nanocomposites is carried out here in a two-step procedure. First, the elastic properties of the silica-nanoparticle-filled polymer are calculated using the Ji and Mori–Tanaka models. Next, the Mori–Tanaka model is employed to predict the overall piezoelectric coefficients and elastic moduli of the piezoelectric fibrous nanocomposite.

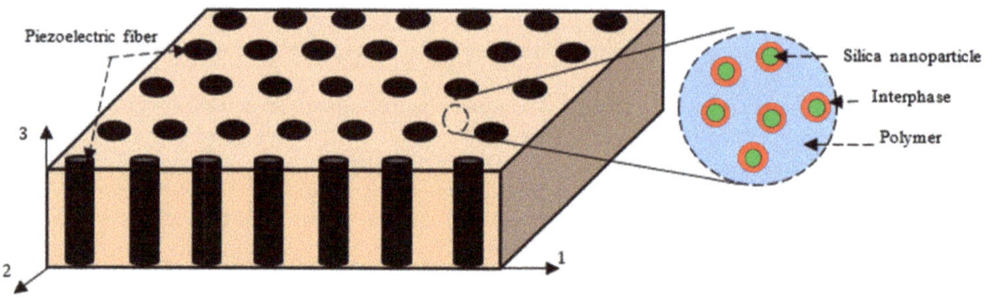

Figure 1. Demonstration of the piezoelectric fibrous nanocomposite with a silica-nanoparticle-filled polymer matrix.

2.1. Silica-Nanoparticle-Filled Polymer

In this sub-section, a micromechanics approach is presented to estimate the effective properties of the polymer matrix containing silica nanoparticles. Micromechanical models predict the composite properties in terms of the volume fraction, geometry of fillers, and material properties of constituents [27–30]. Concerning the nanoparticle-filled polymer systems, it is required to incorporate the interphase between the nanoparticle and the polymer matrix in the simulation [20,29,31–33]. The interfacial region possesses material properties in between those of the polymer matrix and those of the nanoparticle [20,22,31,32,34]. In the micromechanical simulation, the interphase is considered as the third phase, which surrounds the silica nanoparticles, as shown in Figure 1. The Young's modulus of the silica-filled polymer materials with the interphase can be estimated by the Ji model [35,36] as follows:

$$E_{NC} = E_m \left[(1-\lambda) + \frac{\lambda - \beta}{(1-\lambda) + \frac{\lambda(k-1)}{\ln k}} + \frac{\beta}{(1-\lambda) + \frac{(\lambda-\beta)(k+1)}{2} + \beta \frac{E_{NP}}{E_m}} \right]^{-1} \tag{1}$$

in which E_m and E_{NP} are the Young's moduli of the matrix and nanoparticle, respectively. The values of λ, β, and k are given by

$$\lambda = \sqrt{\left(\frac{2t_i}{d_{NP}} + 1\right)^3 \varphi_{NP}},\ \beta = \sqrt{\varphi_{NP}},\ k = \frac{E_i}{E_m} \tag{2}$$

where φ_{NP} and d_{NP} are the volume fraction and diameter of the nanoparticle, and E_i and t_i are the Young's modulus and thickness of the interphase, respectively. The Young's modulus of the silica-filled polymer materials in the absence of the interphase, i.e., a two-phase composite, is [36]

$$E_{NC} = E_m \left[(1-\beta) + \frac{\beta}{(1-\beta) + \beta \frac{E_{NP}}{E_m}} \right]^{-1} \quad (3)$$

Poisson's ratio of a nanoparticle-filled polymer can be estimated by the rule of the mixture as $v_{NC} = v_{NP}\varphi_{NP} + v_m(1-\varphi_{NP})$, in which v_{NP} and v_m are Poisson's ratio of the nanoparticle and matrix, respectively.

Silica nanoparticles tend to form agglomerates, which leads to poor distribution into the polymer matrix during the fabrication process [20,37,38]. The non-uniform distribution and formation of nanoparticle agglomeration may be mostly attributed to their great specific surface area and high surface energy [37,38]. A two-parameter micromechanical technique is employed to investigate the effect of nanoparticle agglomeration on the effective properties of the nanoparticle-filled polymer [39–41]. A number of nanoparticles are supposed to uniformly disperse into the polymer, and other nanoparticles appear in the agglomeration state. The whole volume of the nanoparticles in the RVE of the filled polymer specified by V_{NP} is divided into the following two parts:

$$V_{NP} = V_{NP}^{agg} + V_{NP}^{em} \quad (4)$$

where V_{NP}^{agg} is the volume of nanoparticles inside the agglomerated phase and V_{NP}^{em} denotes the volume of nanoparticles in the polymer matrix and outside the agglomerates. The definition of these two parameters for the agglomeration is as follows:

$$\xi = \frac{V_{agg}}{V},\ \zeta = \frac{V_{NP}^{agg}}{V_{NP}} \quad (5)$$

where V is the volume of the filled polymer RVE and V_{agg} denotes the volume of the agglomerate phase within the RVE [39–41]. Thus, the volume fraction of nanoparticles in the agglomeration phase φ_{NP}^{agg} and in the polymer matrix and outside the agglomerates φ_{NP}^{em} is determined as

$$\varphi_{NP}^{agg} = \frac{V_{NP}^{agg}}{V_{agg}} = \frac{\zeta}{\xi}\varphi_{NP},\ \varphi_{NP}^{em} = \frac{V_{NP} - V_{NP}^{agg}}{V - V_{agg}} = \frac{1-\zeta}{1-\xi}\varphi_{NP} \quad (6)$$

We analyze the nanoparticle-filled polymer as a system consisting of agglomerates of spherical shape embedded in a new matrix. We initially predict Young's modulus and Poisson's ratio of the agglomerate and the new matrix phases by the Ji model and the rule of mixture. The Young's modulus and Poisson's ratio are then used to calculate the bulk modulus and shear modulus. Using the bulk moduli and shear moduli of the agglomerate (K_{agg}, G_{agg}) and the new matrix (K_{em}, G_{em}), the equivalent bulk modulus ($\overline{K}_{NC}$) and shear modulus ($\overline{G}_{NC}$) of the nanoparticle-filled polymer system with the nanoparticle agglomeration are predicted by the Mori–Tanaka model [39–41], respectively,

$$\overline{K}_{NC} = K_{em} + \frac{\xi\left(\frac{K_{agg}}{K_{em}} - 1\right)}{1 + \frac{1-\xi}{1+\frac{4G_{em}}{3K_{em}}}\left(\frac{K_{agg}}{K_{em}} - 1\right)} K_{em}, \quad (7)$$

$$\overline{G}_{NC} = G_{em} + \frac{\xi\left(\frac{G_{agg}}{G_{em}} - 1\right)}{1 + \frac{\left(6+12\frac{G_{em}}{K_{em}}\right)}{\left(15+20\frac{G_{em}}{K_{em}}\right)}(1-\xi)\left(\frac{G_{agg}}{G_{em}} - 1\right)} G_{em}. \quad (8)$$

Thus, the elastic properties of a nanoparticle-filled polymer considering the interphase region and agglomeration of nanoparticles can be achieved by the micromechanical technique developed in this section.

2.2. Piezoelectric Fiber Composites

Now, a micromechanical model can be adopted to estimate the equivalent properties of piezoelectric fibrous nanocomposites. By the means of the conventional indicial notation, the constitutive equations of piezoelectric materials are as follows [42–44]:

$$\begin{aligned} \sigma_{ij} &= C_{ijmn}\varepsilon_{mn} + e_{nij}E_n \\ D_i &= e_{imn}\varepsilon_{mn} - k_{in}E_n \end{aligned} \tag{9}$$

where the repeated sub-scripts are summed over the range of $i,\ j, m, n = 1, 2, 3$, and σ_{ij} and ε_{mn} stand for the stress and strain tensors, respectively. E_n and D_i stand for the electric field and the electric displacement vectors, respectively. C_{ijmn}, e_{nij}, and k_{in} denote the elastic stiffness and piezoelectric and permittivity tensors, respectively. Divergence equations expressing the mechanical equilibrium and Gauss' law can be given by Equation (10) [44], respectively:

$$\begin{aligned} \sigma_{ij,j} &= 0 \\ D_{i,i} &= 0 \end{aligned} \tag{10}$$

Moreover, the gradient equations defining the strain displacement equations and electric field potential are expressed, respectively:

$$\begin{aligned} \varepsilon_{ij} &= \tfrac{1}{2}\left(u_{i,j} + u_{j,i}\right) \\ E_i &= -\phi_{,i} \end{aligned} \tag{11}$$

in which u_i and ϕ stand for the mechanical displacement and electric potential, respectively.

The components of piezoelectric fibrous nanocomposites are the nanoparticle-filled polymer as the matrix phase and the piezoelectric fibers as the reinforcing phase. On the basis of the Mori–Tanaka micromechanical approach and considering v_m and v_r as the volume fraction of the matrix and fiber, respectively, the electro-elastic constants of piezoelectric fibrous nanocomposites can be estimated as

$$\overset{\sim c}{\mathbf{C}} = \overset{\sim m}{\mathbf{C}} + v_r\left(\overset{\sim r}{\mathbf{C}} - \overset{\sim m}{\mathbf{C}}\right)\mathbf{B} \tag{12}$$

in which $\overset{\sim r}{\mathbf{C}}$ and $\overset{\sim m}{\mathbf{C}}$ are the electro-elastic modulus matrices for the reinforcement and matrix, respectively, and the piezoelectric concentration tensor is defined as follows:

$$\mathbf{B} = \mathbf{A}\left[v_m\mathbf{I} + v_r\mathbf{A}\right]^{-1}, \mathbf{A} = \left[\mathbf{I} + \hat{\mathbf{S}}\left(\overset{\sim \mathbf{m}}{\mathbf{C}}\right)^{-1}\left(\overset{\sim \mathbf{r}}{\mathbf{C}} - \overset{\sim \mathbf{m}}{\mathbf{C}}\right)\right]^{-1}, \tag{13}$$

where $\hat{\mathbf{S}}$ is the Eshelby tensor and its components can be found in the literature [18,24,27,44]. It is worth mentioning that all simulation procedures and numerical results obtained in this research have been performed using codes written in MATLAB(R2024b) software. Also, it should be noted that all formulations used for the simulations in MATLAB software are analytical relations, and the finite element method has not been employed.

3. Results and Discussion

In this section, we first present the numerical results of the Young's moduli, shear moduli, and piezoelectric coefficients of PZT-7A fiber-reinforced nanocomposites with the silica-nanoparticle-filled polyimide matrix. Then, some comparisons are made between the present predictions and other results available in the literature [45,46].

3.1. Piezoelectro-Elastic Response of Piezoelectric Fibrous Nanocomposites

The micromechanical model consisting of Ji's model, the rule of mixture, and the Mori–Tanaka model is used to investigate the effective constants of the piezoelectric fibrous nanocomposite. The constituents of the composite are PZT-7A, polyimide, and silica nanoparticles. The

Young's modulus and Poisson's ratio of the polyimide and silica nanoparticles are 3.78 GPa and 0.4 and 73 GPa and 0.23 [20,22,44], respectively. Table 1 lists the properties of the PZT-7A fiber. The piezoelectric fiber volume fraction is considered to be 60%. The diameter of the silica nanoparticles and their volume fraction into the polyimide matrix are 30 nm and 3%, respectively. Also, the Young's modulus, Poisson's ratio, and thickness of the interphase region are taken as $E_i = 10E_m$, 0.4, and $t_i = 0.5d_{NP}$ [20,31], respectively.

Table 1. Material constants of PZT5 fiber, epoxy, and PZT-7A fiber [9,44].

Material	PZT5	Epoxy	PZT-7A
C_{11} (GPa)	121	8	148
C_{12} (GPa)	75.4	4.4	76.2
C_{13} (GPa)	75.2	4.4	74.2
C_{22} (GPa)	121	8	148
C_{23} (GPa)	75.2	4.4	74.2
C_{33} (GPa)	111	8	131
C_{44} (GPa)	21.1	1.8	25.4
C_{66} (GPa)	22.8	1.8	35.9
e_{31} (C/m^2)	−5.4	0	−2.1
e_{33} (C/m^2)	9.5	0	9.5

Figure 2 shows the influence of adding silica nanoparticles on the material constants of the piezoelectric fibrous nanocomposites. The numerical results are presented for two values of nanoparticle volume fraction (NPVF), 3% and 5%. The material properties of piezoelectric fibrous composites without nano-inclusions are also illustrated in the figure. Figure 2a indicates the results of the longitudinal Young's modulus versus piezoelectric fiber volume fraction. Incorporating nano-inclusions into the polymer matrix insignificantly affects the longitudinal Young's modulus. This is attributed to the fact that the longitudinal properties of long fiber-reinforced composites are mostly dominated by the material properties and content of fibers [9,18,47,48]. It can be seen in Figure 2a that Young's modulus in the direction parallel to the fiber direction linearly increases as the piezoelectric fiber volume fraction increases. Figure 2b shows the variation in the transverse Young's modulus of the piezoelectric fibrous nanocomposites with the piezoelectric fiber volume fraction. The transverse Young's modulus notably depends on the nanoparticles in the polyimide matrix. The value of E_T of the silica-nanoparticle-containing nanocomposite is greater than that of the composite without silica nanoparticles. When the piezoelectric fiber volume fraction is 60%, the nanocomposite with a 3% silica-nanoparticle-containing polyimide exhibits a 39% improvement in the transverse Young's modulus in comparison with the composite without nanoparticles. Adding silica nanoparticles within the polyimide provides a relatively stronger matrix that is potentially beneficial for the transverse Young's modulus of the piezoelectric fibrous nanocomposites. The increase in the silica nanoparticle amount aids the nanocomposite in obtaining a higher value of the transverse Young's modulus. A similar trend has been found for other types of nanofillers [9,18,48]. It is shown in Figure 2b that the Young's modulus in the direction perpendicular to the piezoelectric fiber nonlinearly increases as the fiber volume fraction increases. Based on the results observed in Figure 2c,d, adding silica nanoparticles is generally beneficial to shear moduli in the longitudinal and transverse directions. Both shear moduli are enhanced by increasing the piezoelectric fiber volume fraction. The values of the transverse shear modulus of the 60% piezoelectric-fiber-reinforced composite without nanoparticles and with a 3% nanoparticle content are calculated as 6.04 GPa and 7.96 GPa, respectively. Figure 2e depicts the piezoelectric coefficient e_{31} of the piezoelectric fibrous nanocomposites versus the piezoelectric fiber volume fraction. The value of piezoelectric coefficient e_{31} can be significantly improved as a result of the nanoparticle addition into the polyimide matrix. A major contribution to the piezoelectric coefficient e_{31} is from the polymer matrix properties. Compared to the piezoelectric fibrous composite, the piezoelectric coefficient e_{31} of the piezoelectric fibrous nanocomposite containing a 3% silica-nanoparticle-filled polyimide matrix exhibits an upward trend with an approximate 37% improvement. Therefore, incorporating nanofillers into the polymer can enhance the in-plane actuation property of piezoelectric fibrous nanocomposites over that of a traditional piezoelectric fibrous composite without nanofillers. This trend has been reported for other composite systems containing CNTs [9,18,19]. The piezoelectric coefficient e_{31} of piezoelectric fibrous nanocomposites exhibits an improvement through the increase in the piezoelectric fiber volume fraction. Figure 2f displays the piezoelectric coefficient e_{33} versus the fiber volume fraction. There is no effect of silica nanoparticles on the piezoelectric coefficient e_{33} because its value is mostly dominated by the piezoelectric property of the fiber. A linear increase is obtained for the piezoelectric coefficient e_{33} as the piezoelectric fiber volume fraction increases. Because the piezoelectric constant e_{31} and the elastic properties such as E_T and G_{TL} of the piezoelectric fibrous nanocomposite containing silica nanoparticles are improved, this composite has good potential for use as a superior actuator material for intelligent structures with a great in-plane actuation option [9,12].

The micromechanical results for investigating the role of the interfacial zone between the silica nanoparticles and polyimide matrix in the material constants of the piezoelectric fibrous nanocomposites are presented in Figure 3. It is worth pointing out that the interphase does not have a notable contribution to the Young's modulus in the longitudinal direction, as seen in Figure 3a. It may be concluded from Figure 3b–d that the formation of the interfacial region is beneficial to the transverse Young's modulus and both shear moduli. Relative to the composite system without the interphase, an increasing trend is observed for these three elastic moduli of the nanocomposite with the interphase. The results of Figure 3e disclose that the interphase tends to improve the piezoelectric coefficient e_{31}. A literature survey shows that the interphase between the polymer matrix and CNTs can contribute to the improvement of the overall properties of piezoelectric–CNT nanocomposites [18]. As shown in Figure 3f, the piezoelectric coefficient e_{33} exhibits no variation in the presence or absence of the interphase.

Generally, the interphase has properties in between those of the nanoparticle and those of the polymer matrix [20,22,29,31,32,34]. To better evaluate the effect of interphase characteristics on the material constants of piezoelectric fibrous nanocomposites, a micromechanical analysis is conducted with different values of interphase stiffness and thickness. The results of the transverse Young's modulus, longitudinal shear modulus, transverse shear modulus, and piezoelectric coefficient e_{31} with changing the interphase elastic modulus are shown in Figure 4a–d, respectively. The effective properties of the piezoelectric fibrous nanocomposite can be enhanced with increasing the interphase elastic modulus. It is noted that a stiffer interphase can increase the mechanical properties of the polymer matrix [18,20,22,29]. One of the ways to enhance the mechanical properties of the interfacial region may be nanoparticle surface treatment.

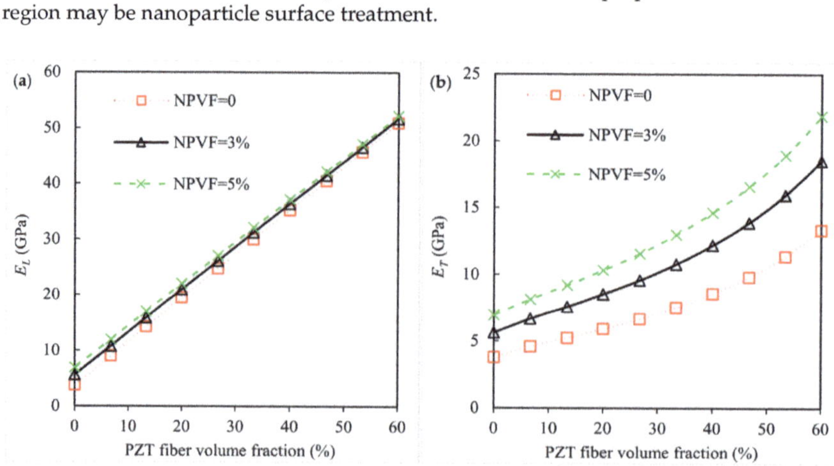

Figure 2. *Cont.*

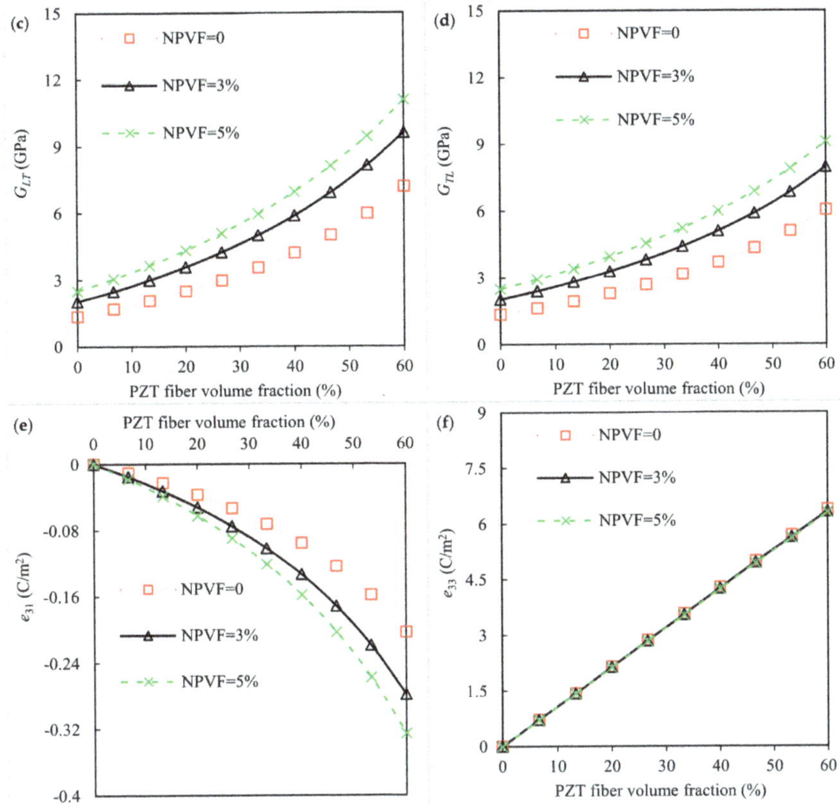

Figure 2. Influence of nanoparticle volume fraction on the (**a**) longitudinal Young's modulus, (**b**) transverse Young's modulus, (**c**) longitudinal shear modulus, (**d**) transverse shear modulus, (**e**) piezoelectric coefficient e_{31}, and (**f**) piezoelectric coefficient e_{33} of the piezoelectric fibrous nanocomposite containing silica nanoparticles.

The influence of changing the interphase thickness on the properties of nanocomposites, including E_T, G_{LT}, G_{TL}, and e_{31}, is depicted in Figure 5a–d, respectively. The increase in interphase thickness significantly improves the elastic and piezoelectric constants. Different methods for nanoparticle functionalization may produce interphases with variable thicknesses. By increasing the interphase thickness from 1 nm to 15 nm, the improvement in the transverse elastic modulus with a 3% silica-nanoparticle-containing polyimide matrix is calculated to be about 20.5%. In turn, for the piezoelectric coefficient e_{31}, the improvement is about 26%. It is worth mentioning that the in-plane actuation caused by the piezoelectric composite is increased by tailoring the piezoelectric constant e_{31} [9,47]. An important conclusion from the above micromechanical studies is the production of a stiff and thick interphase in the nanocomposite fabrication. This is due to the increased stiffness of the polymer matrix by adding nanoparticles, as reported in previous studies [18,20,22,29].

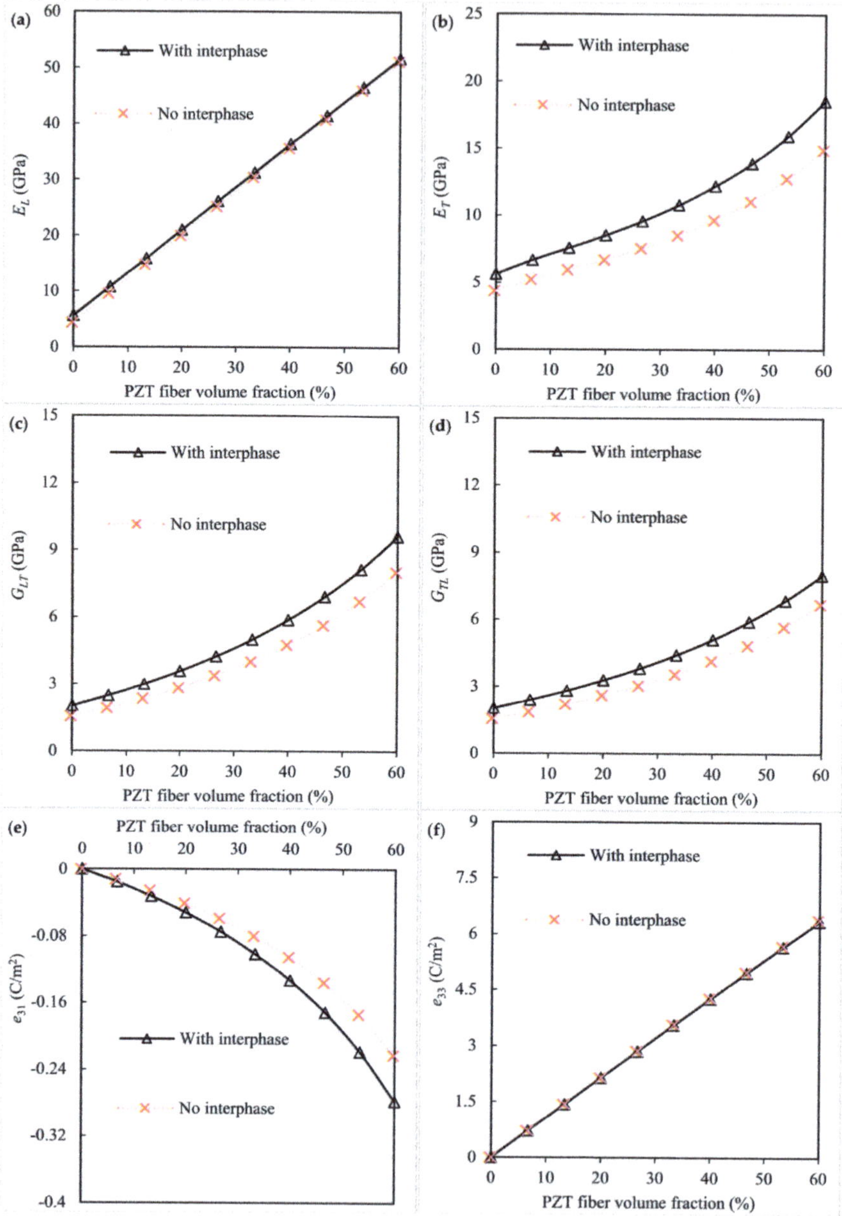

Figure 3. Influence of interphase region on the (**a**) longitudinal Young's modulus, (**b**) transverse Young's modulus, (**c**) longitudinal shear modulus, (**d**) transverse shear modulus, (**e**) piezoelectric coefficient e_{31}, and (**f**) piezoelectric coefficient e_{33} of the piezoelectric fibrous nanocomposite containing silica nanoparticles.

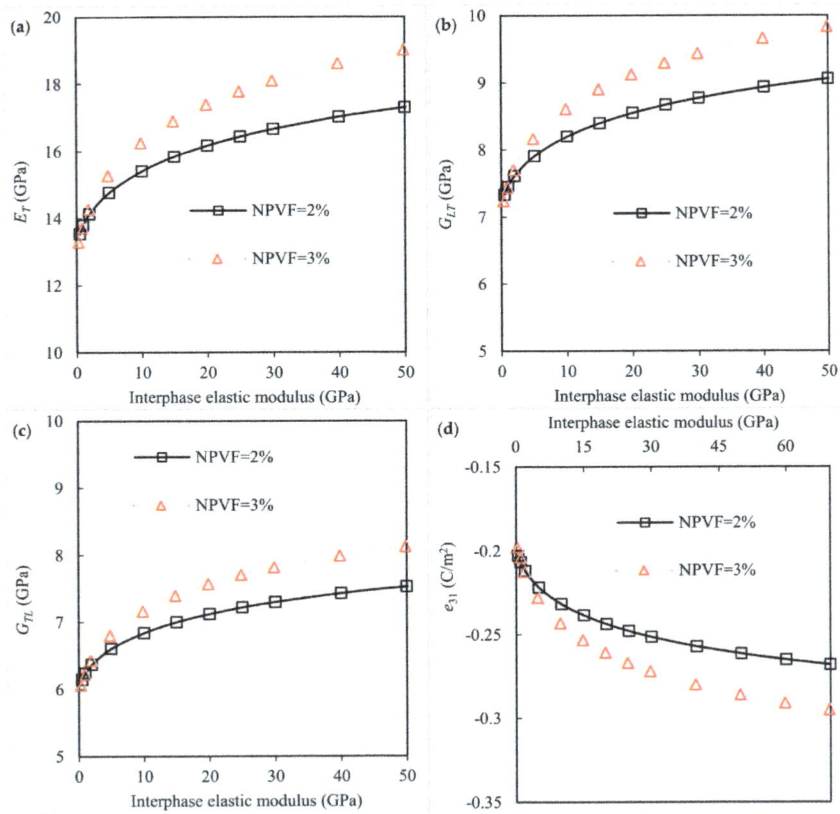

Figure 4. Variation in (**a**) transverse Young's modulus, (**b**) longitudinal shear modulus, (**c**) transverse shear modulus, and (**d**) piezoelectric coefficient e_{31} of piezoelectric fibrous nanocomposite with interphase elastic modulus.

The influence of the silica nanoparticle diameter on the mechanical properties and piezoelectric coefficients of piezoelectric fibrous nanocomposites is studied, and the results are displayed in Figure 6. As seen in Figure 6a, the values for the longitudinal Young's modulus with different nanoparticle diameters are close to each other, indicating the insignificant contribution of nanoparticle size to this elastic property. It is seen from Figure 6b that the addition of uniformly dispersed silica nanoparticles with smaller sizes results in an increase in the Young's modulus in the transverse direction. This may be explained by the interphase contribution to the final properties of nanocomposites becoming more prominent as the nanoparticle size decreases [20,48]. A notable enhancement in the shear moduli along both the longitudinal and transverse directions can be observed by the decrease in nanoparticle size, as shown in Figure 6c,d. The results of Figure 6e disclose that a smaller size of silica nanoparticles is required so as to further improve the piezoelectric coefficient e_{31} of the nanocomposites. In the case of the nanocomposite with a 3% silica-nanoparticle-containing polyimide matrix, the improvement is about 50% by decreasing the nanoparticle diameter from 100 nm to 20 nm. As $d_{NP} > 100$ nm, the change in nanoparticle diameter does not affect the elastic moduli or the piezoelectric coefficient e_{31}. The main reason for this behavior may be the reduced effect of the interphase. As the size of the nanoscale particles increases and goes to the microscale, the role of the interphase in the effective properties of the nanocomposites decreases. According to the outcomes of Figure 6f, the piezoelectric coefficient e_{33} of the nanocomposites does not depend on the nano-inclusion size since the piezoelectric fibers have the main role in this property. Thus, the hybridization of the piezoelectric fibers with smaller nanoparticles induces better elastic moduli E_T, G_{LT}, and G_{TL} and a better piezoelectric coefficient e_{31}.

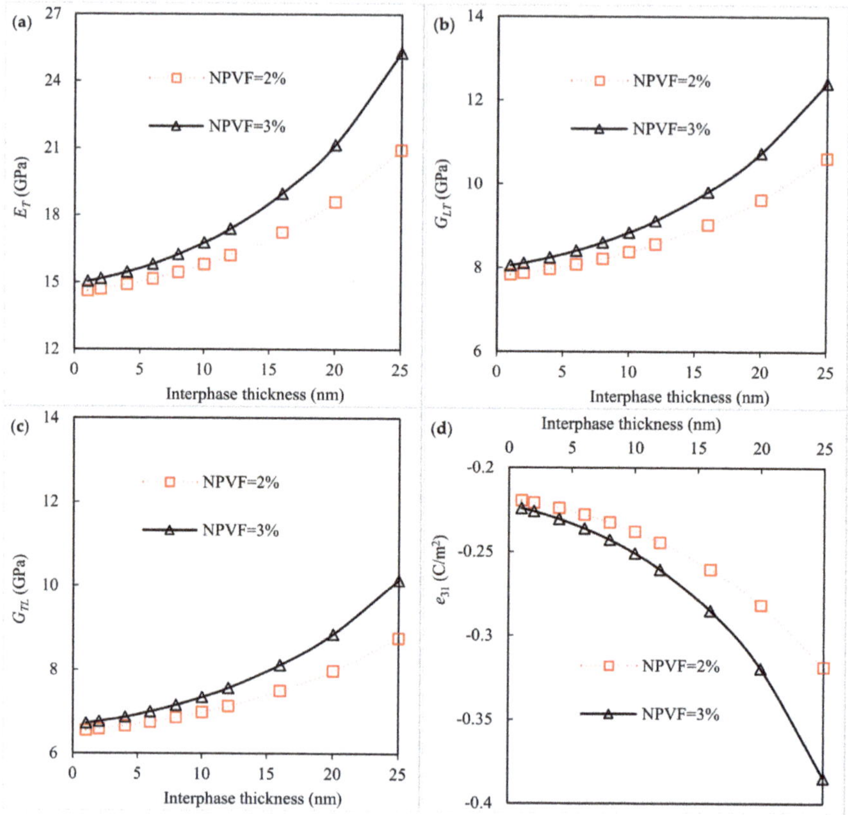

Figure 5. Variation in (**a**) transverse Young's modulus, (**b**) longitudinal shear modulus, (**c**) transverse shear modulus, and (**d**) piezoelectric coefficient e_{31} of piezoelectric fibrous nanocomposite with interphase thickness.

To study the effect of dispersion quality of silica nanoparticles, the elastic moduli and piezoelectric coefficients of the piezoelectric fibrous nanocomposites are calculated for two conditions, including a uniform dispersion and an agglomerated state ($\zeta = 0.9$, $\xi = 0.1$). The numerical results of the micromechanical analysis are presented in Figure 7a–f. In this sensitivity study, the nanoparticle volume fraction is 5%. It is observed from Figure 7a that the longitudinal Young's modulus is minimally affected by the non-uniform dispersion of the nanoparticles. The other three elastic constants, the transverse Young's modulus, longitudinal shear modulus, and transverse shear modulus, appear to significantly decrease due to the formation of silica nanoparticle agglomeration (Figure 7b–d). The agglomeration of the silica nanoparticles produces a negative effect on the piezoelectric coefficient e_{31}. As compared to the uniformly dispersed case, a decrease of about 29.2% in the piezoelectric coefficient e_{31} is observed by forming the agglomeration. As mentioned in previous studies [20,37,38,48], nanoparticle agglomeration leads to a reduction in the mechanical properties of polymer matrix nanocomposites. The nanocomposite containing uniformly dispersed silica nanoparticles exhibits a higher piezoelectric coefficient e_{31} than that containing agglomerated nanoparticles. It is shown in Figure 7f that the dispersion quality does not affect the estimated piezoelectric coefficient e_{33}. Uniformly dispersing and avoiding the agglomeration of nanoparticles into the polymer matrix are critical for advanced composite materials to take the maximum material constants, i.e., E_T, G_{LT}, G_{TL}, and e_{31}.

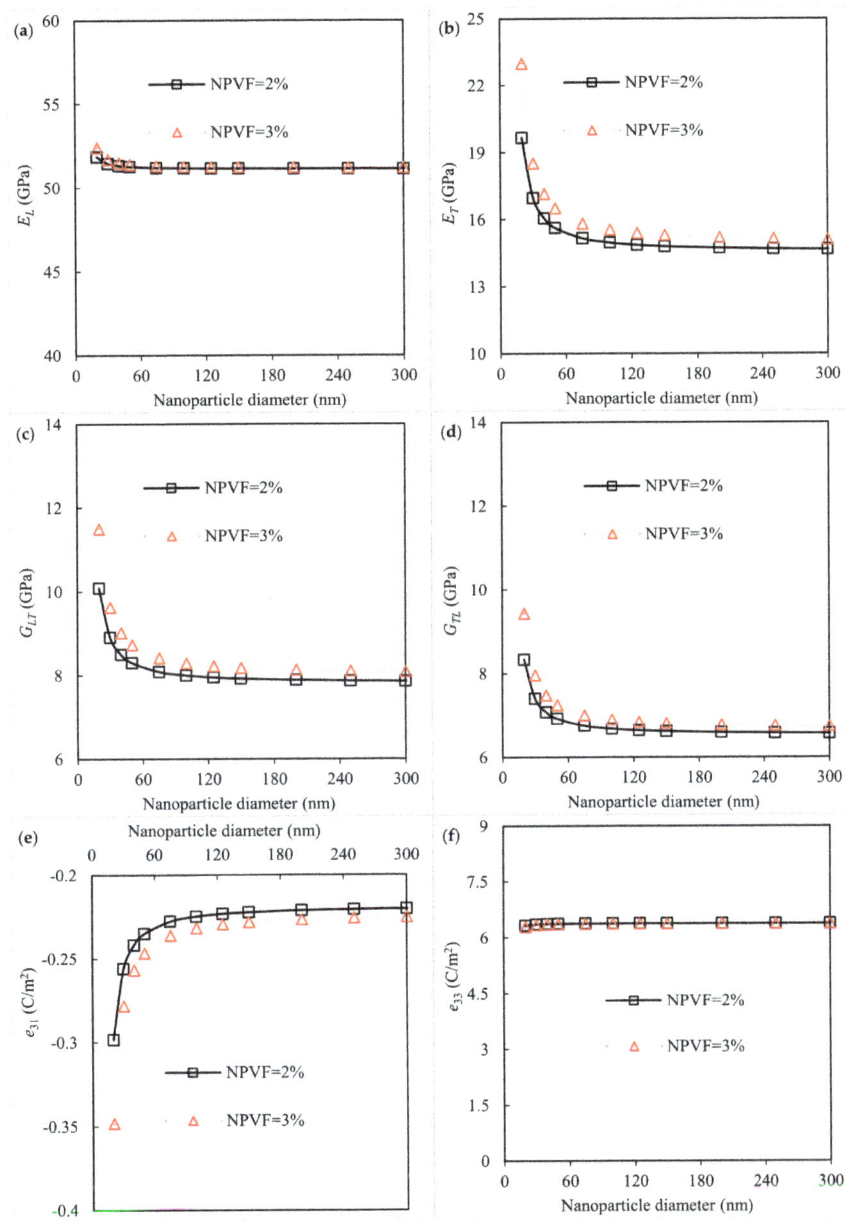

Figure 6. Influence of silica nanoparticle diameter on the (**a**) longitudinal Young's modulus, (**b**) transverse Young's modulus, (**c**) longitudinal shear modulus, (**d**) transverse shear modulus, (**e**) piezoelectric coefficient e_{31}, and (**f**) piezoelectric coefficient e_{33} of the piezoelectric fibrous nanocomposite containing silica nanoparticles.

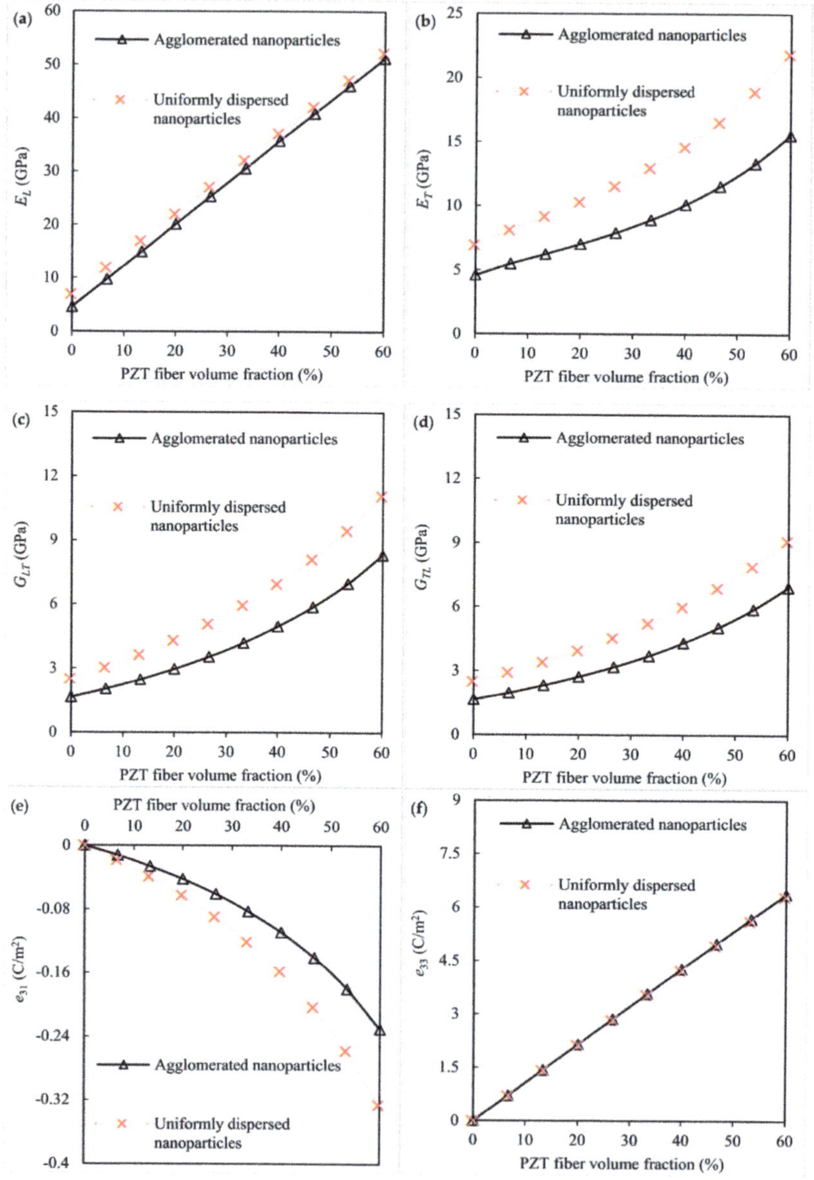

Figure 7. Influence of silica nanoparticle dispersion quality on the (**a**) longitudinal Young's modulus, (**b**) transverse Young's modulus, (**c**) longitudinal shear modulus, (**d**) transverse shear modulus, (**e**) piezoelectric coefficient e_{31}, and (**f**) piezoelectric coefficient e_{33} of the piezoelectric fibrous nanocomposite containing silica nanoparticles.

The results of the micromechanical analysis with silica and alumina nanoparticles individually incorporated into the polyimide matrix are presented in Figure 8. The volume fraction and diameter for both nanoparticles are identical. The influence of the interphase stiffness on the effective properties of the piezoelectric fibrous nanocomposites is also examined. The change in the interphase stiffness is taken in a range from the soft material to the stiff material. The soft interphase ($E_{i,soft}$) can be categorized as the material having very low stiffness in comparison with the reinforcement stiffness [49] as

$$E_{i,soft} = \frac{E_{NP} + E_m}{20} \tag{14}$$

The stiff interphase ($E_{i,stiff}$) can be categorized as the material having an average value of reinforcement and matrix stiffness and is very high in comparison with the matrix stiffness [49] as

$$E_{i,stiff} = \frac{E_{NP} + E_m}{2} \tag{15}$$

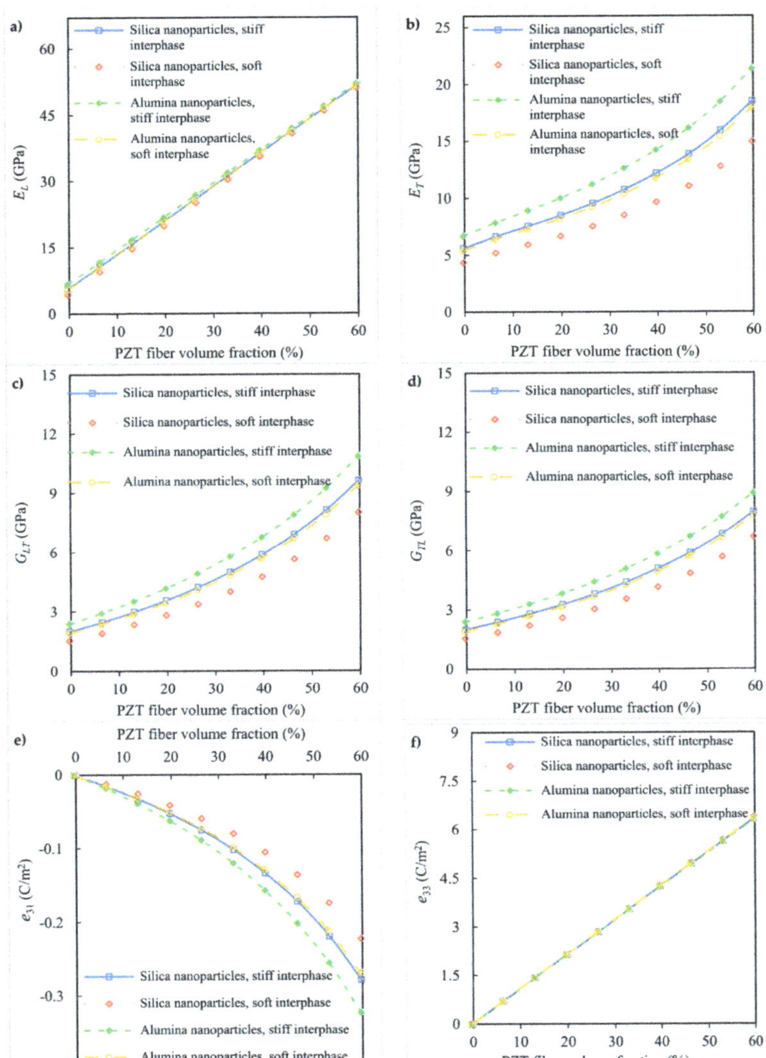

Figure 8. Influence of nanoparticle types on the (**a**) longitudinal Young's modulus, (**b**) transverse Young's modulus, (**c**) longitudinal shear modulus, (**d**) transverse shear modulus, (**e**) piezoelectric coefficient e_{31}, and (**f**) piezoelectric coefficient e_{33} of the piezoelectric fibrous nanocomposite containing nanoparticles.

On the basis of the outcomes shown in Figure 8a,f, the incorporation of different types of nano-inclusions has a negligible effect on the longitudinal Young's modulus and piezoelectric coefficient e_{33}. In contrast, changes in the nano-inclusion type embedded within the matrix affect the elastic moduli E_T, G_{LT}, and G_{TL} and the piezoelectric coefficient e_{31} of the nanocomposites, as shown in Figure 8b–e. Compared to the silica nanoparticles, the use of alumina nanoparticles in the polyimide matrix can further improve these material properties. Again, nanocomposites with a stiff interphase show higher elastic moduli E_T, G_{LT}, and G_{TL} and a higher piezoelectric coefficient e_{31} than those with a soft interphase. Due to the good mechanical and piezoelectric properties, the piezoelectric fiber–nanoparticle–polymer nanocomposites can find various industrial applications, such as in actuators, sensors, and energy-harvesting devices [9,12,26].

3.2. Comparisons with Experimental and Numerical Results

Now, the predictions are compared with the available experimental data of some silica-filled polymer composites for validating the micromechanical model. Figure 9 presents a comparison between the present predictions and experimental data [45] of the Young's modulus of silica-nanoparticle-filled poly(ether-ether-ketone) (PEEK) nanocomposites. Micromechanical tests are carried out in two different states: (1) in the presence of an interphase with $k = 5$, $t_i = 0.25 \times d_{NP}$ and (2) in the absence of an interphase. The silica nanoparticles with a mean diameter ~30 nm are uniformly dispersed in the polymer matrix. The Young's moduli of the silica nanoparticle and PEEK matrix are 73 GPa and 3.9 GPa, respectively. Figure 9 shows that the model without the interphase agrees better with the experiments for lower values of nanoparticle volume fraction. However, at higher nanoparticle contents, the model predictions with the interfacial region between the nanoparticles and polymer matrix give a more reasonable agreement as compared to the experiments [45]. The interphase region with a higher Young's modulus than that of the matrix material increases the Young's modulus of the nanocomposites significantly.

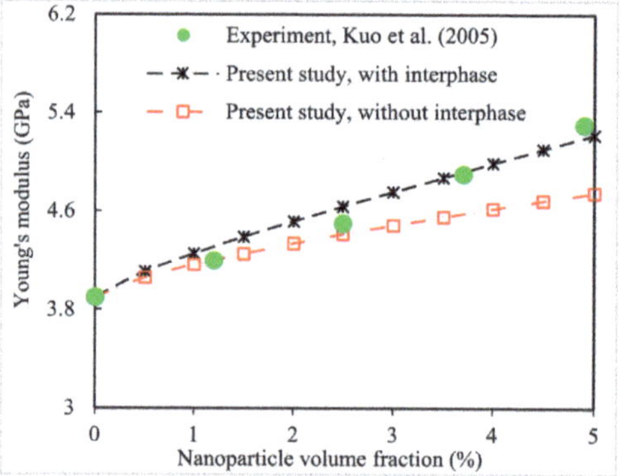

Figure 9. Present predictions for Young's modulus of silica-nanoparticle-filled PEEK nanocomposites compared to experimental data [45].

The Young's modulus of the silica-nanoparticle-filled nylon-6 nanocomposites determined by the present micromechanics method are compared with the experimental data [46]. Figure 10 shows the outcome of this comparison. The effect of considering the interphase in the micromechanical modeling on the final elastic modulus is also examined. It is observed that the two sets of results evaluated by the micromechanical model by taking the interphase region and the experimental route are in a good agreement.

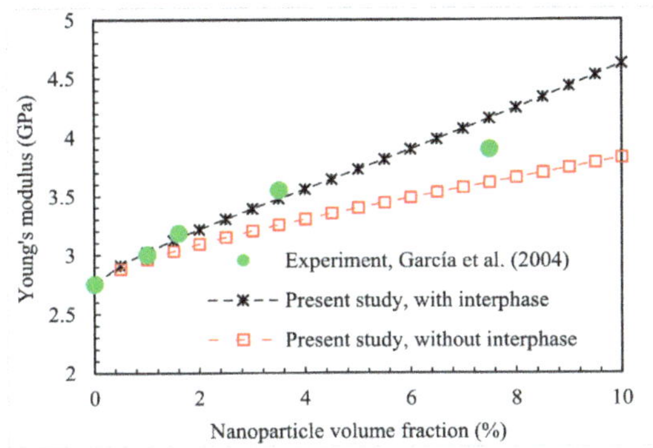

Figure 10. Present predictions for Young's modulus of silica-nanoparticle-filled nylon-6 nanocomposites compared to experimental data [46].

In another comparison, two effective properties of the PZT5-fiber-reinforced epoxy composite including the elastic constant C_{33} and piezoelectric coefficient e_{31} predicted by the present micromechanical model are compared with the Mori–Tanaka predictions carried out in [9]. The material constants of the PZT5 fiber as well as the epoxy are given in Table 1 [9,44]. It is shown in Figure 11a,b that the two sets of results are in a very good agreement for both effective constants.

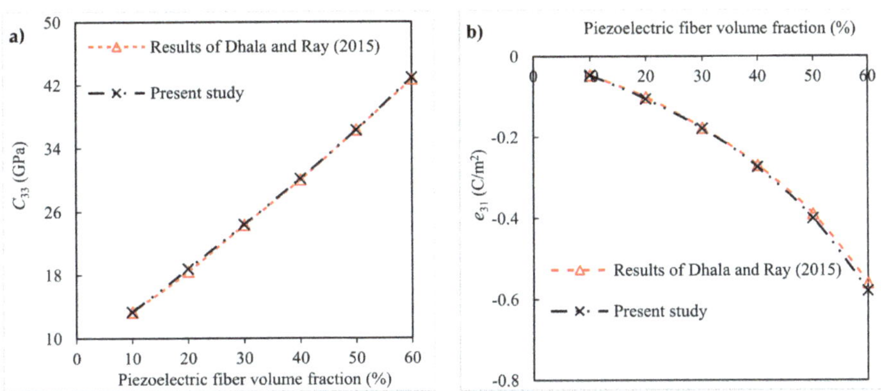

Figure 11. Comparison between the results of the present model and results of [9] for (**a**) elastic constant C_{33} and (**b**) piezoelectric coefficient e_{31} of PZT5-fiber-reinforced epoxy composites.

Wang et al. [50] used silica nanoparticles to produce a polymer nanocomposite: methyl methacrylate (MMA) was chosen as the matrix and copolymerized with a low amount of cationic functional comonomer 2-(methacryloyloxy)ethyltrimethylammonium chloride (MTC). Figure 12 shows another comparison between the present predictions and experimental measurements [50] for the Young's modulus of silica-nanoparticle-filled P(MMA-co-MTC) nanocomposites. Silica nanoparticles have an average diameter of around 20 nm [50]. A good agreement is observed between the model predictions and the experimental measurements at all nanoparticle contents.

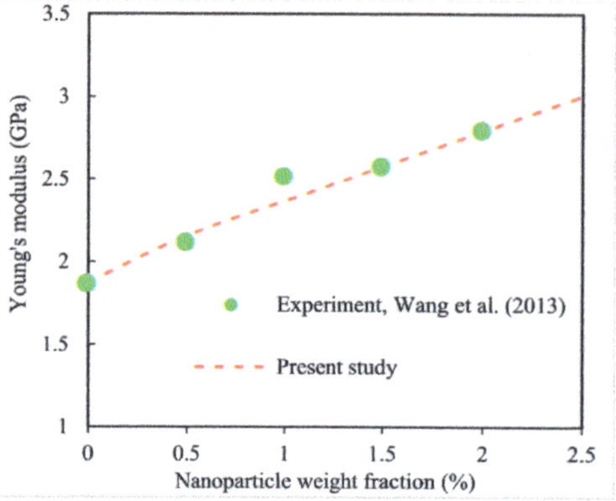

Figure 12. Present predictions for Young's modulus of silica-nanoparticle-filled P(MMA-co-MTC) nanocomposites compared to experimental data [50].

4. Conclusions

In this paper, the piezoelectro-elastic coefficients of PZT-7A-fiber-reinforced nanocomposites with a silica-nanoparticle-filled polyimide matrix were evaluated. First, the Ji and Mori–Tanaka models were hierarchically employed to predict the elastic properties of the silica-nanoparticle-filled polymer. The nanoparticle–polymer interphase and the nanoparticle agglomeration were included in the analysis. Then, considering the nanoparticle-filled polymer as the matrix and the piezoelectric fiber as the reinforcement, the Mori–Tanaka model was employed to predict the elastic and piezoelectric constants of the piezoelectric fibrous nanocomposites. A good agreement was observed between the present predictions and other results available in the literature. The results showed that adding silica nanoparticles into the polyimide matrix improves the elastic and piezoelectric properties (E_T, G_{LT}, G_{TL}, and e_{31}) of the piezoelectric fibrous nanocomposites. As compared to the composite without nanoparticles, 39%, 31.8%, and 37% improvements in the values of E_T, G_{TL}, and the piezoelectric coefficient e_{31} were observed once the volume fractions of the fiber and nanoparticle were 60% and 3%, respectively. More improvement in the elastic moduli E_T, G_{LT}, and G_{TL} and the piezoelectric coefficient e_{31} was found by decreasing the nanoparticle diameter. A thicker and stiffer interphase led to an increase in the elastic moduli E_T, G_{LT}, and G_{TL} and the piezoelectric coefficient e_{31} of the piezoelectric fibrous nanocomposites. However, the nanoparticle agglomeration that formed in the polymer matrix decreased the elastic moduli E_T, G_{LT}, and G_{TL} and the piezoelectric coefficient e_{31}. It was observed that increasing the piezoelectric fiber volume fraction increased the piezoelectro-elastic constants of the piezoelectric fibrous nanocomposites.

Author Contributions: Conceptualization, U.U.; methodology, U.U.; software, U.U., S.H.M. and M.K.A.; validation, U.U.; formal analysis, M.H.A.; investigation, S.H.M.; resources, S.H.M. and F.A.; writing—original draft, U.U.; writing—review and editing, S.H.M. and M.K.A.; visualization, M.H.A.; supervision, F.A. and M.K.A.; project administration, M.H.A., F.A. and M.K.A.; funding acquisition, F.A. All authors have read and agreed to the published version of the manuscript.

Funding: This research was funded through the Research Institute Supporting Program (RICSP-24-2), King Saud University, Riyadh, Saudi Arabia.

Institutional Review Board Statement: Not applicable.

Data Availability Statement: The original contributions presented in the study are included in the article, further inquiries can be directed to the corresponding author.

Acknowledgments: The authors extend their appreciation to the King Saud University for funding this work through the Research Institute Supporting Program (RICSP-24-2), King Saud University, Riyadh, Saudi Arabia.

Conflicts of Interest: The authors declare no conflict of interest.

References

1. Xu, L.F.; Yu, T.C.; Feng, X.; Yang, C.P.; Chen, Y.; Chen, W.; Zhou, J. Dimension dependence of thickness resonance behavior of piezoelectric fiber composites. *Mater. Chem. Phys.* **2018**, *218*, 34–38. [CrossRef]
2. Tan, D.; Yavarow, P.; Erturk, A. Nonlinear elastodynamics of piezoelectric macro-fiber composites with interdigitated electrodes for resonant actuation. *Compos. Struct.* **2018**, *187*, 137–143. [CrossRef]
3. Zhou, B.; Ma, X.; Xue, S. Nonlinear analysis of laminated beams with braided fiber piezoelectric composite actuators. *Int. J. Appl. Mech.* **2020**, *12*, 2050043. [CrossRef]
4. Su, Y.; Li, W.; Yuan, L.; Chen, C.; Pan, H.; Xie, G.; Conta, G.; Ferrier, S.; Zhao, X.; Chen, G.; et al. Piezoelectric fiber composites with polydopamine interfacial layer for self-powered wearable biomonitoring. *Nano Energy* **2021**, *89*, 106321. [CrossRef]
5. He, Z.; Liu, J.; Chen, Q. Higher-order asymptotic homogenization for piezoelectric composites. *Int. J. Solids Struct.* **2023**, *264*, 112092. [CrossRef]
6. Yu, Y.; Narita, F. Evaluation of electromechanical properties and conversion efficiency of piezoelectric nanocomposites with carbon-fiber-reinforced polymer electrodes for stress sensing and energy harvesting. *Polymers* **2021**, *13*, 3184. [CrossRef]
7. Tang, T.; Yu, W. Variational asymptotic micromechanics modeling of heterogeneous piezoelectric materials. *Mech. Mater.* **2008**, *40*, 812–824. [CrossRef]
8. Keramati, Y.; Ansari, R.; Hassanzadeh-Aghdam, M.K. Effect of graphene nano-sheets on the elastic and piezoelectric coefficients of unidirectional PZT-7A/polyimide hybrid composites. *J. Intell. Mater. Syst. Struct.* **2023**, *34*, 1548–1560. [CrossRef]
9. Dhala, S.; Ray, M.C. Micromechanics of piezoelectric fuzzy fiber-reinforced composite. *Mech. Mater.* **2015**, *81*, 1–17. [CrossRef]
10. Al Mahmud, H.; Radue, M.S.; Chinkanjanarot, S.; Pisani, W.A.; Gowtham, S.; Odegard, G.M. Multiscale modeling of carbon fiber-graphene nanoplatelet-epoxy hybrid composites using a reactive force field. *Compos. Part B Eng.* **2019**, *172*, 628–635. [CrossRef]
11. Sahu, R.; Ponnusami, S.A.; Weimer, C.; Harursampath, D. Interface engineering of carbon fiber composites using CNT: A review. *Polym. Compos.* **2024**, *45*, 9–42. [CrossRef]
12. Keramati, Y.; Ansari, R.; Hassanzadeh-Aghdam, M.K.; Umer, U. Micromechanical simulation of thermal expansion, elastic stiffness and piezoelectric constants of graphene/unidirectional $BaTiO_3$ fiber reinforced epoxy hybrid nanocomposites. *Acta Mech.* **2023**, *234*, 6251–6270. [CrossRef]
13. Cui, X.; Liu, J.; Liu, H.; Wu, G. Enhanced interfacial strength and mechanical properties of carbon fiber composites by introducing functionalized silica nanoparticles into the interface. *J. Adhes. Sci. Technol.* **2019**, *33*, 479–492. [CrossRef]
14. Hwayyin, R.N.; Hussien, S.K.; Ameed, A.S. The effect of nano-silica on the mechanical properties of composite polyester/carbon fibers. *J. Mech. Eng. Sci.* **2022**, *16*, 9175–9186. [CrossRef]
15. Zheng, Y.; Ning, R.; Zheng, Y. Study of SiO_2 nanoparticles on the improved performance of epoxy and fiber composites. *J. Reinf. Plast. Compos.* **2005**, *24*, 223–233. [CrossRef]
16. Gang, D.; Chilan, C.; Haobin, T.; Zhenhua, L.; Dingzhong, Z.; Kang, Q. The research on the effect of SiO_2 and CF on the tensile and tribological properties of PI composite. *Proc. Inst. Mech. Eng. Part J J. Eng. Tribol.* **2015**, *229*, 1513–1518. [CrossRef]
17. Tang, Y.; Ye, L.; Zhang, D.; Deng, S. Characterization of transverse tensile, interlaminar shear and interlaminate fracture in CF/EP laminates with 10 wt% and 20 wt% silica nanoparticles in matrix resins. *Compos. Part A Appl. Sci. Manuf.* **2011**, *42*, 1943–1950. [CrossRef]
18. Hasanzadeh, M.; Ansari, R.; Hassanzadeh-Aghdam, M.K. Evaluation of effective properties of piezoelectric hybrid composites containing carbon nanotubes. *Mech. Mater.* **2019**, *129*, 63–79. [CrossRef]
19. Godara, S.S.; Mahato, P.K. Micromechanical technique based prediction of effective properties for hybrid smart nanocomposites. *Mech. Adv. Mater. Struct.* **2022**, *29*, 2065–2073. [CrossRef]
20. Boutaleb, S.; Zaïri, F.; Mesbah, A.; Nait-Abdelaziz, M.; Gloaguen, J.M.; Boukharouba, T.; Lefebvre, J.M. Micromechanics-based modelling of stiffness and yield stress for silica/polymer nanocomposites. *Int. J. Solids Struct.* **2009**, *46*, 1716–1726. [CrossRef]
21. Kontou, E.; Christopoulos, A.; Koralli, P.; Mouzakis, D.E. The effect of silica particle size on the mechanical enhancement of polymer nanocomposites. *Nanomaterials* **2023**, *13*, 1095. [CrossRef]
22. Sun, L.; Gibson, R.F.; Gordaninejad, F. Multiscale analysis of stiffness and fracture of nanoparticle-reinforced composites using micromechanics and global–local finite element models. *Eng. Fract. Mech.* **2011**, *78*, 2645–2662. [CrossRef]
23. Ray, M.C. Micromechanics of piezoelectric composites with improved effective piezoelectric constant. *Int. J. Mech. Mater. Des.* **2006**, *3*, 361–371. [CrossRef]
24. Dinzart, F.; Sabar, H. Magneto-electro-elastic coated inclusion problem and its application to magnetic-piezoelectric composite materials. *Int. J. Solids Struct.* **2011**, *48*, 2393–2401. [CrossRef]
25. Koutsawa, Y. Overall thermo-magneto-electro-elastic properties of multiferroics composite materials with arbitrary heterogeneities spatial distributions. *Compos. Struct.* **2015**, *133*, 764–773. [CrossRef]
26. Ferreira, P.M.; Machado, M.A.; Vidal, C.; Carvalho, M.S. Modelling electro-mechanical behaviour in piezoelectric composites: Current status and perspectives on homogenisation. *Adv. Eng. Softw.* **2024**, *193*, 103651. [CrossRef]
27. Mishra, N.; Das, K. A Mori–Tanaka based micromechanical model for predicting the effective electroelastic properties of orthotropic piezoelectric composites with spherical inclusions. *SN Appl. Sci.* **2020**, *2*, 1206. [CrossRef]

28. Pan, J.; Bian, L. A physics investigation for influence of carbon nanotube agglomeration on thermal properties of composites. *Mater. Chem. Phys.* **2019**, *236*, 121777. [CrossRef]
29. Mortazavi, B.; Bardon, J.; Ahzi, S. Interphase effect on the elastic and thermal conductivity response of polymer nanocomposite materials: 3D finite element study. *Comput. Mater. Sci.* **2013**, *69*, 100–106. [CrossRef]
30. Ghasemi, A.R.; Hosseinpour, K. The SWCNTs roles in stress/strain distribution of three-phase multilayered nanocomposite cylinder under combined internal pressure and thermo-mechanical loading. *J. Braz. Soc. Mech. Sci. Eng.* **2018**, *40*, 391. [CrossRef]
31. Cannillo, V.; Bondioli, F.; Lusvarghi, L.; Montorsi, M.; Avella, M.; Errico, M.E.; Malinconico, M. Modeling of ceramic particles filled polymer–matrix nanocomposites. *Compos. Sci. Technol.* **2006**, *66*, 1030–1037. [CrossRef]
32. Snipes, J.S.; Robinson, C.T.; Baxter, S.C. Effects of scale and interface on the three-dimensional micromechanics of polymer nanocomposites. *J. Compos. Mater.* **2011**, *45*, 2537–2546. [CrossRef]
33. Nematollahi, H.; Mohammadi, M.; Munir, M.T.; Zare, Y.; Rhee, K.Y. Two-Step Method for Predicting Young's Modulus of Nanocomposites Containing Nanodiamond Particles. *J. Mater. Res. Technol.* **2024**, *33*, 2343–2352. [CrossRef]
34. Pakseresht, M.; Ansari, R.; Hassanzadeh-Aghdam, M.K. An efficient homogenization scheme for analyzing the elastic properties of hybrid nanocomposites filled with multiscale particles. *J. Braz. Soc. Mech. Sci. Eng.* **2021**, *43*, 3. [CrossRef]
35. Ji, X.L.; Jing, J.K.; Jiang, W.; Jiang, B.Z. Tensile modulus of polymer nanocomposites. *Polym. Eng. Sci.* **2002**, *42*, 983–993. [CrossRef]
36. Zare, Y. Assumption of interphase properties in classical Christensen–Lo model for Young's modulus of polymer nanocomposites reinforced with spherical nanoparticles. *RSC Adv.* **2015**, *5*, 95532–95538. [CrossRef]
37. Zhang, H.; Zhang, Z.; Friedrich, K.; Eger, C. Property improvements of in situ epoxy nanocomposites with reduced interparticle distance at high nanosilica content. *Acta Mater.* **2006**, *54*, 1833–1842. [CrossRef]
38. Naito, K. Tensile properties of polyacrylonitrile-and pitch-based hybrid carbon fiber/polyimide composites with some nanoparticles in the matrix. *J. Mater. Sci.* **2013**, *48*, 4163–4176. [CrossRef]
39. Ji, X.Y.; Cao, Y.P.; Feng, X.Q. Micromechanics prediction of the effective elastic moduli of graphene sheet-reinforced polymer nanocomposites. *Model. Simul. Mater. Sci. Eng.* **2010**, *18*, 045005. [CrossRef]
40. Dastgerdi, J.N.; Marquis, G.; Salimi, M. The effect of nanotubes waviness on mechanical properties of CNT/SMP composites. *Compos. Sci. Technol.* **2013**, *86*, 164–169. [CrossRef]
41. Shi, D.L.; Feng, X.Q.; Huang, Y.Y.; Hwang, K.C.; Gao, H. The effect of nanotube waviness and agglomeration on the elastic property of carbon nanotube-reinforced composites. *J. Eng. Mater. Technol.* **2004**, *126*, 250–257. [CrossRef]
42. Mallik, N.; Ray, M.C. Effective coefficients of piezoelectric fiber-reinforced composites. *AIAA J.* **2003**, *41*, 704–710. [CrossRef]
43. Aboudi, J.; Arnold, S.M.; Bednarcyk, B.A. *Micromechanics of Composite Materials: A Generalized Multiscale Analysis Approach*; Butterworth-Heinemann: Oxford, UK, 2013.
44. Odegard, G.M. Constitutive modeling of piezoelectric polymer composites. *Acta Mater.* **2004**, *52*, 5315–5330. [CrossRef]
45. Kuo, M.C.; Tsai, C.M.; Huang, J.C.; Chen, M. PEEK composites reinforced by nano-sized SiO_2 and Al_2O_3 particulates. *Mater. Chem. Phys.* **2005**, *90*, 185–195. [CrossRef]
46. García, M.; van Vliet, G.; ten Cate, M.G.; Chavez, F.; Norder, B.; Kooi, B.; van Zyl, W.E.; Verweij, H.; Blank, D.H.A. Large-scale extrusion processing and characterization of hybrid nylon-6/SiO_2 nanocomposites. *Polym. Adv. Technol.* **2004**, *15*, 164–172. [CrossRef]
47. Ray, M.C. The concept of a novel hybrid smart composite reinforced with radially aligned zigzag carbon nanotubes on piezoelectric fibers. *Smart Mater. Struct.* **2010**, *19*, 035008. [CrossRef]
48. Mahmoodi, M.J.; Hassanzadeh-Aghdam, M.K.; Safi, M. Effects of nano-sized silica particles on the off-axis creep performance of unidirectional fiber-reinforced polymer hybrid composites. *J. Compos. Mater.* **2021**, *55*, 1575–1589. [CrossRef]
49. Joshi, P.; Upadhyay, S.H. Effect of interphase on elastic behavior of multiwalled carbon nanotube reinforced composite. *Comput. Mater. Sci.* **2014**, *87*, 267–273. [CrossRef]
50. Wang, X.; Wang, L.; Su, Q.; Zheng, J. Use of unmodified SiO_2 as nanofiller to improve mechanical properties of polymer-based nanocomposites. *Compos. Sci. Technol.* **2013**, *89*, 52–60. [CrossRef]

polymers

MDPI

Article

A Multi-Phase Analytical Model for Effective Electrical Conductivity of Polymer Matrix Composites Containing Micro-SiC Whiskers and Nano-Carbon Black Hybrids

Usama Umer [1,*], Mustufa Haider Abidi [1], Zeyad Almutairi [1,2] and Mohamed K. Aboudaif [1]

1 Advanced Manufacturing Institute, King Saud University, P.O. Box 800, Riyadh 11421, Saudi Arabia
2 Mechanical Engineering Department, College of Engineering, King Saud University, P.O. Box 800, Riyadh 11421, Saudi Arabia
* Correspondence: uumer@ksu.edu.sa

Abstract: Multifunctional polymer composites containing micro/nano hybrid reinforcements have attracted intensive attention in the field of materials science and engineering. This paper develops a multi-phase analytical model for investigating the effective electrical conductivity of micro-silicon carbide (SiC) whisker/nano-carbon black (CB) polymer composites. First, CB nanoparticles are dispersed within the non-conducting epoxy to achieve a conductive CB-filled nanocomposite and its electrical conductivity is predicted. Some critical microstructures such as volume percentage and size of nanoparticles, and interphase characteristics surrounding the CB are micromechanically captured. Next, the electrical conductivity of randomly oriented SiC-containing composites in which the nanocomposite and whisker are considered as the matrix and reinforcement phases, respectively, is estimated. Influences of whisker aspect ratio and volume fraction on the effective electrical conductivity of the SiC/CB-containing polymer composites are explored. Some comparison studies are performed to validate the accuracy of the model. It is observed before the percolation threshold that the addition of nanoparticles with a uniform dispersion can improve the electrical conductivity of the polymer composites containing SiC/CB hybrids. Moreover, the results show that the electrical conductivity is more enhanced by the decrease in nanoparticle size. Interestingly, the composite percolation threshold is significantly reduced when SiC whiskers with a higher aspect ratio are added. This work will be favorable for the design of electro-conductive polymer composites with high performances.

Keywords: electrical property; polymer composite; hybrid fillers; aspect ratio; multi-phase model

1. Introduction

Polymer composites containing micro/nano hybrid reinforcements have increasingly attracted attention in materials science, particularly in applications requiring both lightweight and high electrical conductivity [1,2]. Polymer composite materials, despite their widespread use in structural applications, suffer from inherently low electrical conductivity, particularly in non-conductive polymer matrices ($\sim 10^{-15}$–10^{-9} S/m), which limits their functionality in aerospace and electronics industries [1,2]. This necessitates the addition of conductive nanofillers to significantly enhance electrical conductivity while retaining the mechanical advantages of the composite structure [3,4].

Silicon carbide whiskers and carbon black nanoparticles are emerging as effective hybrid fillers due to their potential to enhance electrical properties, making them ideal for advanced applications in electronics, energy storage, and aerospace. Zhang et al. fabricated

 https://doi.org/10.3390/polym17020128

a composite with copper matrix and SiC whiskers as the reinforcing phase by means of the co-electrodeposition technology [5]. They concluded that the manufactured composite displays improved shear strength and enhanced thermal conductivity. Fan and colleagues presented a comprehensive review of the application of carbon black as the reinforcing phase [6]. First, the physical and chemical characteristics of CB were characterized in conjunction with the mixing procedure, and afterward, eight reinforcing mechanisms of CB were listed and summarized. The inclusion of CB nanoparticles into a non-conductive polymer matrix can transform the material into a conductive composite, primarily through the formation of conductive networks at the percolation threshold [7,8]. However, a critical challenge with CB is that achieving percolation often requires a high volume fraction of CB, which can compromise the mechanical integrity of the composite [9]. Mazaheri et al. scrutinized the electrical conductivity and percolation characteristics of polymer nanocomposites (PNCs) consisting of CB spherical particles as the reinforcing phase [10]. Their analytical modeling considered the influence of interphase region, quantum electron tunneling, volume fraction, radius, and filler/interphase/matrix conductivities. Despite the high percolation threshold, CB remains a popular choice due to its cost-effectiveness and superior electrical properties compared to other nanofillers such as carbon nanotubes (CNTs) and graphene nanoplatelets (GNPs). CB exhibits a rapid increase in electrical conductivity near the percolation point, primarily driven by electron tunneling between particles [11,12]. However, the need for large amounts of CB can result in filler aggregation and reduced processability, limiting its performance in polymer nanocomposites [13,14].

To address these limitations, hybrid systems combining micro- and nano-sized fillers have been explored to optimize electrical performance while reducing filler content. Kim et al. proposed an analytical method for composites containing multiple heterogeneities with conductive coated layers to estimate the percolation threshold influence, the tunneling influence by hard/soft core concept as well as the electrical conductivity of polymeric composites with ellipsoidal fillers [12]. Saberi et al. explored the electrical conductivity of short carbon fiber-reinforced nanocomposites enhanced with GNPs [15]. To this end, a two-phase micromechanical modeling method was used to predict the effective electrical conductivity of composites. They concluded that the hybrid nanocomposite displayed better conductivity by increasing the volume fraction and aspect ratio of the GNP and reducing its thickness. They also deduced that increasing the carbon fiber aspect ratio results in higher conductivity. Previously, an analytical modeling approach was proposed to investigate the electrical conductivity of hybrid CNT/CB/polymer nanocomposites [16]. The model incorporated cylindrical nanotubes and spherical CB fillers covered by an interphase region serving as an electron hopping duct. Various materials may be used as the reinforcing phase to enhance the electrical conductivity of polymeric composites which require further investigation. In particular, SiC whiskers, known for their high aspect ratio, mechanical strength, and good electrical properties, may provide a complementary scale to CB, enhancing the formation of multi-scale conductive networks [17]. This synergistic interaction between SiC whiskers and CB nanoparticles may significantly lower the percolation threshold while improving electrical conductivity, as the whiskers serve as bridges within the composite, facilitating more efficient electron transport [18–20].

In this work, an analytical micromechanics method with some important microstructures for predicting the effective electrical conductivity of polymer composites reinforced with micro-SiC whiskers and CB nanoparticles is proposed which, to date, has not been studied. This model captures critical microstructural parameters, including the volume fraction and size of the CB nanoparticles, the interphase characteristics, and the aspect ratio of the SiC whiskers. Initially, the electrical conductivity of the CB-filled nanocomposite is predicted, followed by an analysis of the hybrid SiC/CB system, where the SiC whiskers act as reinforcement within the conductive nanocomposite matrix. The percolation behavior and the influence of whisker geometry and content are explored using a micromechanics-based approach.

2. Analytical Modeling

The analytical procedure to estimate the equivalent electrical conductivity of the polymer composite containing SiC/CB hybrids starts with the estimation of the electrical conductivity of the polymeric nanocomposite material containing CB nanoparticles a priori. For this purpose, incorporating the interaction between the nanoparticles and the polymer may be a significant issue [12–16]. An equivalent solid continuum interphase region is captured among the carbon-based nanoparticles and the polymer characterizing the interaction. Subsequently, incorporating the polymeric nanocomposite as the matrix phase and the SiC whisker as the reinforcements, the electrical conductivity of the hybrid composite can be calculated. Figure 1a presents a schematic representation of the initial configuration of the composite, depicted within a global XYZ coordinate system. The figure illustrates the dispersion of CB nanoparticles, represented as red spheres, within a non-conducting polymer matrix, shown in a purple hue. The coordinate axes are labeled with the X, Y, and Z directions indicating the spatial arrangement of the representative volume element (RVE).

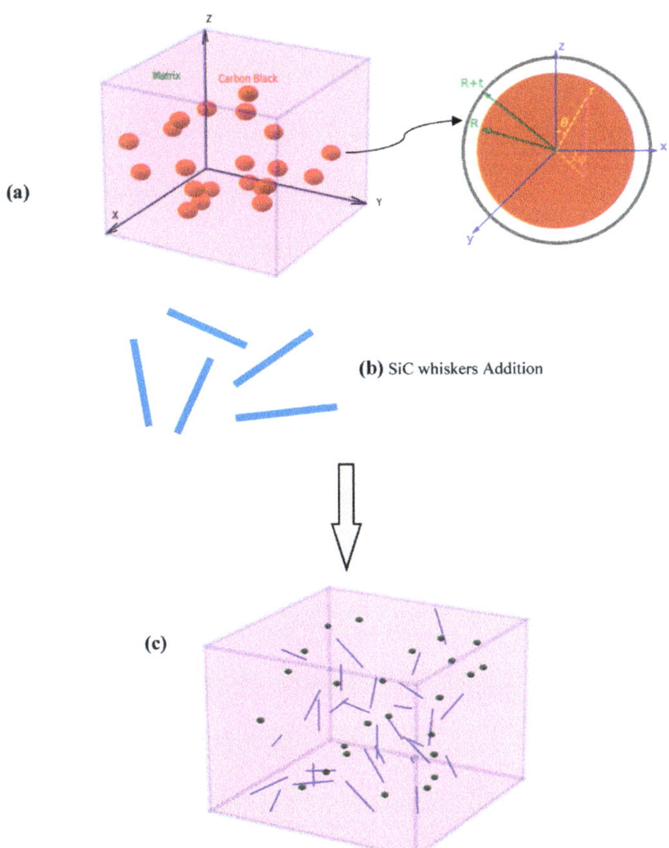

Figure 1. (**a**) RVE representing the random dispersion of CB nanoparticles, (**b**) SiC whiskers and (**c**) RVE for the CB/SiC/polymer ternary hybrid composite.

Additionally, the inset diagram highlights the geometric parameters associated with the CB particles, including their radius R and the thickness of the surrounding interphase t, which collectively contribute to the effective electrical properties of the composite. Figure 1b shows the addition of SiC whiskers. Subsequently, Figure 1c illustrates the final configuration of the hybrid composite, where the SiC whiskers, represented as blue lines, are randomly oriented and dispersed throughout the CB-filled polymer matrix. This arrangement is crucial for understanding the interactions between the micro and nano reinforcements and their collective impact on the electrical conductivity of the hybrid composite. The overarching goal of this representation is to facilitate a comprehensive analysis of the structural and electrical characteristics of the CB/SiC hybrid composite, ultimately leading to enhanced performance in electro-conductive applications.

2.1. *Effective Electrical Conductivity of CB-Reinforced Nanocomposites*

The initial step to determine the electrical conductivity of the polymeric nanocomposite depicted in Figure 1a is the calculation of the electrical conductivity of the effective CB nanofiller consisted of CB nanoparticle and interphase region. Then, the integration of CB nanofiller with the polymer matrix is taken into account accordingly.

2.1.1. Electrical Conductivity of Effective CB Nanofillers

The electrical potential and conductivity of effective spherical CB nanoparticles under an applied electric field can be derived by solving Laplace's equation in spherical coordinates with appropriate boundary conditions. The equation governing the electric potential Φ in spherical coordinates is given as follows [10,16,21]:

$$\nabla^2\Phi = \frac{1}{r^2}\frac{\partial}{\partial r}\left(r^2\frac{\partial\Phi}{\partial r}\right) + \frac{1}{r^2 sin\theta}\frac{\partial}{\partial\theta}\left(sin\theta\frac{\partial\Phi}{\partial\theta}\right) = 0 \tag{1}$$

where Φ depends on the radial distance r and the polar angle θ. To solve for the electric potential, boundary conditions are applied at the interfaces between the CB core, the surrounding interphase layer, and the polymer matrix. Denoting the electric potentials in the CB core, the interphase, and the matrix as Φ_{CB}, Φ_{int}, and Φ_m, respectively, the boundary conditions at R and $R+t$ are as follows [10,16]:

$$\Phi_{CB}|_{r=R} = \Phi_{int}|_{r=R} \tag{2}$$

$$-\sigma_{CB}\frac{\partial\Phi_{CB}}{\partial r}|_{r=R} = -\sigma_{int}\frac{\partial\Phi_{int}}{\partial r}|_{r=R} \tag{3}$$

$$\Phi_{int}|_{r=R+t} = \Phi_m|_{r=R+t} \tag{4}$$

$$-\sigma_{int}\frac{\partial\Phi_{CB}}{\partial r}|_{r=R+t} = -\sigma_m\frac{\partial\Phi_{int}}{\partial r}|_{r=R+t} \tag{5}$$

where σ_{CB}, σ_{int}, σ_m denote the electrical conductivities of the CB core, the interphase, and the polymer matrix, sequentially. The azimuthal symmetry of the system implies that the potential solution is independent of the azimuthal angle, reducing Laplace's equation to the following general solution for each region [10,16,22]:

$$\Phi_{CB}(r,\theta) = \sum_{n=0}^{\infty}(A_n r^n + \frac{B_n}{r^{n+1}})P_n(cos\theta), \quad r \le R \tag{6}$$

$$\Phi_{int}(r,\theta) = \sum_{n=0}^{\infty}(A'_n r^n + \frac{B'_n}{r^{n+1}})P_n(cos\theta), \quad R \le r \le R+t \tag{7}$$

$$\Phi_m(r,\theta) = \sum_{n=0}^{\infty} (A''_n r^n + \frac{B''_n}{r^{n+1}}) P_n(cos\theta), \ \ r \geq R+t \tag{8}$$

By applying the boundary conditions, the coefficients are determined, and the electric potential simplifies to the first-order terms (since higher-order terms vanish). Thus, the electric potentials in the three regions become the following:

$$\Phi_{CB}(r,\theta) = A_1 r cos\theta, \ \ r \leq R \tag{9}$$

$$\Phi_{int}(r,\theta) = (A'_1 r + \frac{B'_1}{r^2}) cos\theta, \ \ R \leq r \leq R+t \tag{10}$$

$$\Phi_m(r,\theta) = \left(-E_0 r + \frac{B''_1}{r^2}\right) cos\theta, \ \ r \geq R+t \tag{11}$$

The coefficients of the above equation are obtained from the boundary conditions as follows:

$$A_1 = \frac{-9\sigma_{int}\sigma_m E_0}{(\sigma_{int}+2\sigma_m)(2\sigma_{int}+\sigma_{CB}) - \frac{2R^3}{(R+t)^3}(\sigma_{int}-\sigma_m)(\sigma_{int}-\sigma_{CB})} \tag{12}$$

$$A'_1 = \frac{2\sigma_{int}+\sigma_{CB}}{3\sigma_{int}} A_1 \tag{13}$$

$$B'_1 = \frac{\sigma_{int}-\sigma_{CB}}{3\sigma_{int}} R^3 A_1 \tag{14}$$

$$B''_1 = (R+t)^3 (A'_1 + \frac{B'_1}{(R+t)^3} + E_0) \tag{15}$$

In addition, the electric fields ($E = -\nabla\Phi$) within the CB core and the interphase region are as follows:

$$E^i_{CB} = -\frac{\partial\Phi_{CB}}{\partial j} = -\frac{1}{cos\theta}\frac{\partial\Phi_{CB}}{\partial r} = -A_1 \tag{16}$$

$$E^i_{int} = -\frac{\partial\Phi_{int}}{\partial j} = -\frac{1}{cos\theta}\frac{\partial\Phi_{int}}{\partial r} = -A'_1 + \frac{2B'_1}{r^3}, \tag{17}$$

where $i \in \{x, y, z\}$. The equivalent electrical conductivity of the spherical CB nanoparticle (σ_{CB}) is derived by solving the current density relation $J = \sigma_{CB}E$, where J and E are the spatially averaged current densities and electric fields, in sequence. This yields the following expression for the effective conductivity of a CB nanoparticle [10,16]:

$$\sigma^i_{CB} = \left[\frac{3\sigma_{CB}\sigma_{int}R^3 + \sigma_{int}(2\sigma_{int}+\sigma_{CB})\left[(R+t)^3 - R^3\right] - 6\sigma_{int}(\sigma_{int}-\sigma_{CB})\ln[(R+t)/R]}{3\sigma_{int}R^3 + (2\sigma_{int}+\sigma_{CB})\left[(R+t)^3 - R^3\right] - 6(\sigma_{int}-\sigma_{CB})\ln[(R+t)/R]}\right] \tag{18}$$

2.1.2. Electrical Conductivity of CB/Polymer Nanocomposites

In a composite system where CB nanoparticles are dispersed within a polymer matrix, the electrical conductivity σ_t is derived using a mean-field approximation. According to Ohm's law, the current density $\langle J \rangle = \sigma_t \langle E \rangle$ relates to the external electric field E_0. The internal current densities of the CB nanoparticles J^{in}_{CB} and the matrix J^{in}_m are given as follows [10,16,23,24]:

$$J^{in}_{CB} = \sum_{i=x,\,y,\,z} \sigma^i_{CB} \frac{\sigma_t}{\sigma_t + B^i_{CB}(\sigma^i_{CB} - \sigma_t)} E_0 \tag{19}$$

$$J_m^{in} = \sum_{k=X,Y,Z} \sigma_m^k \frac{\sigma_t}{\sigma_t + B_m^k(\sigma_m^k - \sigma_t)} E_0 \tag{20}$$

Substituting these expressions into the overall current density equation yields the following:

$$\sum_{i=x,\,y,\,z} \frac{V_{CB}}{\alpha_{CB}} \frac{\sigma_t - \sigma_{CB}^i}{\sigma_t + B_{CB}^i(\sigma_{CB}^i - \sigma_t)} + \sum_{k=X,Y,Z} \left(1 - \frac{V_{CB}}{\alpha_{CB}}\right) \frac{\sigma_t - \sigma_m^k}{\sigma_t + B_m^k(\sigma_m^k - \sigma_t)} = 0 \tag{21}$$

where V_{CB} is the volume fraction of CB nanoparticles, and $\alpha_{CB} = \left(\frac{R}{R+t}\right)^3$ accounts for the ratio of the CB core to the effective nanoparticle volume. The depolarization factors B_{CB} and B_m are equal to $1/3$ for the spherical particles.

Finally, the overall electrical conductivity of the CB/polymer nanocomposite is obtained from the following equation:

$$\left[\frac{V_{CB}}{3\alpha_{CB}}\left[\frac{\sigma_t - \sigma_{CB}}{\sigma_t + B_{CB}(\sigma_{CB} - \sigma_t)} + \frac{4(\sigma_t - \sigma_{CB})}{2\sigma_t + (1 - B_{CB})(\sigma_{CB} - \sigma_t)}\right] + 3\left(1 - \frac{V_{CB}}{\alpha_{CB}}\right)\left[\frac{\sigma_t - \sigma_m}{2\sigma_t + \sigma_m}\right]\right] = 0 \tag{22}$$

In this model, the electrical conductivity of the interphase region can be further expressed as $\sigma_{int} = \frac{d_c}{A_c R_{int}}$, where d_c is the tunneling distance, R_c is the contact area, and R_{int} is the tunneling resistance as reported in [10,16,25].

$$R_{int}(d_c) = \frac{d_c h^2}{A_c e^2 \sqrt{2m\lambda_l}} exp\left(\frac{4\pi d_c \sqrt{2m\lambda_l}}{h}\right) \tag{23}$$

where h, e, m, and λ_l are the Planck constant, electron charge, electron mass, and potential barrier height, sequentially. This relation shows the impact of tunneling distance and potential barrier height on the probability of electron transport across the interphase.

2.2. *Effective Electrical Conductivity of SiC/CB Hybrid Composites*

Dispersing CB nanoparticles into a non-conductive polymer such as epoxy transforms the polymer into a conductive matrix. This modification enhances the general electrical properties of the material, specifically improving the electrical conductivity of random SiC composites. The electrical conductivity of the new matrix is computed using Equation (22). Herein, a closed-form model designed to evaluate the electrical conductivity of the ternary CB/SiC/polymer composite is presented. Assume an arbitrarily dispersion of SiC whiskers within the conductive nanocomposite matrix. When the volume content of SiC surpasses a certain critical threshold, the electrical conductivity of SiC composites increases significantly. The percolation threshold (V_c) assuming a random distribution of SiC whiskers, can be determined using the following expression [15,16,26]:

$$V_c(H) = \frac{9H(1-H)}{2 + 15H - 9H^2} \tag{24}$$

where H is obtained from the following equation:

$$H(AR) = \frac{1}{AR^2 - 1}\left[\frac{AR}{\sqrt{AR^2 - 1}} \ln\left(AR + \sqrt{AR^2 - 1}\right) - 1\right] \tag{25}$$

where AR represents the aspect ratio of SiC, defined as $AR = l/D$, and l is the length and D is the diameter of the SiC whisker. The percolation threshold represents the critical volume fraction of SiC at which there is a noticeable and rapid increase in the electrical conductivity.

The electrical conductivity of a microfiber-filled composite can be described by the following equation [2,15]:

$$\frac{\sigma_c}{\sigma_{m.con}} = 1 + \frac{V_{SiC}}{3}\left[\xi\frac{\sigma_{SiC}}{\sigma_{m.con}} + (1-\xi)\frac{1}{\frac{\sigma_{m.con}}{\sigma_{SiC}} + H}\right] \tag{26}$$

where $\sigma_{m.con}$ is the electrical conductivity of the conductive matrix. Additionally, σ_{SiC} and V_{SiC} represent the electrical conductivity and the volume fraction of SiC, respectively. The variable ξ is the fraction of percolated SiC whiskers, and it can be roughly calculated as follows [15,27]:

$$\xi = \frac{V_{SiC}^{1/3} - V_c^{1/3}}{1 - V_c^{1/3}} \tag{27}$$

It is worth mentioning that V_{SiC} ranges from V_C to 1, and the value of ξ varies from 0 to 1.

3. Results and Discussion

3.1. Validation

In this section, the proposed modeling approach is validated by comparing its predictions with the available results from the literature. Two sets of validation are performed: first, the electrical conductivity of CB-reinforced polyurethane (PUR) nanocomposites is compared with experimental data from the study by Rebeque et al. [28]; second, the electrical conductivity of short carbon fiber (SCF) reinforced polymer-based composites is validated against the results provided by Pal and Kumar [29].

First, the experimental data related to CB/PUR nanocomposites is scrutinized to validate the developed formulation herein [28]. For the sake of comparison, the Planck constant h, the electron charge e, electron mass m, and potential barrier height λ_l are considered as 6.626068×10^{-34} m^{2}kg/s, $-1.602176565 \times 10^{-19}$C, $9.10938291 \times 10^{-31}$ kg, and 1 eV, in sequence. The specifications of the CB, the interphase region, and the matrix are as follows: radius of the CB $R = 17.5$ nm, thickness of the interphase $t = 15$ nm, conductivity of the CB and PUR $\sigma_{CB} = 200$ S/m and $\sigma_m = 1.87 \times 10^{-11}$S/m, respectively, and the tunneling distance $d_c = 2$ nm. The model's predictions are compared with the experimental data for different CB volume fractions to assess its accuracy in capturing the percolation behavior and effective electrical conductivity. Figure 2 presents the comparison between the predicted and experimental values of electrical conductivity for the CB/PUR nanocomposites. The model accurately captures the rapid increase in conductivity as the CB volume fraction exceeds the percolation threshold, showing good agreement with experimental data [28].

The second validation is performed using the results of the electrical conductivity for the SCF-reinforced polymer-based composites from Pal and Kumar [29]. The results provide the relationship between composite electrical conductivity and SCF volume fraction dispersed into the matrix. The electrical conductivities of SCF (σ_{SCF}) and the matrix σ_m, and the fiber aspect ratio AR are considered as 66,250 S/m, 1 S/m, and 100, respectively. Present predictions of the electrical conductivity for the SCF-reinforced composites are compared with the results reported in Ref. [29], and depicted in Figure 3. The comparison indicates that the model successfully predicts the electrical conductivity behavior over a range of SCF volume fractions, with a good alignment between the two sets of results across the entire range. This suggests that the model can accurately represent the conductive network formation within SCF/polymer composite systems, taking into account the aspect ratio and volume fraction of the SCF reinforcements.

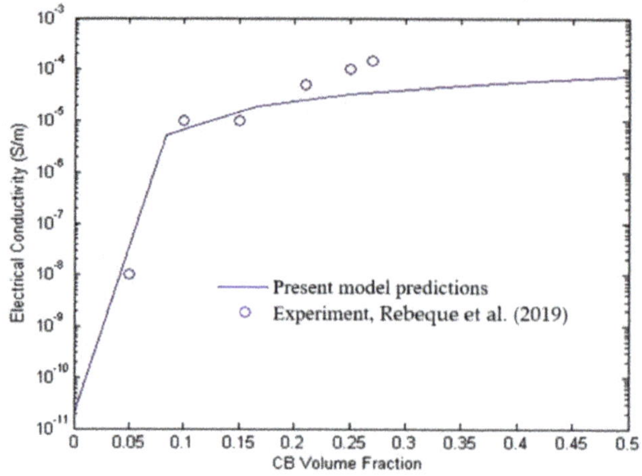

Figure 2. Comparison between the present model predictions and the experimental results obtained from the CB/PUR nanocomposite specimen [28].

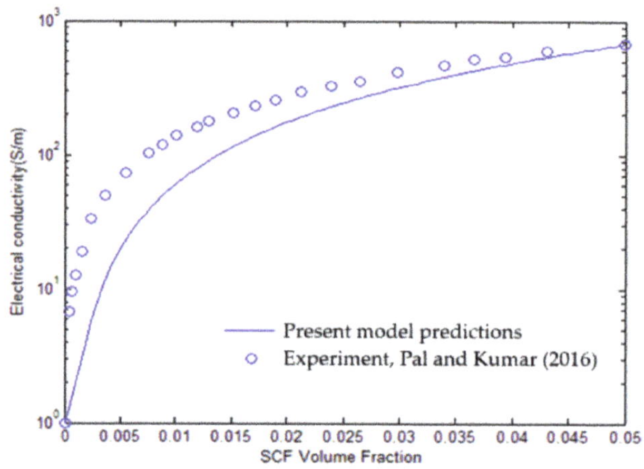

Figure 3. Comparison between the present model predictions and the results of Pal and Kumar for the SCF/polymer-based composite specimen [29].

3.2. Parametric Studies

In this section, a comprehensive investigation on the multiscale modeling of ternary CB/SiC/polymer hybrid composites is presented. To this end, a detailed parametric study is conducted on the electrical conductivity of CB/SiC whisker-reinforced polymer composites. The analysis focuses on the effect of several key parameters, such as the volume fraction of CB and SiC, CB nanoparticle size, interphase thickness, SiC whisker conductivity, and aspect ratio on the effective electrical conductivity of the composites. The specifications of the polymer matrix, CB, and SiC whiskers are provided in Table 1 [15,16,29–31]. These values are used as the base parameters for the simulation, and any variations or adjustments to these parameters will be explicitly mentioned in the corresponding figure.

Table 1. Specifications of the matrix, CB and SiC fillers.

CB	$R = 5\,\text{nm}$	$\sigma_{CB} = 10^4\,\frac{\text{S}}{\text{m}}$	$\sigma_m = 10^{-12}\,\frac{\text{S}}{\text{m}}$	$\lambda_l = 1.1\,\text{eV}$	$d_c = 1\,\text{nm}$	$t = 10\,\text{nm}$
SiC	$D_{SiC} = 2\,\mu\text{m}$	$l_{SiC} = 200\,\mu\text{m}$	$\sigma_{SiC} = 3 \times 10^4\,\frac{\text{S}}{\text{m}}$	$AR = 100$	-	-

In Figure 4, the influence of varying CB volume fractions on the electrical conductivity of CB/SiC/polymer composites is depicted. Three different volume fractions of CB are compared: 0.005, 0.015, and 0.03. The figure shows that, before reaching the percolation threshold, there is a significant difference in the electrical conductivity between the curves. The lowest CB volume fraction (0.005, represented by the blue line) yields the least conductivity, indicating the scarcity of conductive pathways within the composite. In contrast, the composites with 0.015 and 0.03 CB volume fractions, show higher conductivity even before the percolation threshold is reached. This is due to the increased likelihood of conductive network formation as the CB content rises. Thus, the formation of more conductive networks can be an important parameter that governs the electrical conductivity of the hybrid composites before percolation. After the percolation threshold; however, all curves converge, demonstrating that beyond this point, the volume fraction of CB has a little additional impact on conductivity, and the system achieves a saturated level of conductivity.

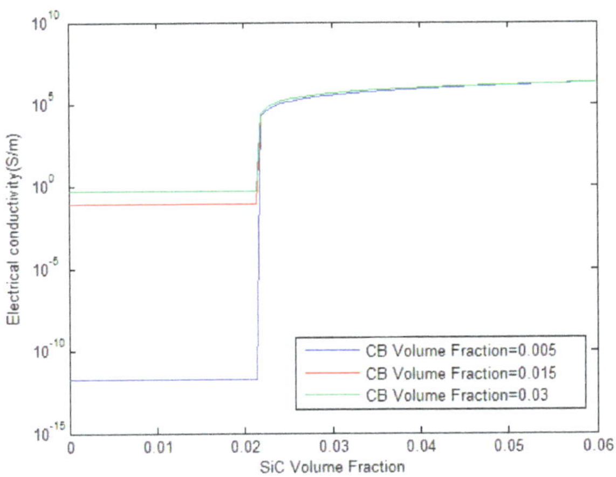

Figure 4. Influence of CB volume fraction on the electrical conductivity of CB/SiC/polymer hybrid composites.

In the present study, a uniform dispersion of nanoparticles without any aggregation is considered in the polymer matrix. Also, it is noticed that the model assumes uniform tunneling distances without considering potential variations due to particle aggregation, which could impact the conductivity predictions. It has been reported that uniform dispersion without any aggregation of nanoparticles may be desirable to improve the electrical properties of composites [23,32,33]. But, other experimental works indicate that a less than uniform dispersion with carbon-based nanofiller aggregations may favor the formation of conductive networks in the composite materials [23,34,35].

Figure 5 demonstrates the effect of CB dispersion into the polymer matrix on the electrical conductivity of the hybrid composites. Two cases are presented: one without any CB (blue line) and one with a CB volume fraction of 0.05 (red line). The figure clearly shows that the presence of CB drastically enhances the electrical conductivity of the hybrid composites. The blue line, representing the absence of CB, shows a low baseline conductivity across the range of SiC whisker volume fractions. In comparison, the red line with 0.05 CB exhibits significantly higher conductivity throughout the entire range. Interestingly, both curves exhibit the same percolation threshold, indicating that while the addition of CB improves the overall conductivity, it does not shift the threshold.

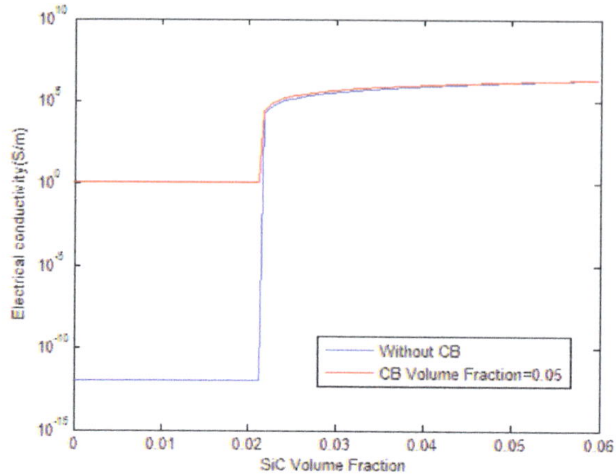

Figure 5. Influence of CB dispersion on the electrical conductivity of CB/SiC/polymer hybrid composites.

In Figure 6, the influence of CB nanoparticle radius on the electrical conductivity is examined. Three radii are considered: 5 nm, 7.5 nm, and 10 nm. The results indicate that smaller CB particles (blue line, 5 nm) provide significantly higher conductivity before the percolation threshold compared to larger particles. This may be attributed to the greater number and larger surface area of smaller particles, which increases the probability of forming conductive paths via tunneling mechanisms. The figure shows that the blue line exhibits the highest conductivity before percolation, followed by the red and green lines for 7.5 nm and 10 nm CB radii, sequentially. After percolation, the differences between the curves are less pronounced, though the green line for 10 nm shows a slight increase in conductivity compared to the others. It has been shown that small size functionalized nanoparticles can increase the conductivity drastically and improve the overall physical properties of composites [36–44]. Yaduvanshi et al. observed that DC conductivity of HAT4 has enhanced by two orders of magnitude due to copper nano-particles [42]. Therefore, a reason for the enhancement of the electrical conductivity before the percolation threshold may be described by the existence of more conductive networks for nanoparticles with smaller size for the same percentage.

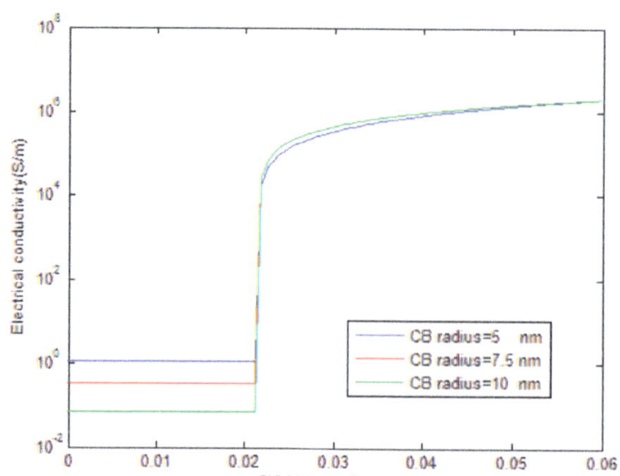

Figure 6. Influence of CB radius on the electrical conductivity of CB/SiC/polymer hybrid composites.

During the micromechanical modeling of carbon nanoparticle-polymer nanocomposites, an interphase is captured among the nanofiller and the polymer matrix which characterizes the non-bonded van der Waals interaction between them. Such an interphase can be regarded as an equivalent solid continuum with a definite thickness [45–48]. The effect of CB interphase thickness on the effective electrical conductivity of polymer composites containing SiC whisker/CB hybrids is illustrated in Figure 7. Three interphase thicknesses (7 nm, 9 nm, and 12 nm) are compared, and the results show that thicker interphases lead to higher conductivity before and after the percolation threshold. The thinnest interphase (7 nm) shows the lowest conductivity before percolation, while the thickest interphase (12 nm) results in the highest conductivity across the entire range. This is because a thicker interphase provides a larger region for conductive electron transport, enhancing the overall conductivity of the composite. After the percolation threshold, the green line continues to exhibit higher conductivity than the others, demonstrating the continued benefit of a thicker interphase in facilitating electron transport.

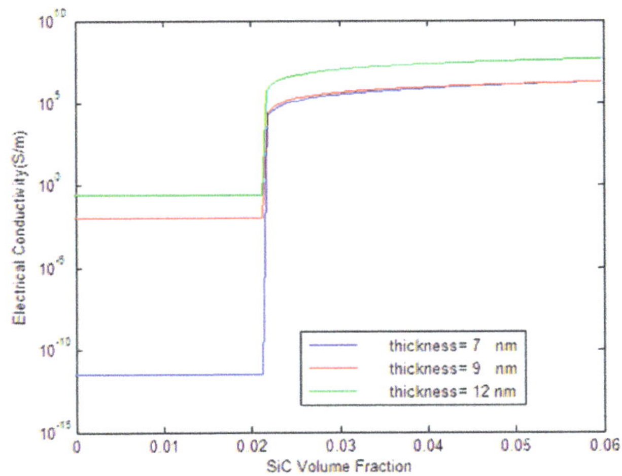

Figure 7. Influence of CB interphase thickness on the electrical conductivity of CB/SiC/polymer hybrid composites.

Figure 8 explores the influence of SiC whisker conductivity on the overall electrical conductivity of the hybrid composites. Three different SiC conductivities are considered: 10×10^3 S/m, 30×10^3 S/m, and 60×10^3 S/m. The figure shows that there is little difference in conductivity between the three cases before the percolation threshold, as the SiC whiskers do not significantly contribute to the conductive network in this region. However, after the percolation threshold, the green line (60×10^3 S/m SiC conductivity) demonstrates the highest conductivity, followed by the red line (30×10^3 S/m), and the blue line (10×10^3 S/m). This suggests that higher SiC conductivities enhance the capability of conductive pathways, thereby improving the overall electrical performance of the composite.

Figure 9 presents the effect of the SiC whisker aspect ratio on the electrical conductivity of the polymer composites containing micro/nano hybrids. Three aspect ratios are compared: 80, 100, and 120. The results indicate that the aspect ratio has a significant impact on both the percolation threshold and the post-percolation conductivity. The yellow line, representing the highest aspect ratio (120), reaches the percolation threshold at a lower SiC volume fraction compared to the red and blue lines, indicating that higher aspect ratios lead to earlier percolation. This is because longer whiskers are more efficient at forming conductive networks at lower volume fractions. After percolation, the yellow line remains higher than the others, showing that a higher aspect ratio not only reduces the percolation threshold but also enhances conductivity post-percolation. The outcomes offer that the electrical conductivity of hybrid composites is improved by the increase in the SiC whisker aspect ratio. Thus, fabricating a hybrid composite with a high-length whisker can be more effective in its electrical conductivity.

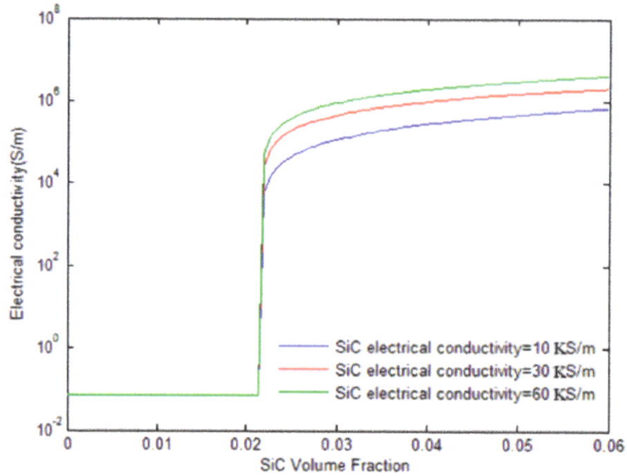

Figure 8. Influence of SiC whisker's conductivity on the electrical conductivity of CB/SiC/polymer hybrid composites.

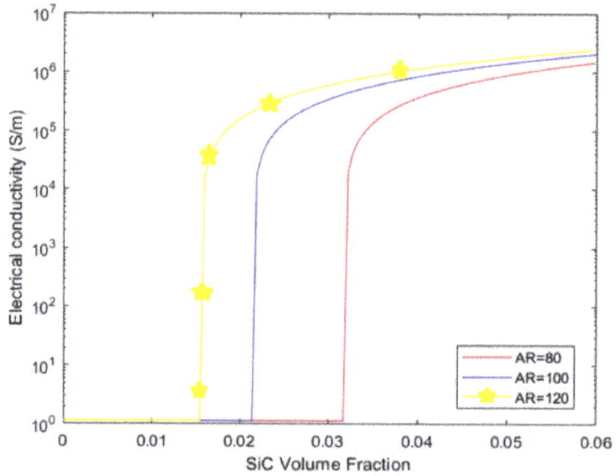

Figure 9. Influence of SiC whisker's aspect ratio on the electrical conductivity and the percolation threshold of CB/SiC/polymer hybrid composites.

4. Conclusions

In this study, an analytical micromechanics method was developed to investigate the effective electrical conductivity of hybrid polymer composites reinforced with SiC whiskers and CB nanoparticles. A two-phase modeling approach was employed, wherein CB nanoparticles were first dispersed in a non-conductive polymer matrix to form a conductive nanocomposite, followed by the incorporation of randomly oriented and dispersed SiC whiskers as a secondary reinforcement to further enhance the composite's electrical performance. The model captured critical microstructural parameters such as the volume fraction and size of CB nanoparticles, as well as the aspect ratio and volume fraction of SiC whiskers. The model predictions were validated using available results from the literature for different types of composite systems, demonstrating a good agreement between the two sets of results. The findings revealed that increasing the volume fraction of CB significantly enhances electrical conductivity before the percolation threshold, while the influence diminishes after percolation is achieved. Furthermore, the radius and interphase thickness of CB nanoparticles were shown to have a critical impact on the composite's conductivity, with smaller radii and thicker interphase regions contributing to the improved electrical conductivity of the CB/SiC/polymer composites. The study also highlights the role of SiC whisker aspect ratio, with higher aspect ratios resulting in significantly lower percolation thresholds and higher electrical conductivity in the post-percolation regime. This work provides a robust framework for understanding and predicting the electrical conductivity of SiC/CB hybrid polymer composites. The results indicate that optimizing the microstructural characteristics of both CB nanoparticles and SiC whiskers is essential for enhancing the electrical performance of the composite.

Author Contributions: Conceptualization, U.U. and M.H.A.; methodology, U.U. and M.H.A.; software, U.U.; validation, U.U., M.H.A. and Z.A.; formal analysis, U.U. and M.H.A.; investigation, U.U.; resources, Z.A. and M.K.A.; data curation, M.H.A. and M.K.A.; writing—original draft preparation, U.U. and M.H.A.; writing—review and editing, M.H.A. and Z.A.; visualization, M.H.A. and Z.A.; supervision, Z.A. and M.K.A.; project administration, Z.A. and M.K.A.; funding acquisition, Z.A. and M.K.A. All authors have read and agreed to the published version of the manuscript.

Funding: The research was funded through the Research Institute/Center Supporting Program (RICSP-25-2), King Saud University, Riyadh, Saudi Arabia.

Institutional Review Board Statement: Not applicable.

Data Availability Statement: Could be made available on request.

Acknowledgments: The authors extend their appreciation to the King Saud University for funding this work through the Research Institute/Center Supporting Program (RICSP-25-2), King Saud University, Riyadh, Saudi Arabia.

Conflicts of Interest: The authors declare no conflict of interest.

References

1. Al-Saleh, M.H.; Saadeh, W.H. Hybrids of conductive polymer nanocomposites. *Mater. Des.* **2013**, *52*, 1071–1076. [CrossRef]
2. Chanda, A.; Sinha, S.K.; Datla, N.V. Electrical conductivity of random and aligned nanocomposites: Theoretical models and experimental validation. *Compos. Part A Appl. Sci. Manuf.* **2021**, *149*, 106543. [CrossRef]
3. Jin, J.; Lin, Y.; Song, M.; Gui, C.; Leesirisan, S. Enhancing the electrical conductivity of polymer composites. *Eur. Polym. J.* **2013**, *49*, 1066–1072. [CrossRef]
4. Allaoui, A.; Bai, S.; Cheng, H.M.; Bai, J.B. Mechanical and electrical properties of a MWNT/epoxy composite. *Compos. Sci. Technol.* **2002**, *62*, 1993–1998. [CrossRef]
5. Zhang, Y.; Lai, L.; Cui, D.; Zhu, Y.; Cai, H.; Yan, B.; Li, Y.; Yang, Z.; Ding, G. Hybrid effect on mechanical and thermal performance of copper matrix composites reinforced with SiC whiskers. *Ceram. Int.* **2024**, *50*, 16553–16563. [CrossRef]
6. Fan, Y.; Fowler, G.D.; Zhao, M. The past, present and future of carbon black as a rubber reinforcing filler—A review. *J. Clean. Prod.* **2020**, *247*, 119115. [CrossRef]
7. Xu, H.P.; Dang, Z.M.; Yao, S.H.; Jiang, M.J.; Wang, D. Exploration of unusual electrical properties in carbon black/binary-polymer nanocomposites. *Appl. Phys. Lett.* **2007**, *90*, 152912. [CrossRef]
8. Wu, D.; Lv, Q.; Feng, S.; Chen, J.; Chen, Y.; Qiu, Y.; Yao, X. Polylactide composite foams containing carbon nanotubes and carbon black: Synergistic effect of filler on electrical conductivity. *Carbon* **2015**, *95*, 380–387. [CrossRef]
9. Wu, X.; Lu, C.; Zhang, X.; Zhou, Z. Conductive natural rubber/carbon black nanocomposites via cellulose nanowhisker templated assembly: Tailored hierarchical structure leading to synergistic property enhancements. *J. Mater. Chem. A* **2015**, *3*, 13317–13323. [CrossRef]

10. Mazaheri, M.; Payandehpeyman, J.; Jamasb, S. Modeling of Effective Electrical Conductivity and Percolation Behavior in Conductive-Polymer Nanocomposites Reinforced with Spherical Carbon Black. *Appl. Compos. Mater.* **2022**, *29*, 695–710. [CrossRef]
11. Alemour, B.; Yaacob, M.H.; Lim, H.N.; Hassan, M.R. Review of electrical properties of graphene conductive composites. *Int. J. Nanoelectron. Mater.* **2018**, *11*, 371–398.
12. Kim, S.Y.; Noh, Y.J.; Yu, J. Prediction and experimental validation of electrical percolation by applying a modified micromechanics model considering multiple heterogeneous inclusions. *Compos. Sci. Technol.* **2015**, *106*, 156–162. [CrossRef]
13. Hashemi, R.; Weng, G.J. A theoretical treatment of graphene nanocomposites with percolation threshold, tunneling-assisted conductivity and microcapacitor effect in AC and DC electrical settings. *Carbon* **2016**, *96*, 474–490. [CrossRef]
14. He, B.; Mortazavi, B.; Zhuang, X.; Rabczuk, T. Modeling Kapitza resistance of two-phase composite material. *Compos. Struct.* **2016**, *152*, 939–946. [CrossRef]
15. Saberi, M.; Ansari, R.; Hassanzadeh-Aghdam, M.K. Predicting the electrical conductivity of short carbon fiber/graphene nanoplatelet/polymer composites. *Mater. Chem. Phys.* **2023**, *309*, 128324. [CrossRef]
16. Saberi, M.; Moradi, A.; Ansari, R.; Hassanzadeh-Aghdam, M.K.; Jamali, J. Developing an efficient analytical model for predicting the electrical conductivity of polymeric nanocomposites containing hybrid carbon nanotube/carbon black nanofillers. *Compos. Part A Appl. Sci. Manuf.* **2024**, *185*, 108374. [CrossRef]
17. Townsend, J.; Burtovyy, R.; Aprelev, P.; Kornev, K.G.; Luzinov, I. Enhancing Mechanical and Thermal Properties of Epoxy Nanocomposites via Alignment of Magnetized SiC Whiskers. *ACS Appl. Mater. Interfaces* **2017**, *9*, 22927–22940. [CrossRef]
18. Chen, Y.; Pan, F.; Wang, S.; Liu, B.; Zhang, J. Theoretical estimation on the percolation threshold for polymer matrix composites with hybrid fillers. *Compos. Struct.* **2015**, *124*, 292–299. [CrossRef]
19. Sumfleth, J.; Adroher, X.C.; Schulte, K. Synergistic effects in network formation and electrical properties of hybrid epoxy nanocomposites containing multi-wall carbon nanotubes and carbon black. *J. Mater. Sci.* **2009**, *44*, 3241–3247. [CrossRef]
20. Szeluga, U.; Kumanek, B.; Trzebicka, B. Synergy in hybrid polymer/nanocarbon composites. *A Rev. Compos. Part A Appl. Sci. Manuf.* **2015**, *73*, 204–231. [CrossRef]
21. Griffiths, D.J.; Inglefield, C. *Introduction to Electrodynamics*; Cambridge University Press: Cambridge, UK, 2005; Volume 73. [CrossRef]
22. Geim, A.K.; Novoselov, K.S. The rise of graphene. In *Nanoscience and Technology: A Collection of Reviews from Nature Journals*; World Scientific: Singapore, 2009; pp. 11–19. [CrossRef]
23. Feng, C.; Jiang, L. Micromechanics modeling of the electrical conductivity of carbon nanotube (CNT)–polymer nanocomposites. *Compos. Part A Appl. Sci. Manuf.* **2013**, *47*, 143–149. [CrossRef]
24. Yan, K.Y.; Xue, Q.Z.; Zheng, Q.B.; Hao, L.Z. The interface effect of the effective electrical conductivity of carbon nanotube composites. *Nanotechnology* **2007**, *18*, 255705. [CrossRef]
25. Simmons, J.G. Generalized Formula for the Electric Tunnel Effect between Similar Electrodes Separated by a Thin Insulating Film. *J. Appl. Phys.* **1963**, *34*, 1793–1803. [CrossRef]
26. Gao, L.; Li, Z. Effective medium approximation for two-component nonlinear composites withshape distribution. *J. Phys. Condens. Matter* **2003**, *15*, 4397. [CrossRef]
27. Deng, F.; Zheng, Q.S. An analytical model of effective electrical conductivity of carbon nanotube composites. *Appl. Phys. Lett.* **2008**, *92*, 071902. [CrossRef]
28. Rebeque, P.V.; Silva, M.J.; Cena, C.R.; Nagashima, H.N.; Malmonge, J.A.; Kanda, D.H.F. Analysis of the electrical conduction in percolative nanocomposites based on castor-oil polyurethane with carbon black and activated carbon nanopowder. *Polym. Compos.* **2019**, *40*, 7–15. [CrossRef]
29. Pal, G.; Kumar, S. Multiscale modeling of effective electrical conductivity of short carbon fiber-carbon nanotube-polymer matrix hybrid composites. *Mater. Des.* **2016**, *89*, 129–136. [CrossRef]
30. Motaghi, A.; Hrymak, A.; Motlagh, G.H. Electrical conductivity and percolation threshold of hybrid carbon/polymer composites. *J. Appl. Polym. Sci.* **2015**, *132*, 41744. [CrossRef]
31. Bystrický, R.; Škrátek, M.; Rusnák, J.; Precner, M.; Ťapajna, M.; Hnatko, M.; Šajgalík, P. Electrical and magnetic properties of silicon carbide composites with titanium and niobium carbide as sintering aids. *Ceram. Int.* **2023**, *49*, 5319–5326. [CrossRef]
32. Song, Y.S.; Youn, J.R. Influence of dispersion states of carbon nanotubes on physical properties of epoxy nanocomposites. *Carbon* **2005**, *43*, 1378–1385. [CrossRef]
33. Golosova, A.A.; Adelsberger, J.; Sepe, A.; Niedermeier, M.A.; Lindner, P.; Funari, S.S.; Jordan, R.; Papadakis, C.M. Dispersions of polymer-modified carbon nanotubes: A small-angle scattering investigation. *J. Phys. Chem. C* **2012**, *116*, 15765–15774. [CrossRef]
34. Li, J.; Ma, P.C.; Chow, W.S.; To, C.K.; Tang, B.Z.; Kim, J.K. Correlations between percolation threshold, dispersion state, and aspect ratio of carbon nanotubes. *Adv. Funct. Mater.* **2007**, *17*, 3207–3215. [CrossRef]
35. Martin, C.A.; Sandler, J.K.W.; Shaffer, M.S.P.; Schwarz, M.K.; Bauhofer, W.; Schulte, K.; Windle, A.H. Formation of percolating networks in multi-wall carbon-nanotube–epoxy composites. *Compos. Sci. Technol.* **2004**, *64*, 2309–2316. [CrossRef]

36. Mahesh, P.; Shah, A.; Swamynathan, K.; Singh, D.P.; Douali, R.; Kumar, S. Carbon dot-dispersed hexabutyloxytriphenylene discotic mesogens: Structural, morphological and charge transport behavior. *J. Mater. Chem. C* **2020**, *8*, 9252–9261. [CrossRef]
37. Iqbal, A.; Urbanska, M.; Dąbrowski, R.S.; Kumar, S.; Dhar, R. Impact of carbon quantum dots on self-assembly and dielectric relaxation modes of a room temperature tri-component fluorinated antiferroelectric liquid crystal mixture. *Soft Matter* **2023**, *19*, 9293–9307. [CrossRef] [PubMed]
38. Singh, P.K.; Dhar, R.; Dabrowski, R. Enhancement of dielectric and electro-optical characteristics of liquid crystalline material 4′-octyl-4-cyano-biphenyl with dispersed functionalized and nonfunctionalized multiwalled carbon nanotubes. *Phys. Rev. E* **2023**, *107*, 044704. [CrossRef]
39. Ji, X.; Lu, Z.; Wang, J.; Ye, N.; Zhang, H.; Zhou, L.; Li, J.; Lu, Y. Construction of micro-nano hybrid structure based on carbon nanotube whisker and alumina for thermally conductive yet electrically insulating silicone rubber composites. *Compos. Sci. Technol.* **2024**, *249*, 110495. [CrossRef]
40. Verma, R.; Mishra, M.; Dhar, R.; Dabrowski, R. Enhancement of electrical conductivity, director relaxation frequency and slope of electro-optical characteristics in the composites of single-walled carbon nanotubes and a strongly polar nematic liquid crystal. *Liq. Cryst.* **2017**, *44*, 544–556. [CrossRef]
41. Tripathi, P.; Mishra, M.; Kumar, S.; Dabrowski, R.; Dhar, R. Dependence of physical parameters on the size of silver nano particles forming composites with a nematic liquid crystalline material. *J. Mol. Liq.* **2018**, *268*, 403–409. [CrossRef]
42. Yaduvanshi, P.; Mishra, A.; Kumar, S.; Dhar, R. Enhancement in the thermodynamic, electrical and optical properties of hexabutoxytriphenylene due to copper nanoparticles. *J. Mol. Liq.* **2015**, *208*, 160–164. [CrossRef]
43. Lal, A.; Verma, H.; Chirra, S.; Dhar, R.; Dabrowski, R.; Pandey, K.L. Gold nanorod-induced effects in a mesogenic compound 4-(trans-4-n-Hexylcyclohexyl) isothiocyanatobenzene. *ACS Omega* **2023**, *8*, 29012–29024. [CrossRef] [PubMed]
44. Tang, W.; Luo, L.; Chen, Y.; Li, J.; Dai, Y.; Xie, Y.; Ma, Y.; Zhang, J.; Zhang, Y. Noble-metal-free Bi-OZIS nanohybrids for sacrificial-agent-free photocatalytic water splitting: With long-lived photogenerated electrons. *Sep. Purif. Technol.* **2025**, *357*, 130047. [CrossRef]
45. Kundalwal, S.I.; Meguid, S.A. Micromechanics modelling of the effective thermoelastic response of nano-tailored composites. *Eur. J. Mech.-A/Solids* **2015**, *53*, 241–253. [CrossRef]
46. Hassanzadeh-Aghdam, M.K. Evaluating the effective creep properties of graphene-reinforced polymer nanocomposites by a homogenization approach. *Compos. Sci. Technol.* **2021**, *209*, 108791. [CrossRef]
47. Tsai, J.L.; Tzeng, S.H.; Chiu, Y.T. Characterizing elastic properties of carbon nanotubes/polyimide nanocomposites using multi-scale simulation. *Compos. Part B Eng.* **2010**, *41*, 106–115. [CrossRef]
48. Hassanzadeh-Aghdam, M.K.; Mahmoodi, M.J.; Ansari, R. Creep performance of CNT polymer nanocomposites-An emphasis on viscoelastic interphase and CNT agglomeration. *Compos. Part B Eng.* **2019**, *168*, 274–281. [CrossRef]

polymers

MDPI

Article

Multiscale Numerical Study of Enhanced Ductility Ratios and Capacity in Carbon Fiber-Reinforced Polymer Concrete Beams for Safety Design

Moab Maidi [1,2,*], Gili Lifshitz Sherzer [3] and Erez Gal [1]

1 Department of Civil and Environmental Engineering, Ben-Gurion University of the Negev, Beer Sheva 8410501, Israel; erezgal@bgu.ac.il
2 Department of Civil Engineering, Braude College of Engineering, Karmiel 2161002, Israel
3 Department of Civil Engineering, Ariel University, Ariel 4070000, Israel; gilil@ariel.ac.il
* Correspondence: moab@post.bgu.ac.il; Tel.: +972-523597109

Abstract: Rigid reinforced concrete (RC) frames are generally adopted as stiff elements to make the building structures resistant to seismic forces. However, a method has yet to be fully sought to provide earthquake resistance through optimizing beam and column performance in a rigid frame. Due to its high corrosion resistance, the integration of CFRP offers an opportunity to reduce frequent repairs and increase durability. This paper presents the structural response of CFRP beams integrated into rigid frames when subjected to seismic events. Without any design provision for CFRP systems in extreme events, multiscale simulations and parametric analyses were performed to optimize the residual state and global performance. Macroparameters, represented by the ductility ratio and microfactors, have been analyzed using a customized version of the modified compression field theory (MCFT). The main parameters considered were reinforcement under tension and compression, strength of concrete, height-to-width ratio, section cover, and confinement level, all of which are important to understand their influence on seismic performance. The parametric analysis results highlight the increased ductility and higher load-carrying capacity of the CFRP-reinforced tested component compared to the RC component. These results shed light on the possibility of designing CFRP-reinforced concrete components that could improve ductile frames with increased energy dissipation and be suitable for applications in non-corrosive seismic-resistant buildings. This also shows reduced brittleness and enhancement in the failure mode. Numerical simulations and experimental results showed a strong correlation with a deviation of about 8.3%, underlining the reliability of the proposed approach for designing seismic-resistant CFRP-reinforced structures.

Keywords: ductility ratio enhancement; seismic performance; CFRP; high corrosion resistance; energy dissipation; performance; limited displacement; structural capacity; ductility index

1. Introduction

Seismic events pose significant risks to buildings and their occupants. Conventional construction, which often relies heavily on steel reinforcement, consumes substantial natural resources and has a notable environmental impact [1–10]. In contrast, using CFRP for reinforcement presents a more sustainable alternative, reducing resource use and environmental footprint. Studies comparing these materials highlight CFRP's advantages in terms of both efficiency and ecological impact [11]. The search for alternatives to steel bars in reinforced concrete is increasingly pertinent, yet the construction sector will unlikely abandon reinforced concrete technology anytime soon. Reinforced concrete remains a cornerstone of construction due to its numerous advantages. This material's durability, cost-effectiveness, and versatility make it indispensable for various structural applications, from residential buildings to large infrastructures. Reinforced concrete's ability to be molded into different shapes and its excellent fire resistance further solidify its status as a preferred construction material. However, innovations such as Carbon Fiber Reinforced Polymer (CFRP) and other advanced materials are being explored to address some of the limitations of steel reinforcement, particularly its susceptibility to corrosion, which can compromise the longevity and safety of concrete structures. Durability issues can be avoided by using fiber reinforced polymer (FRP) instead of steel. Fibers contribute to the tensile capacity of CFRP, as demonstrated in the studies by Maidi et al. [12].

Reducing the amount of CFRP in the compressive fiber can enhance the ductility of the component, while reducing the section depth may increase the curvature in the ultimate state of the section. Over-reinforcing the tensional fiber to support the compressive area can further improve the ductility of the section. The transition from deeper sections to flatter ones enhances ductility by increasing the component's capacity for plastic deformation. However, increasing the percentage of tensioned reinforcement beyond 2% may create a yield strain in the compressed area, which can adversely affect the overall ductility of the section. Their work showed that higher tolerance and curvature can be achieved by decreasing the section depth and increasing its width, thereby creating an alternate compression area. These adjustments, appropriate reinforcement ratios, and concrete strength must be carefully considered when designing sections requiring high ductility to ensure optimal performance.

Providing shear and compressive capacities to overcome the limitations of CFRP material. Moreover, Maidi et al. [13] pointed out that corrosion-induced degradation in concrete and reinforced concrete (RC) structures often starts within the first few decades of their lifespan, and this remains a significant challenge to seismic resistance. Existing research tools can evaluate performance, but they usually fail to predict changes in seismic resistance resulting from alterations in the core properties of RC structures. To this gap, the research addresses the substitution of steel reinforcement with CFRP [14]. CFRP is a lightweight, high-strength material that enhances load-carrying capacity, stiffness, and ductility to create effective CFRP-reinforced concrete (CRC). However, under high stress or strain, CFRP can experience brittle failure, potentially compromising the structural integrity of the reinforced concrete member.

According to the model presented in Section 3, ductility is influenced by the curvature and failure mechanism, which ensures considerable strain in the compressed fiber. The increase in ductility is achieved by effective confinement and the use of relatively low concrete strengths. Therefore, improvements are necessary to enhance beam ductility and compare performance with traditional reinforced concrete (RC). Although CFRP does not corrode in the same manner as steel, prolonged water exposure, high temperatures, and high long-term stress can cause durability issues [15–17]. An alkali concrete environment can also attenuate the durability of some types of fiber, but CFRP is usually made of alkali-resistant materials. When compared to steel, CFRP has several disadvantages in structural applications, including lack of ductility and complicated bending of bars [18]. Although bending is possible during production, the strength of bent bars is lower and construction options are limited. CFRP bars also have poor performance when loaded in compression [19], due to stability issues.

In seismically active areas, material ductility is a property of greater importance than strength, for instance, seismic events [12,13,20–23]. Although ductility enables the design of safer structures, a nonlinear analysis complicates calculations, and design values [22]. On a structural level, capacity design is very important because it aims to avoid non-ductile types of failure such as shear. This is implemented in the current Croatian (and European) design code HRN EN 1998-1 [23,24], as well as in most other codes across the globe [25–28]. On an elemental level, detailing provisions can significantly influence element behavior. For example, the amount of compressive and confinement reinforcement can improve the behavior of elements [21], while increasing the tensile reinforcement can decrease their ductility. Elements that fail by crushing concrete before the steel reinforcement yields have a very low ductility, as in the case of over-reinforced beams.

One significant gap in the existing literature is the absence of an analysis that considers all relevant parameters collectively and assesses the extent of their impact on structural behavior. Such comprehensive studies are crucial for developing more accurate predictive models and enhancing construction materials' effectiveness in real-world applications.

Research on the ductility of elements with CFRP rebars is limited. Most works have focused on the use of CFRP for structural repair and strengthening [29], rather than in new structures. In addition, several methods used to design/construct new elements are analyzed, as they relate well to CFRP-reinforced elements and provide crucial insight into the behavior of concrete elements [30]. This paper discusses the ductility of new structures, with emphasis on new beam structures, by examining parameters on ductility and shear ductility. The need to explore alternatives to steel as the primary reinforcement, such as various fibers, was highlighted in a study by Lu et al. (2022) [31]. Furthermore, as discussed in [12], recommendations for optimizing fiber performance were provided, focusing on reinforcement ratios, cross-sectional dimensions, concrete strengths, and concrete cover. These recommendations aim to inform and enhance design codes. When properly designed, combining reinforced concrete with CFRP offers a promising solution to bridging these gaps in structural performance and durability.

2. Methodology

This research focuses on analyzing the energy dissipation of flexural components under bending, investigating various macro, meso, and micro parameters through Push-Over Analysis (POA). The methodology employed is based on the approach presented in the study by Maidi and Shufrin [32]. The simulations are conducted using the Response-2000 program (v1.0), which incorporates the Modified Compression Field Theory (MCFT) model developed by Bentz in 2000 [33].

The macro parameters are characterized by ductility ratio (μ), or ductility index (*DI*) predict the residual state and are defined by two meso parameters: shear force (*V*) and displacement (*D*). The study also examines how micro parameters, such as tensile CFRP reinforcement (*Af*) and compressive CFRP reinforcement (*A′f*), depth (*d*), concrete strength (*f′c*), and its cover thickness (*df*), influence the macro parameters (*V*) and (*D*). A parametric study is conducted to analyze the behavior of the CFRP-reinforcement section, considering the interaction between macro, meso, and micro parameters. The study ultimately demonstrates improvements in flexural performance and energy dissipation, highlighting potential enhancements in seismic performance. The methodology framework is illustrated in Figure 1.

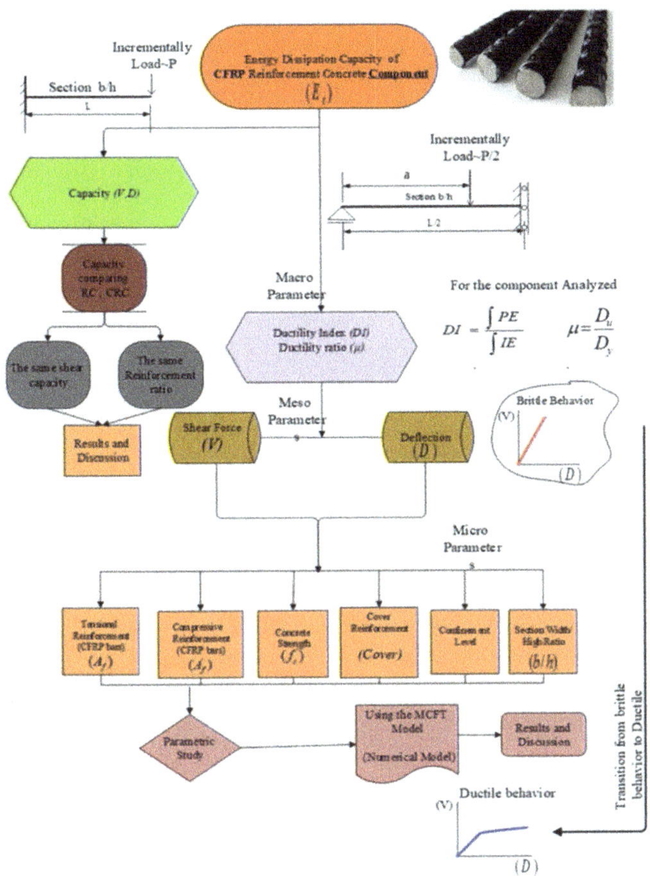

Figure 1. Methodology framework.

3. Multiscale Study

3.1. Macroscale Study (MASS)

The ductility ratio, which is considered a macroparameter, reflects the capacity of the tested bending component to dissipate maximum plastic energy, and is calculated as the ratio of elastic to plastic dissipation energies. The amount of dissipated energy is expressed using Equation (1):

$$E_t = \int_0^{D_u} EPB(D)d(D) = E_e^{epb} + E_p^{epb} \quad (1)$$

where (EPB) represents the elastic-plastic behavior, (EB) represents the elastic behavior and ${E_e}^{eb}$ is the total elastic energy in the elastic behavior. The following characteristics are presented in Figure 2, V_e, De are the base shear and top displacement, respectively, in the elastic state. D_y, D_u correspond to the yield and ultimate state displacements, respectively, with V_y denoting the shear force capacity in the ductile behavior. E_e^{epb}, E_p^{epb} represent the elastic and plastic energies, respectively, in the ductile behavior.

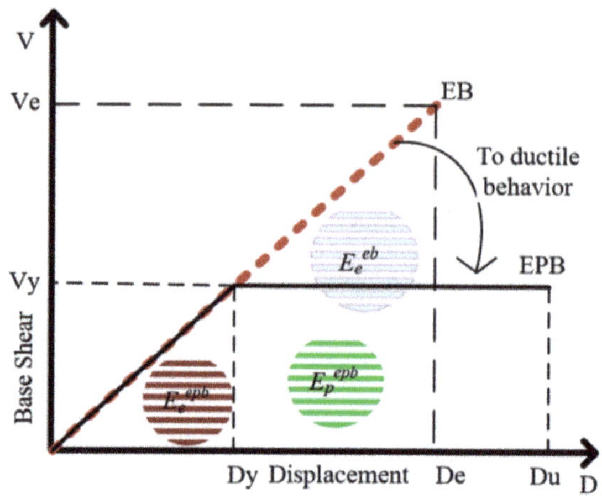

Figure 2. From Transition from brittle to ductile force–displacement behaviors.

The macroparameter, characterized as the ductility ratio μ, can then be expressed by Equation (2) [34]:

$$\mu = \frac{D_u}{D_y} \tag{2}$$

3.2. Mesoscale Study (MESS)

The correlation between the yield state strain and the displacement D_y in bending can be determined with Equations (3) and (4) [21]:

$$\varphi_y = \frac{\varepsilon_{cy} + \varepsilon_{fy}}{d_y = d_f} \tag{3}$$

$$D_y = \int_0^L x \cdot \varphi_y dx \tag{4}$$

where ε_{cy} denotes the strain ratio in concrete at yield state, ε_{fy} denotes the strain corresponding to the yield strength of CFRP reinforcement, defined as f_{fy}/E_f, and d denotes the distance from the extreme compression fiber to the centroid of tension reinforcement. Similarly, the ultimate state can be obtained by Equations (5)–(7):

$$\varphi_u = \frac{\varepsilon_{cu} + \varepsilon_{fu}}{d_u} \tag{5}$$

$$d_u = d_f - c \tag{6}$$

$$D_u = \int_0^L x \cdot \varphi_u dx \tag{7}$$

where ε_{cu} denotes the strain in concrete at the ultimate state, and ε_{fu} denotes the strain corresponding to the ultimate strength of CFRP reinforcement, defined as f_{fu}/E_f.

Note: For the CFRP material, the strain value and the strengths at both the yield and ultimate states are equal, as illustrated in Section 4.

3.3. Microscale Study (MISS)

This section describes the method used to analyze the impact of six microparameters on the performance of a doubly reinforced section (DRS) and a single reinforced section (SRS) under a theoretical moment-displacement representation framework. The DRS featured reinforcement bars at both the top and bottom of the beam, while the SRS had a single reinforcement section at the bottom. The analysis incorporated the displacement approach proposed by Kišiček et al. [22], the balanced stress–strain diagrams by Kiši, Sori, and Gali, the effects of confinement methods on the ductility of over-reinforced RC sections, as reported by Ahmed et al. [35] and the Concrete Compression Failure Principle (CCFP) outlined by Paulay and Park. Furthermore, the tensile failure limitation proposed by Bazan and Fernandez-Davila [34] was incorporated to evaluate failure in the compressive zone and facilitate the cross-section transition from brittle to ductile behavior. Additionally, the residual state of the CFRP was assessed through a four-point bending test, focusing on the influence of six microparameters, as outlined in Sections 3.3.1–3.3.6 below.

3.3.1. Tensional CFRP Reinforcement Effect

The failure mode in CFRP-reinforced concrete structures is significantly influenced by the amount of reinforcement in the tensioned area. Increasing the amount of CFRP reinforcement in this zone can lead to failure at the compressed fiber, which is generally a more desirable failure than brittle failure in the tension zone. Increasing the reinforcement in the compressed fiber can enhance the structure's flexural and bending capacity. However, if the tensile reinforcement ratio exceeds the balanced failure ratio, further increases in tensile reinforcement will not yield additional benefits. The balanced reinforcement ratio was determined by varying the CFRP reinforcement areas.

3.3.2. Compressive CFRP Reinforcement Effect

The ultimate displacement and moment were analyzed to understand the effect of double reinforced in both tension and compression sides.

3.3.3. Effect of Concrete Strength

The ductility of a section is influenced by the strain level in the compressive stressed area, which is directly affected by the characteristic strength of the concrete. Consequently, this chapter examines the impact of concrete strength on the bending behavior of the section and the strain levels under different stress conditions. High concrete strength is expected to delay the onset of maximum strain in the concrete, which may negatively impact the overall ductility. This preliminary conclusion is supported by a work that showed that different types of concrete require different stress levels to achieve the same strain, as shown in Table 1 [21].

Table 1. Mechanical properties of tested concrete.

Concrete Type	Tension Strength (MPa)	Peak Strain (mm/m)	Cylinder Strength (MPa)	Modulus of Elasticity (MPa)
1	1.69 (2.83)	1.87	20.7	21,946.2
2	1.61	1.87	15	19,550.9
3	1.68	1.86	20	21,678.8
4	1.75	1.9	25	23,476.9
6	1.82	1.96	30	29,826.8
7	1.96	2.1	40	30,006.6
8	2.1	2.25	50	30,375.8
9	2.24	2.39	60	32,616.6
10	2.52	2.37	80	36,594.9

3.3.4. Effect of Concrete Cover on Concrete Cracking

Concrete covers protect steel reinforcement from corrosion and fire resistance, and significantly influence the cracking behavior of concrete under various loads. Numerical analyses were conducted to investigate the impact of the concrete cover on cracking behavior. Specimens were subjected to four-point bending tests, and the relationship between crack width and spacing was analyzed. The Response-2000 numerical model used in this study is based on equations from both the American Concrete Institute [36,37] and Eurocode 2 [38], which establish a correlation between the spacing and width of concrete cracks, as shown in Equation (8):

$$w = \varepsilon_{cm} s_m \tag{8}$$

where ε_{cm} denotes the average strain due to stress at a certain level, and sm denotes the crack spacing, which is calculated by Equation (6):

$$s_m = 3c_{max} \tag{9}$$

where c_{max} denotes the distance from the furthest tension fiber to the nearest reinforcement, which is directly related to the cover thickness.

Instead of increasing the distance c_{max}, a multiplying factor γ can be applied to cmax, which is then incorporated into Equation (5) as follows: $w = \varepsilon_{cm}\gamma s_m$. Then, the factor strain is multiplied by the factor γ, plugged into Equation (2), to enhance ductility. Therefore, increasing the cover thickness not only leads to a greater crack width but also improves ductility, as illustrated in Figure 3.

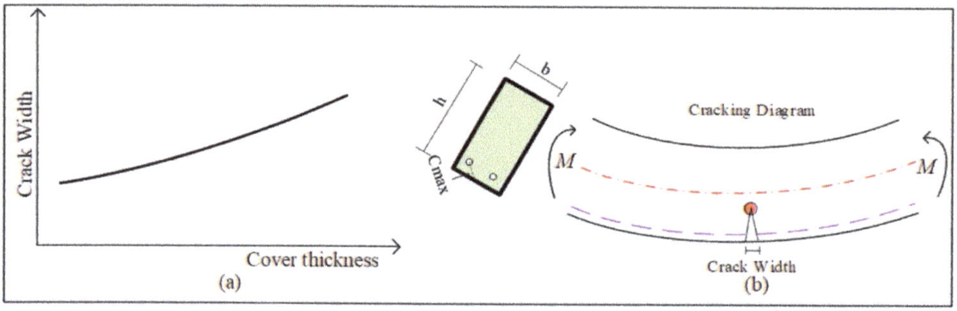

Figure 3. Typical relationships between the concrete cover thickness and crack width. (**a**) Schematic of crack width versus cover thickness curve. (**b**) Schematic diagram of crack opening in a beam.

3.3.5. Effect of Confinement Concrete Level

The strength and ductility of concrete are significantly influenced by the level of confinement provided by the vertical reinforcement within the beam's structure. The pivotal roles played by the stiffness and behavior of the confining reinforcement, ranging from elastic to elastoplastic, are crucial in determining the overall performance of concrete [39]. This research incorporates a concrete confinement model to enhance the shear capacity and ductility ratios.

In the analysis framework, the axial and lateral directions are referred to as directions 1 and 3, respectively, assuming that stresses and strains are equivalent in both lateral directions ($\sigma_2 \approx \sigma_3$). Compressive stresses and strains are considered positive, while volume expansion is negative. When confined by constant lateral pressure, the axial stress–strain response of concrete is characterized at three distinct points on the stress–strain curve, as outlined in the referenced study [40]. This methodological approach is visually represented in Figure 4, clearly depicting the model's application to enhance our understanding of confined concrete behavior under various loading conditions.

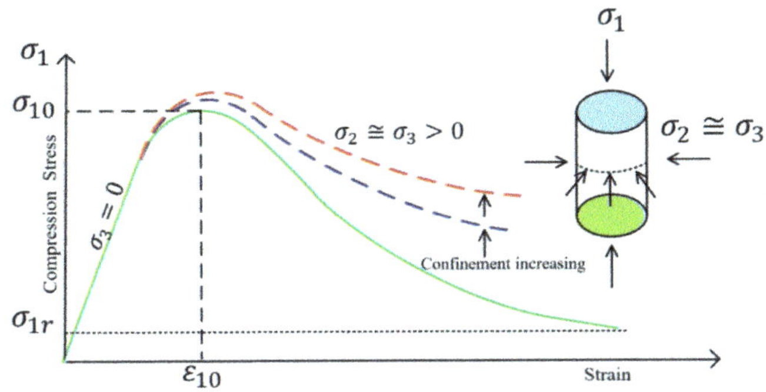

Figure 4. Constitutive model of confined concrete under uniaxial compression.

The ultimate strength of the concrete, denoted as (σ_{10}), is directly influenced by the lateral pressure applied, while the residual capacity, (σ_{1r}), is determined by the internal friction remaining after initial stress effects. The analysis employs the Leon–Pramono [40], criterion as defined in Equation (10) to precisely characterize these critical points on the stress–strain curve:

$$\left[(1-k)(\sigma_3/f_c')^2 + \frac{\sigma_1 - \sigma_3}{f_c'}\right]^2 + k^2 m(\sigma_3/f_c') - k^2 c = 0 \tag{10}$$

where k denotes the hardening parameter, defined as 0.1 at the elastic limit. The confinement modulus, m, as described in Equation (11), is calculated as follows:

$$m = \frac{f_c'^2 \cdot f_r^2}{f_c' \cdot f_r} \tag{11}$$

This equation simplifies the understanding of confinement effects on the compressive strength of concrete, where ϕ denoted as the confinement ratio, is calculated as follows:

$$\phi = \sigma_3/f_c' \tag{12}$$

Finally, the increase in strength due to confinement, denoted as i, is calculated using Equation (13) and is illustrated in Figure 5:

$$i = \sigma_{10}/f_c' \tag{13}$$

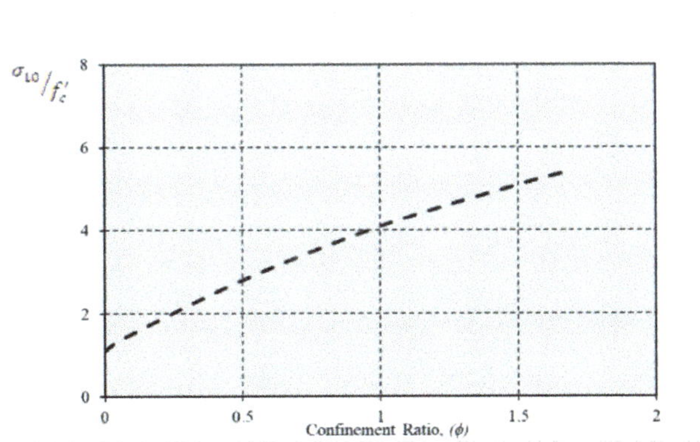

Figure 5. The increase ratio (i) in the strength vs. confinement ratios.

3.3.6. Effect of Section Dimensions Ratio

In this section, we demonstrate how the sensitivity of reinforcement amounts, concrete strength, and b/h ratios to determine the optimal dimension ratio with specific reinforcement ratios. The primary goal is to identify the optimum ratio between the height and width of the tested section, thereby establishing the ideal b/h ratio.

3.4. Modeling of Moment Displacement Curve

The effect of the ductility behavior is displayed in Figure 5, the moment-displacement curve can be schematically divided into three straight segments, which influence the ductility behavior. The key points on the moment-displacement curve are (*Dcr*, *Mcr*), (*Dy*, *My*), and (*Du*, *Mu*). The symbols used in deriving the moment-displacement model are illustrated in Figure 6.

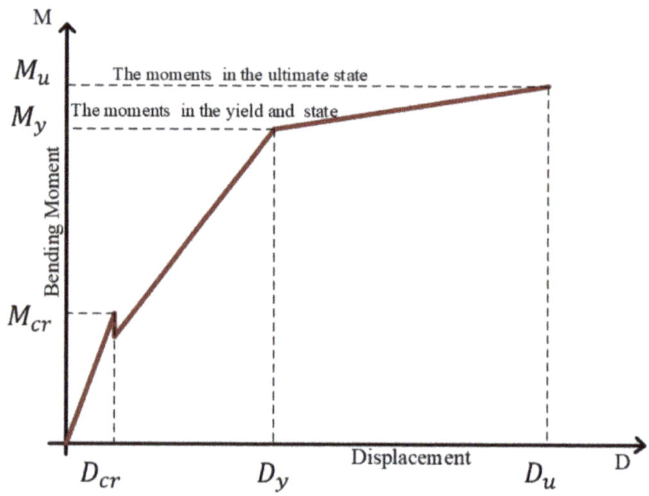

Figure 6. Schematic model of a moment displacement curve.

The symbols used in deriving the moment-displacement model are illustrated in Figure 7.

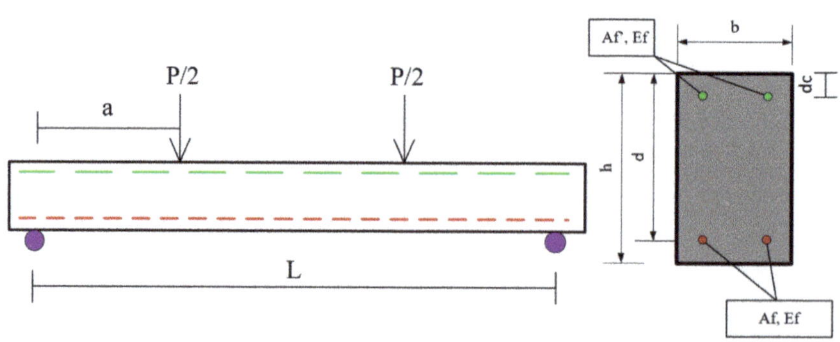

Figure 7. Beam dimensions and cross-section for four-point bending analyses.

The three main stages to obtain a capacity curve with the three classic control points are presented below.

Stage I: Elastic Stage (0 to Cracking Moment, M_{cr}), This stage begins with no initial displacement and zero moment, and progresses to the cracking moment (M_{cr}). The mid-span displacement increases from 0 to (D_{cr}), corresponding to the cracking displacement. At this stage, the displacement can be computed using a linear elastic relationship up to M_{cr}. Often this analysis is based on beam theory or specific structural behavior Equations (14)–(16). The general presentation is as follows:

$$M_{cr} = \frac{I_g \cdot f_r}{y} \tag{14}$$

when the section is semitic at y axial:

$$M_{cr} = \frac{2I_g \cdot f_r}{h} \tag{15}$$

$$D_{cr} = \frac{M_{cr}}{24 \cdot E_c \cdot I_g}\left\langle 3 \cdot L^2 - 4 \cdot a^2 \right\rangle \tag{16}$$

where I_g denotes the moment of inertia of the uncracked section, calculated using $I_g = b \times h$ 3/12, where b denotes the width and h the height of the section. The flexural tensile strength of concrete, fr can be determined as $fr = 0.7(f'_c)$ (MPa) for normal-weight concrete, where $f'c$ is the concrete's compressive strength and y is the distance between the natural axis and the extreme of the tensional zone.

Stage II: The maximum moment increases from the cracking moment M_{cr} to the yielding moment, corresponding to concrete yielding M_y, the mid-span displacement increases from D_{cr} to D_y.

A generally accepted constitutive model used for concrete in compression Figure 6 assumes that the strain distribution along the depth of the beam section remains linear. Concrete contribution to tension is negligible in the derivation, which simplifies the analysis. According to Figure 7, the total compressive force from the concrete can be given by Equations (17) and (18).

$$C_c = \int_0^{\varepsilon_c} f_c(\varepsilon) \cdot b \cdot c / \varepsilon_c d\varepsilon \tag{17}$$

where ε_c denotes the concrete strain at the extreme compression fiber and c denotes the height of the concrete compression block. The solution to Equation (17) is presented in Equation (18) as a simplified expression for the total compressive force C_c.

$$C_c = k_1 \cdot f'_c \cdot b \cdot c \tag{18}$$

where k_1 is a coefficient that accounts for the shape of the stress distribution in the compression zone and c is the depth of the compression block as presented in Equation (19) and Figures 8 and 9.

$$f_c(\varepsilon) = f'_c\left[2 \cdot \left(\frac{\varepsilon}{\varepsilon_0}\right) - \left(\frac{\varepsilon}{\varepsilon_0}\right)^2\right] \tag{19}$$

The total bending moment, M_c, provided by the concrete with respect to the neutral axis is expressed by Equations (20)–(22).

$$M_c = \frac{\int_0^{\varepsilon_c} f_c(\varepsilon) \cdot b \cdot c^2 \cdot \varepsilon}{\varepsilon_c^2 d\varepsilon} = k_1 \cdot k_2 \cdot f'_c \cdot b \cdot c^2 \tag{20}$$

$$k_1 = \frac{1}{f'_c \cdot \varepsilon_c}\int_0^{\varepsilon_c} f_c(\varepsilon) d\varepsilon = \frac{\varepsilon_c}{\varepsilon_0} \cdot \left(1 - \frac{1}{3} \cdot \frac{\varepsilon_c}{\varepsilon_0}\right) \tag{21}$$

$$k_2 = \frac{1}{\varepsilon_c}\frac{\int_0^{\varepsilon_c} f_c(\varepsilon) \cdot \varepsilon d\varepsilon}{\int_0^{\varepsilon_c} f_c(\varepsilon) d\varepsilon} = \frac{2}{3}\left(\frac{1 - \frac{3}{8} \cdot \frac{\varepsilon_c}{\varepsilon_0}}{1 - \frac{1}{3} \cdot \frac{\varepsilon_c}{\varepsilon_0}}\right) \tag{22}$$

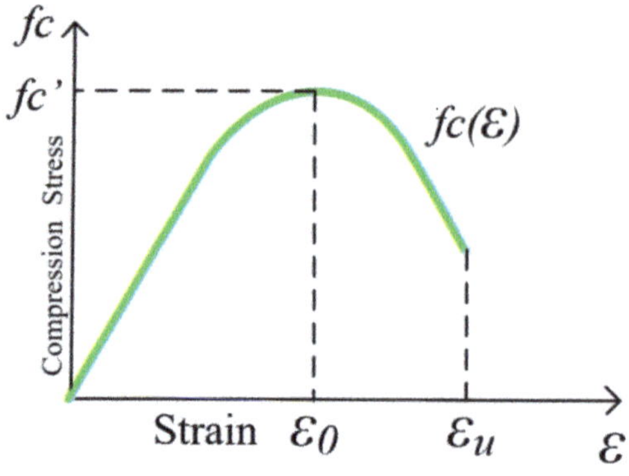

Figure 8. Constitutive model of concrete subjected to uniaxial compression for a non-confined condition.

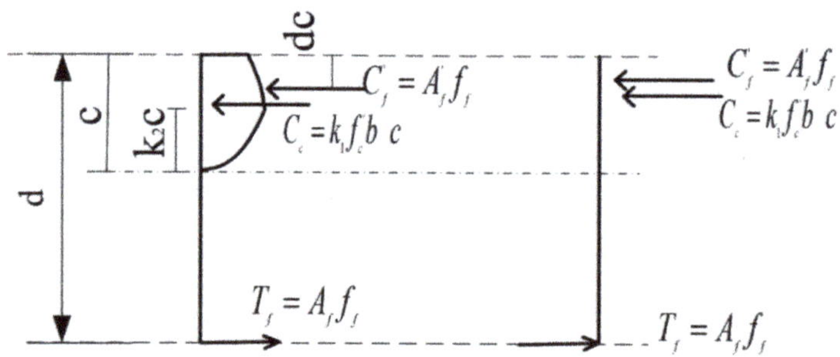

Figure 9. Stresses block and force distribution in the beam cross-section.

Typically, $\varepsilon_0 = 0.002$, which allows k_1 and k_2 to be presented as follows:

$$k_1 = 500{\cdot}\varepsilon_c\langle 1 - 166.7{\cdot}\varepsilon_c\rangle \tag{23}$$

$$k_2 = \frac{\langle 2 - 375{\cdot}\varepsilon_c\rangle}{\langle 3 - 500{\cdot}\varepsilon_c\rangle} \tag{24}$$

Thus, the yielding moment M_y is given by Equation (25).

$$M_y = k_1{\cdot}k_2{\cdot}f'_c{\cdot}b{\cdot}c^2 + A'_f{\cdot}E_f{\cdot}\varepsilon'_f(c - d_c) + A_f{\cdot}f_f{\cdot}(d - c) \tag{25}$$

where A'_f, E_f, ε'_f, d_c, A_f, and ff represent the CFRP area at the top fiber, modulus of elasticity, top fiber strain, the active depth, CFRP area at the bottom fiber, and forces associated with reinforcement.

The mid-span displacement at yielding, Dy, can then be calculated by the following:

$$D_y = \frac{M_y}{24{\cdot}E_c{\cdot}I_{eff,y}}\left\langle 3{\cdot}L^2 - 4{\cdot}a^2\right\rangle \tag{26}$$

or in terms of curvature φ_y by the following:

$$D_y = \frac{\varphi_y}{24}\left\langle 3{\cdot}L^2 - 4{\cdot}a^2\right\rangle \tag{27}$$

The effective moment of inertia at yielding ($I_{eff,y}$) is given by the following:

$$I_{eff,y} = \left(\frac{M_{cr}}{M_y}\right)^3{\cdot}I_{unc} + \left(1 - \left(\frac{M_{cr}}{M_y}\right)^3\right){\cdot}I_{cr} \tag{28}$$

where I_{unc} denotes the uncracked second moment of area, calculated by the following:

$$I_{unc} = \left(\frac{1}{12} * b * h^3\right) + b \cdot h\left(y' - \frac{h}{2}\right)^2 + \left(\frac{E_f}{E_c} - 1\right) \cdot A_f \cdot (d - y')^2 \quad (29)$$

The moment of inertia I_{cr} for the cracked section can be derived using the transformed area method, as follows:

$$I_{cr} = \frac{1}{3} b \cdot c^3 + \frac{E_f}{E_c} \cdot A_f \langle d - c \rangle^2 \quad (30)$$

where $E_c = 4730\sqrt{f_c'}$ (MPa), as specified by ACI 318-05 [36]. In calculating displacements, the stress–strain relationship for concrete in compression is assumed to be linearly elastic, so long as the concrete's extreme fiber remains within the ascending branch of the curve shown in Figure 6.

Stage III: As the maximum moment increases from the yielding moment M_y to the ultimate moment M_u, the mid-span displacement increases from D_y to D_u calculated by:

$$M_u = k_1 \cdot k_2 \cdot f_c' \cdot b \cdot c^2 + A_f \cdot E_f \cdot \varepsilon_{ff}(d - c) \quad (31)$$

where k_1, k_2, and c are determined by the same equations as in stage II, and εff is the CFRP strain at beam failure. If failure occurs due to CFRP rupture, εff is taken as the ultimate CFRP strain, εfu, so $\varepsilon ff = \varepsilon fu$.

After CFRP failure, the displacement Du can be given by the following:

$$D_u = \frac{M_u}{24 \cdot E_c \cdot I_{eff,u}} \langle 3 \cdot L^2 - 4 \cdot a^2 \rangle \quad (32)$$

where the effective moment of inertia at ultimate ($I_{eff,u}$) is as follows:

$$I_{eff,u} = \left(\frac{M_{cr}}{M_u}\right)^3 \cdot I_{unc} + \left(1 - \left(\frac{M_{cr}}{M_u}\right)^3\right) \cdot I_{cr} \quad (33)$$

The following are used to calculate the ultimate displacement:

$$\varphi_u = \frac{M_u}{E_c \cdot I_{eff,u}} \quad (34)$$

$$D_u = \frac{\varphi_u}{24 \cdot E_c} \langle 3 \cdot L^2 - 4 \cdot a^2 \rangle \quad (35)$$

The moment can be converted to an equivalent shear force using the following:

$$P = \frac{2M_{max}}{a} \quad (36)$$

This procedure enables the construction of a shear force-displacement curve for the selected beam, allowing for detailed examination of the beam's behavior under the specified parameters.

4. Materials Characterization

This study incorporates a steel rebar force equivalent to that of the compressed section. The concrete properties and steel bars were selected based on the guidelines outlined in (CSA S806-12 2012) [27], while the properties of CFRP were based on those reported in studies by Suriyati et al. and Zhou et al. [41–43].

4.1. Concrete

Concrete properties significantly influence the behavior and performance of reinforced structures under various loading conditions. To ensure accurate modeling, different types of concrete were selected based on established studies. The mechanical properties of these concrete types are summarized in Table 1, as reported by Suriyati et al. and Zhou et al. [41–43].

The stress–strain relationship for different types of concrete is typically represented by a standard curve, as provided in ACI 318 (1995) [36].

4.2. Steel Bars

The properties of the steel reinforcement bars incorporated into the structural models were selected based on standardized specifications. The specific properties of the steel bars used in this study are detailed in Table 2.

Table 2. Mechanical properties of the steel bars.

Properties	Ultimate Strain (m/mm)	Yield Strength (MPa)	Ultimate Strength (MPa)	Elastic Modulus (GPa)
Steel bars	7	276	414	200

4.3. CFRP Bars

CFRP is a high-strength, lightweight reinforcement material known for enhancing the load-carrying capacity, stiffness, and ductility of concrete structures. The mechanical properties of the CFRP used in this study are presented in Figure 10.

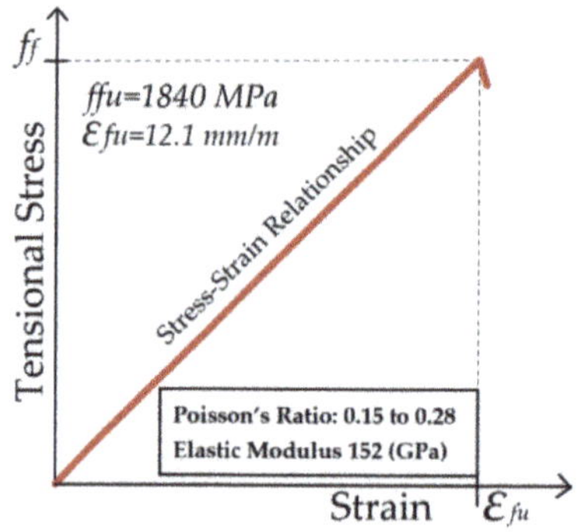

Figure 10. Mechanical properties of the CFRP reinforcement material.

In addition to its high strength and lightweight properties, CFRP has a low density of approximately 1.60×10^{-4} N/mm^3 and a thermal expansion coefficient of 40×10^{-6} °C. In this study, the diameter of the CFRP bars was selected to match that of the steel rebar. The brittle behavior of CFRP bars is illustrated in Figure 8. The properties of the CFRP used in this study are presented in Figure 10.

5. Results

5.1. Load-Bearing Capacity of CFRP-Reinforced Concrete Elements

Two cantilever beam configurations (see Figure 11), both with the same cross-section and concrete strength, where one beam was reinforced with a steel bar and the other with CFRP, were compared.

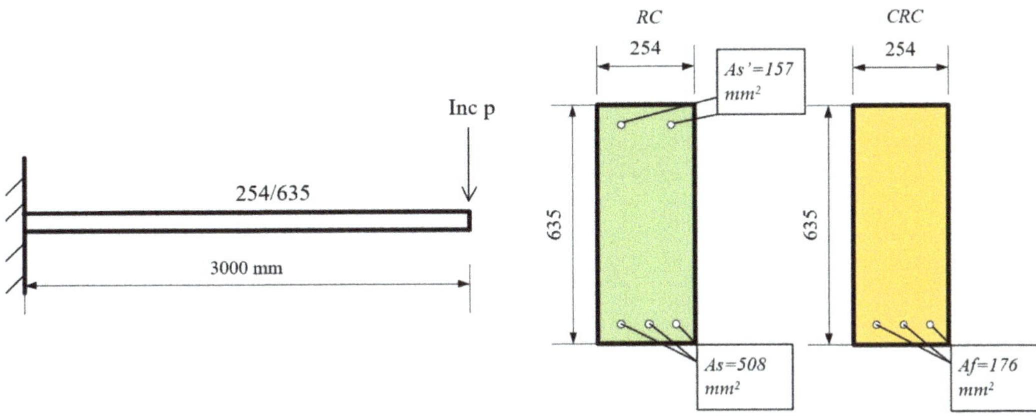

Figure 11. Schematic of RC and CRC Components Under Loading.

First, a test was performed to determine the equivalent CFRP area value in the CRC component required to achieve the same capacity as the RC component. A second test compared the capacities of the two components with the same preset reinforcement values. In both scenarios, the ductility and effective stiffness, as defined by Equations (2) and (37), were calculated as follows:

$$k_{eff} = \frac{V_u}{D_u} \tag{37}$$

For the first component, the cantilever span was 3000 mm, with a cross-sectional depth of 635 mm and a width of 254 mm. The concrete strength was $fc' = 30\ \text{N/mm}^2$, and the cover dimensions were $ds, ds', df, df' = 51$ mm. The POA indicated that the reinforcement areas were suitable, with the top reinforcement areas measuring $As' = 157\ \text{mm}^2$ and $Af' = 0\ \text{mm}^2$, and the bottom reinforcement areas measuring $As = 508\ \text{mm}^2$, $Af = 176\ \text{mm}^2$. The reinforcement area was determined by comparing the capacities of the two numerical models. The POA calculated capacity under monotonic loading. While these values provided equal capacities for the two components, they resulted in different ductility levels, as illustrated in Figure 12.

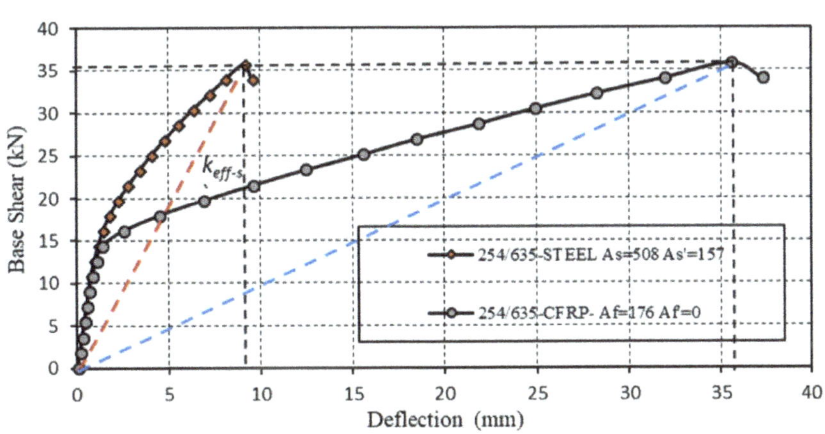

Figure 12. RC and CRC components equal capacity in the ultimate state, the blue and the red line are the effective stiffnesses.

The ductility of the CRC component was significantly higher than that of the RC component (15.33 and 8.77, respectively). The load-bearing capacities of the cross-sections were similar, with the steel-reinforced section achieving Vus = 35.634 kN and the CFRP-reinforced section achieving Vuf = 35.766 kN. However, the effective stiffness differed considerably; the steel-reinforced section had an effective stiffness of $keff\text{-}s$ = 3.90 kN/m, while the CFRP-reinforced section had an effective stiffness of $keff\text{-}f$ = 1.0065 kN/m. In conclusion, with the same capacity in both components, the concrete element with steel reinforcement exhibited greater effective stiffness but lower ductility. In contrast, the ductility parameter was relatively high with CFRP reinforcement. To achieve the same capacity as the RC component, the required amount of CFRP reinforcement in the CRC component was 176 mm^2, compared to 665 mm^2 steel reinforcement in the RC component. Namely, the amount of reinforcement with CFRP required constitutes 74% lower of the steel reinforcement area. Of note, when the carbon fibers were only introduced into the bottom area, the level of ductility in both situations was in favor of the CRC component, as presented in Figure 12. Additionally, it had better ductility than the same section with both bottom and top reinforcement. The ductility level in the RC component (8.77) was significantly lower than in the CRC component (15.33). The load-bearing capacities of the cross-sections were similar, with the steel-reinforced section achieving $Vu{,}s$ = 35.634 kN and the CFRP-reinforced section achieving $Vu{,}f$ = 35.766 kN. However, the effective stiffness differed considerably; the steel-reinforced section had an effective stiffness of $keff\text{-}s$ = 3.90 kN, while the CFRP-reinforced section had an effective stiffness of $keff\text{-}f$ = 1.0065 kN.

In the following situation, the cantilever span was also 3000 mm, with a cross-sectional depth of 635 mm, a width of 254 mm, and concrete strength of fc' = 30 N/mm^2. The cover dimensions were ds, ds′, df, df′ = 51 mm, the top reinforcement areas were As', Af' = 157 mm^2, and the bottom reinforcement areas were As, Af = 508 mm^2. POA demonstrated different capacities and ductility levels for the RC and CRC components, as shown in Figure 13.

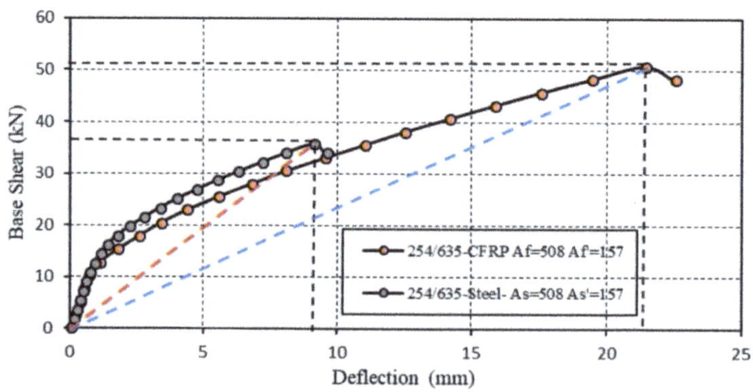

Figure 13. RC and CRC components with the same reinforcement ratios, the blue and the red line are the effective stiffnesses.

Figure 13 shows the force-displacement curves for a beam with a 3000 mm span, where the orange curve represents the CRC cross-section, and the gray curve represents the RC section. The ductility level at the component was defined as the ratio between the displacement in the yielding state and the displacement in the failure state. The ductility level of the RC component was 8.77, compared to 15.87 for the CRC. Shear force at failure (ultimate state) was Vus = 33.853 kN for the RC component and Vuf = 50.663 kN for the CRC component. The effective stiffness, defined as the ratio between the shear force and horizontal displacement in the ultimate state, was *keff-s* = 3.53 kN/m for the steel-reinforced beam compared to *keff-f* = 2.365 kN/m for the same section reinforced with CFRP.

5.2. Sensitivity Analysis

Six microparameters were projected to significantly impact on the ductility of a beam section under bending: the amount of CFRP used in the tension and compression zones, the geometry of the beam cross-section, concrete cover, the concrete strength and confinement ratio as presented in Figure 14. A single beam section with a b/h ratio of 200/400 mm and with different cross-sectional geometries was used to evaluate the effects of the first three microparameters. To identify an optimal ductile section for shear loading, a series of beam sections, starting with a b/h ratio of 200/600 mm and progressing to a section with a b/h ratio of 600/200 mm, was examined.

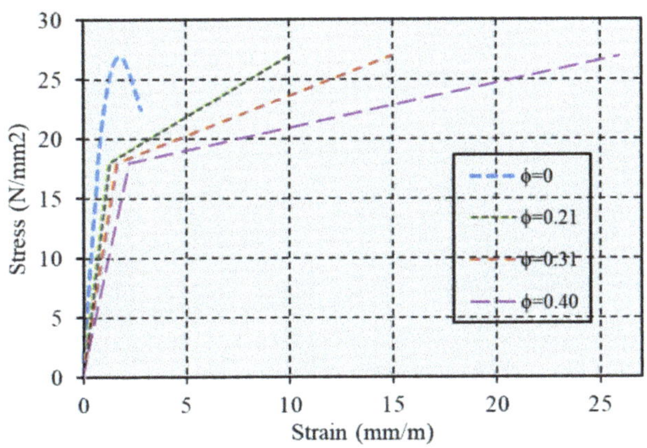

Figure 14. Stress–strain of variable confined concrete with a compressive strength of (fc' = 27).

5.2.1. Single Reinforcement

The effect of increasing amounts of CFRP in the tension zone on displacement was tested on a single beam with constant CFRP compressive reinforcement Af' = 0 and concrete compressive strength fc' = 30 N/mm^2, but with varied bottom reinforcement. Figure 15a–c in the results section present the outcomes of this analysis.

5.2.2. Double Reinforcement

The effect of increasing amounts of CFRP in the compressive zone on bending behavior was measured in a beam with tensile reinforcement areas held constant at Af = 300,700 mm^2 and Af = 1050 mm^2, varied Af′ in the compressive zone, and a constant concrete strength of fc' = 30 N/mm^2. The results of the analysis are shown in Figure 16a–c.

5.2.3. The Role of Concrete Strength

The impact of concrete strength on the ductility ratio specifically in cases with no compressive reinforcement (Af' = 0) and with tensile reinforcement (Af = 150 mm^2) was tested while varying the compressive strength fc'. Figure 17a–c presents the observations and corresponding data.

5.2.4. Role of Cover Thickness in Enhancing the Durability of CFRP RC Structures

To explore the relationship between the distance separating strain centers and deflection under pure flexure in the inspected section, the effect of cover thickness on a rectangular section with dimensions of 200 mm by 800 mm was characterized. The concrete strength was $fc' = 30$ MPa, with reinforcement at the top ($Af' = 900$ mm^2) and bottom ($Af = 1500$ mm^2). The primary variable under investigation was the cover thickness, as outlined in Equations (5) and (6). This parameter influences the moment-displacement relationship and the ductility ratio, as illustrated in Figure 18a–c.

5.2.5. Effect of Confinement on Concrete Behavior

The effect of increasing the passive confinement ratio (ϕ) in the compressive zone of the bending component on ductile behavior was tested, with confinement parameters in the compressive zone applied at four levels, labeled from 1 to 4, as presented in Figure 14.

A rectangular section with dimensions of 200 mm by 500 mm was used, with Af = 800 mm^2, Af′ = 0, and concrete compressive strength of fc′ = 27 N/mm^2 held constant, while the stress–strain models for concrete were varied. The results of the analysis are presented in Figure 19a–c in the Results Section.

5.2.6. Contribution of Section Dimensions Ratio

Considering Equations (1) and (2), it can be inferred that the depth of the cross-section being tested may impact both the curvature and ductility ratio. Therefore, the effect of the b/h ratio of the tested section was assessed while maintaining a constant reinforcement ratio of 0.10 mm^2 and a concrete compressive strength of fc′ = 30 MPa. Figure 20a–c showcases the results of this analysis.

5.3. Sensitivity Analysis Results

This section summarizes all results from Section 5, including shear force-displacement curves and ductility ratios, as functions of the tested parameters (see Figures 15–20). Figure 15 presents the single reinforcement effective ductility.

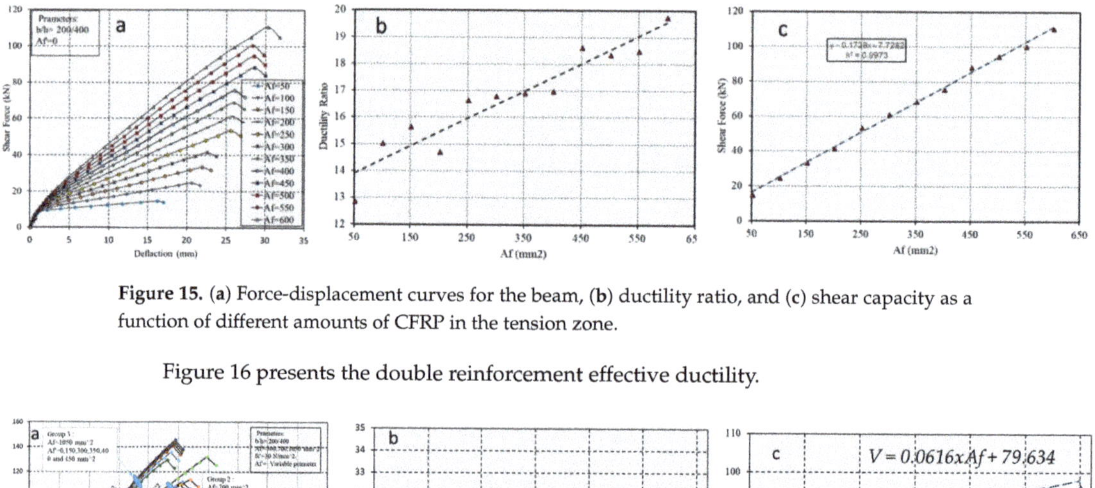

Figure 15. (**a**) Force-displacement curves for the beam, (**b**) ductility ratio, and (**c**) shear capacity as a function of different amounts of CFRP in the tension zone.

Figure 16 presents the double reinforcement effective ductility.

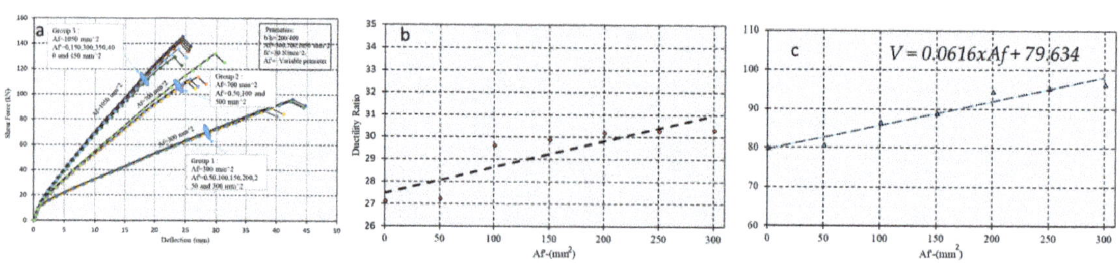

Figure 16. (**a**) Influence of varying amounts of CFRP on the force-displacement curve for a beam in the compressive zone. (**b**) Ductility ratio as a function of different amounts of tensioned CFRP. (**c**) Shear capacity variation in response to different amounts of compressive CFRP.

Figure 17 presents the role of concrete strength effective ductility.

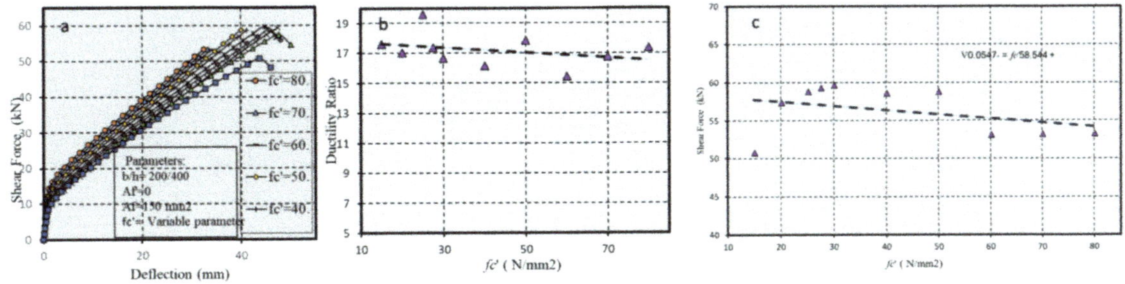

Figure 17. (**a**) Force-displacement curves with different of concrete strength, (**b**) ductility ratio, and (**c**) shear capacity for beams with varying concrete strengths.

Figure 18 presents the role of cover thickness in enhancing the durability of CFRP and RC structures effective ductility.

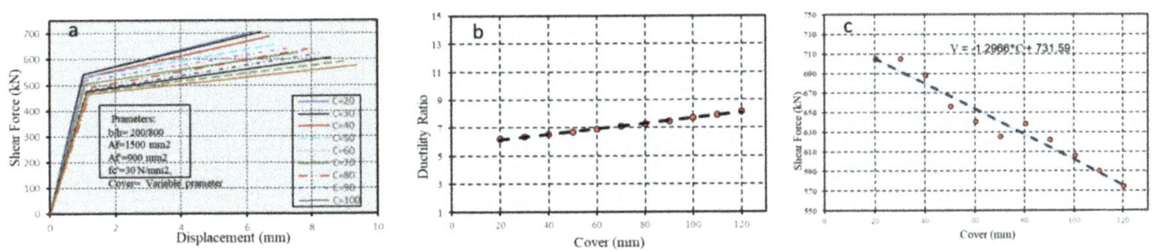

Figure 18. (**a**) Force-displacement curves with different of concrete cover, (**b**) ductility ratio, and (**c**) shear capacity as a function of different cover thickness.

Figure 19 presents the effect of confinement on concrete behavior effective ductility.

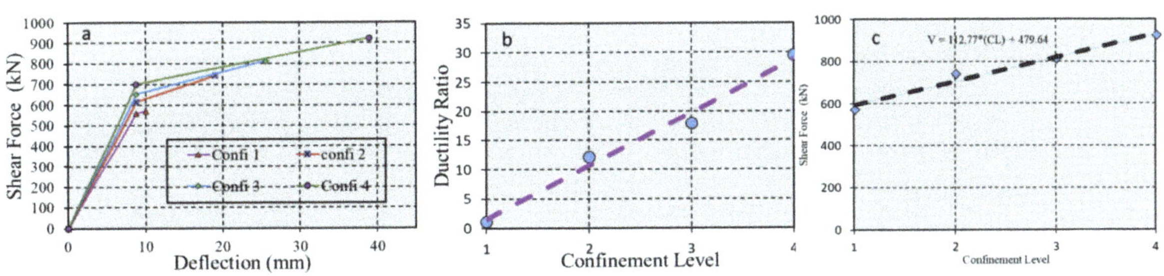

Figure 19. (**a**) Force-displacement curve, (**b**) ductility ratio, and (**c**) shear capacity as a function of confinement.

Figure 20 presents the section dimensions ratio effective ductility.

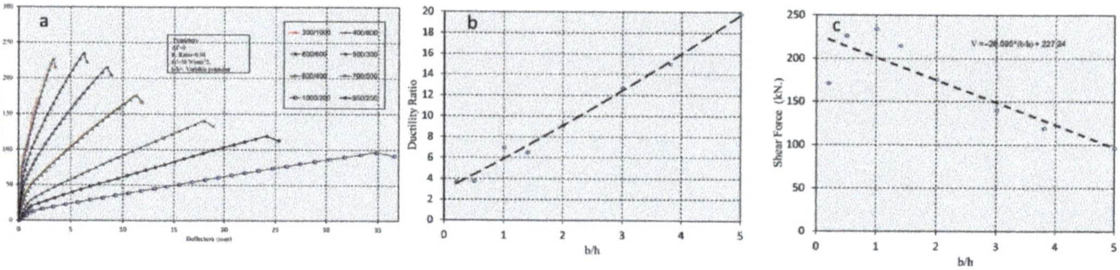

Figure 20. (**a**) Force-displacement curve, (**b**) ductility ratio, and (**c**) shear capacity for varying confinement ratios.

Figures 15–20 present summarized simulation results representing various tests, illustrating shear force-deflection and ductility as functions of the tested parameters. Figure 15a mainly shows how increasing the amount of CFRP in the tensional fibers of a concrete member significantly enhances its ductility. Notably, while improving ductility, CFRP is susceptible to brittle failure under high levels of stress or strain, potentially compromising the structural integrity of the concrete element. To prevent such brittle failure, it is advisable to use more CFRP than typically required to achieve the concrete's ultimate strength. This approach ensures that the CFRP reinforcement remains within its elastic and elastic-plastic ranges, preventing abrupt failures and enhancing the structure's resilience. Adopting this over-reinforcement strategy ensures that failures if they occur, are more likely in the compressed fibers.

Additionally, the high stiffness of CFRP reduces displacement in the compressed area, further decreasing the likelihood of brittle failure. Figure 15b demonstrates that the amount of tensioned CFRP increases with increasing ductility of the concrete. Figure 15c shows a tendency for increased shear capacity with more tensioned CFRP. For the typical stress–strain behavior of CFRP, refer to Figure 10.

In Figure 16, the reduction in CFRP in the compressive fiber modestly enhances both ductility and shear capacity. Figure 17 highlights that higher concrete strengths increase yielding stress but predispose structures to more brittle failure modes than those with lower strengths. The least strong concrete showed the lowest yielding stress but was more ductile.

Figures 18 and 19 reveal an interesting dynamic: as the cover thickness increases, the ductility ratio increases, but shear capacity decreases. Enhanced concrete confinement, as shown in Figures 19 and 20, leads to higher ductility ratios and shear capacities. These enhancements suggest a critical balance between confinement and cover adjustments to optimize ductility and strength in seismic conditions.

Finally, Figure 20 underscores that an increase in the breadth-to-height (b/h) ratio not only improves ductility but also enhances the structure's ability to absorb seismic energy, thereby boosting overall durability. The structural adjustments that enhance the b/h ratio lead to increased curvature at yield and ultimate states, significantly amplifying the ductility.

This comprehensive analysis is summarized in Table 3.

Table 3. Four-point bending of parameters affecting ductility and their level of influence.

Variable Parameter	Ductility Ratio	Shear Capacity
Single reinforcement	↑↑↑	↑↑↑

Table 3. *Cont.*

Variable Parameter	Ductility Ratio	Shear Capacity
Double reinforcement	↑	↑
Concrete Strengthening	↓	↓
Cover thickness	↑	↓↓
Confinement concrete ratio	↑↑↑	↑↑↑
Section dimensions' ratio	↑↑↑	↓↓↓

An arrow pointing up indicates improvement and a downward arrow indicates decline. The number of arrows indicates the intensity of the change.

6. Validation

A specific concrete section reinforced with CFRP (Table 4), identified as (C82) and studied by Wang et al. [44], was chosen for validation of the numerical simulations performed with the Response-2000 software, developed by Bentz (2000) [33]. A critical aspect of the validation involved analysis of the load-deflection curve (LDC) associated with the section, as depicted in Figure 21. The systematic approach used for validation ensured that the simulations accurately mirror the behavior of CFRP-reinforced concrete sections under various loading conditions, thereby boosting the reliability and practical utility of the Response-2000 software for structural engineering analyses.

Table 4. Properties of CFRP RC Sections.

Symbol	(C8)	(C10)	(C12)
Span L (mm)	2100	2100	2100
a (mm)	700	700	700
b (mm)	120	120	120
h (mm)	250	250	250
Af (bars)	2D8	2D10	2D10
Af' (bars)	2D6	2D6	2D6
Ef (MPa)	106.4	106.4	106.4
Ffu (MPa)	1628	1628	1628
$f'c$ (GPa)	42.6	42.6	42.6
Ec (GPa)	31.6	31.6	31.6
Cover (mm)	25	25	25
ρf	0.37%	0.58%	0.84%

Figure 22 presents the Load-Deflection Curves (LDC) for CFRP-Reinforced Concrete Section: Illustrated is the comparison and validation of the performance under load of the concrete section studied by Wang et al. (2021) [44]. The red line denotes numerical results, and the magenta line represents experimental outcomes.

Figure 22 shows that the numerical analyses agreed well with the test results of the CFRP-strengthened segment. Such agreement of the numerical and experimental data verified the accuracy and applicability of Response-2000 in engineering practice, especially in design and analysis involving reinforced concrete members. Discrepancies may arise due to inaccuracies in the specimen's size, support conditions, loading velocities, or other relevant factors.

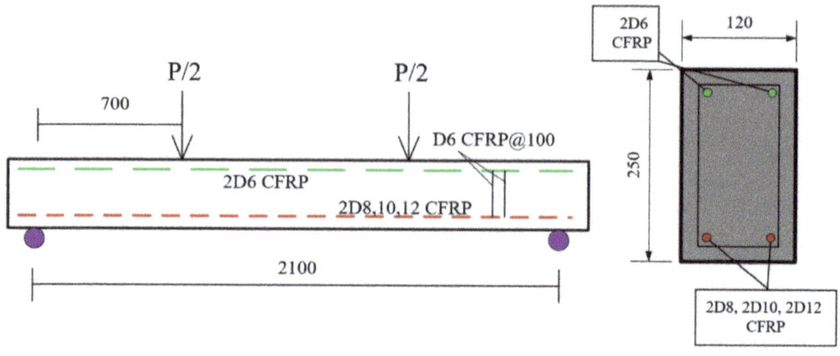

(**a**) Four-point bending scheme (**b**) cross-section

Figure 21. Section geometry and schemes: Illustrated are the dimensions of the CFRP-reinforced concrete sections, where D denotes the diameter of the CFRP bars. This specific section geometry was adapted from Wang et al. [44] study.

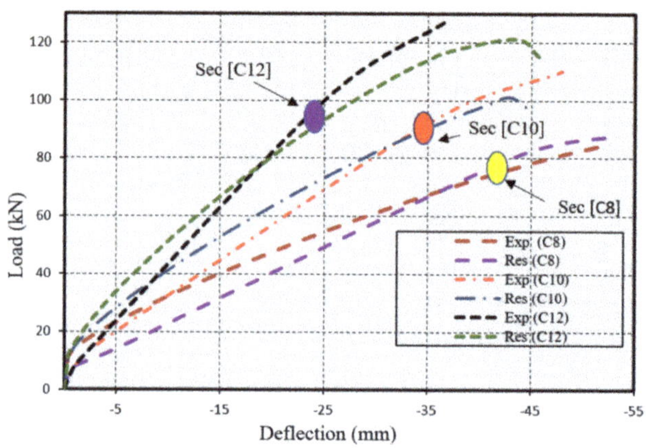

Figure 22. Load-deflection curves comparison using Response-2000 for the CFRP-reinforced section by Wang et al.

7. Conclusions and Practical Recommendations

This research proposed a strategy to enhance structural durability against seismic forces by incorporating CFRP for concrete reinforcement. Through comprehensive multi-scale simulations and evaluations, the study demonstrated that CFRP, despite its brittleness, can significantly improve the ductility of beam components in concrete structures. Key findings from this investigation can be summarized as follows:

- Increasing the amount of CFRP in the tensional zone in CRC components boosts both shear capacity and ductility ratios.
- Additional CFRP in the compressive zone, when matched with an equivalent amount in the tensional zone, does not significantly affect ductility or shear capacity.
- Lower concrete compressive strength can increase ductility and reduce brittleness and capacity.
- Increasing cover thickness enhances ductility ratios while reducing capacity.

- Decreasing the section depth can lead to increased curvature in the ultimate state of the section, and over-reinforcing the tensional fiber ensures that the compressive area improves the section's ductility while decreasing its capacity.
- Enhancing the confinement ratio in concrete compression zones increases ductility and capacity while mitigating brittleness.
- The ductility of CRC cross-section is twice that of RC with identical concrete strength and reinforcement area.
- The shear capacity of CRC and RC structures is similar when the CFRP-to-steel ratio is 26.4%.

This study contributes to the ongoing effort to develop a user-friendly and accessible design methodology for practicing engineers. The recommendations are presented below as follow:

- Increase the amount of CFRP reinforcement in the tensional zone of CRC components to improve both shear capacity and ductility ratios.
- Avoid unnecessary CFRP in the compressive zone except when matched by equivalent amounts in the tensional zone, as this has minimal impact on ductility and capacity.
- Use lower concrete compressive strength to improve ductility and reduce brittleness, recognizing the trade-off with capacity.
- Consider using reduced section depth to increase curvature at the ultimate state, which improves ductility when over-reinforcing the tensional fiber.
- The compressive area should be adequately reinforced to balance ductility with capacity.
- Increase the thickness of the concrete cover provided, which increases the ductility ratios but most likely reduces the capacity.
- Increase confinement ratios in concrete compression zones to reduce brittleness and increase ductility and capacity.
- Design the CRC component with a CFRP-to-steel ratio of about 26.4% for a similar shear capacity to the RC structure.
- Make use of the higher ductility of the CRC sections, up to twice that of the identical concrete strength and area of reinforcement of RC sections.

Author Contributions: Conceptualization, M.M., G.L.S., and E.G.; methodology, M.M., G.L.S., and E.G.; software, M.M.; validation, M.M., G.L.S., and E.G.; formal analysis, M.M.; investigation, M.M.; resources, G.L.S., and E.G.; data curation, M.M., G.L.S., and E.G., writing—original draft preparation, M.M., G.L.S., and E.G.; writing—review and editing, G.L.S., and E.G.; visualization, M.M.; supervision, G.L.S., and E.G.; project administration, G.L.S., and E.G. All authors have read and agreed to the published version of the manuscript.

Funding: This research received no external funding.

Institutional Review Board Statement: Not applicable.

Informed Consent Statement: Not applicable.

Data Availability Statement: The original contributions presented in the study are included in the article, further inquiries can be directed to the corresponding author.

Conflicts of Interest: The authors declare no conflicts of interest.

References

1. Urlainis, A.; Ornai, D.; Levy, R.; Vilnay, O.; Shohet, I.M. Loss and Damage Assessment in Critical Infrastructures Due to Extreme Events. *Saf. Sci.* **2022**, *147*, 105587. [CrossRef]
2. Urlainis, A.; Shohet, I.M.; Levy, R.; Ornai, D.; Vilnay, O. Damage in Critical Infrastructures Due to Natural and Man-Made Extreme Events—A Critical Review. *Procedia Eng.* **2014**, *85*, 529–535. [CrossRef]

3. Wei, H.-H.; Skibniewski, M.J.; Shohet, I.M.; Yao, X. Lifecycle Environmental Performance of Natural-Hazard Mitigation for Buildings. *J. Perform. Constr. Facil.* **2016**, *30*, 04015042. [CrossRef]
4. Wei, H.H.; Skibniewski, M.J.; Shohet, I.M.; Shapira, S.; Aharonson-Daniel, L.; Levi, T.; Salamon, A.; Levy, R.; Levi, O. Benefit-Cost Analysis of the Seismic Risk Mitigation for a Region with Moderate Seismicity: The Case of Tiberias, Israel. *Procedia Eng.* **2014**, *85*, 536–542. [CrossRef]
5. Urlainis, A.; Shohet, I.M. Development of Exclusive Seismic Fragility Curves for Critical Infrastructure: An Oil Pumping Station Case Study. *Buildings* **2022**, *12*, 842. [CrossRef]
6. Urlainis, A.; Shohet, I.M. Seismic Risk Mitigation and Management for Critical Infrastructures Using an RMIR Indicator. *Buildings* **2022**, *12*, 1748. [CrossRef]
7. Urlainis, A.; Shohet, I.M.; Levy, R. Probabilistic Risk Assessment of Oil and Gas Infrastructures for Seismic Extreme Events. *Procedia Eng.* **2015**, *123*, 590–598. [CrossRef]
8. Lifshitz Sherzer, G.; Urlainis, A.; Moyal, S.; Shohet, I.M. Seismic Resilience in Critical Infrastructures: A Power Station Preparedness Case Study. *Appl. Sci.* **2024**, *14*, 3835. [CrossRef]
9. Urlainis, A.; Lifshitz Sherzer, G.; Shohet, I.M. Multi-Scale Integrated Corrosion-Adjusted Seismic Fragility Framework for Critical Infrastructure Resilience. *Appl. Sci.* **2024**, *14*, 8789. [CrossRef]
10. Urlainis, A.; Shohet, I.M. A Comprehensive Approach to Earthquake-Resilient Infrastructure: Integrating Maintenance with Seismic Fragility Curves. *Buildings* **2023**, *13*, 2265. [CrossRef]
11. Al-Lami, A.; Hilmer, P.; Sinapius, M. Eco-Efficiency Assessment of Manufacturing Carbon Fiber Reinforced Polymers (CFRP) in Aerospace Industry. *Aerosp. Sci. Technol.* **2018**, *79*, 669–678. [CrossRef]
12. Maidi, M.; Sherzer, G.L.; Gal, E. Enhancing Ductility in Carbon Fiber Reinforced Polymer Concrete Sections: A Multi-Scale Investigation. *Comput. Concr.* **2024**, *33*, 385–398. [CrossRef]
13. Maidi, M.; Lifshitz Sherzer, G.; Shufrin, I.; Gal, E. Seismic Resilience of CRC- vs. RC-Reinforced Buildings: A Long-Term Evaluation. *Appl. Sci.* **2024**, *14*, 11079. [CrossRef]
14. Clyne, T.W.; Hull, D. *An Introduction to Composite Materials*; Cambridge University Cambridge: Cambridge, UK, 2019; p. 345.
15. Starkova, O.; Aniskevich, K.; Sevcenko, J. Long-Term Moisture Absorption and Durability of FRP Pultruded Rebars. *Mater. Today Proc.* **2021**, *34*, 36–40. [CrossRef]
16. Benmokrane, B.; Brown, V.L.; Ali, A.H.; Mohamed, K.; Shield, C. Reconsideration of the Environmental Reduction Factor C_E for GFRP Reinforcing Bars in Concrete Structures. *J. Compos. Constr.* **2020**, *24*, 6020001. [CrossRef]
17. Ceroni, F.; Cosenza, E.; Gaetano, M.; Pecce, M. Durability Issues of FRP Rebars in Reinforced Concrete Members. *Cem. Concr. Compos.* **2006**, *28*, 857–868. [CrossRef]
18. Imjai, T.; Garcia, R.; Guadagnini, M.; Pilakoutas, K. Strength Degradation in Curved Fiber-Reinforced Polymer (FRP) Bars Used as Concrete Reinforcement. *Polymers* **2020**, *12*, 1653. [CrossRef]
19. Bai, Y.; Keller, T. Shear Failure of Pultruded Fiber-Reinforced Polymer Composites under Axial Compression. *J. Compos. Constr.* **2009**, *13*, 234–242. [CrossRef]
20. Paulay, T. The Philosophy and Application of Capacity Design. *Sci. Iran.* **1995**, *20*, 117–136.
21. Paulay, T.; Park, R. *Reinforcd Concrete Structure*; John Wiley & Sons: Hoboken, NJ, USA, 1974; Volume 776, pp. 1–776. [CrossRef]
22. Renić, T.; Kišiček, T. Ductility of Concrete Beams Reinforced with Frp Rebars. *Buildings* **2021**, *11*, 21. [CrossRef]
23. Ivanković, A.M.; Radić, J.; Savor, Z. New Seismic Design Requirements for Bridge Structures in Croatia. In Proceedings of the FIB Symposium Dubrovnik 2007, Dubrovnik, Croatia, 20–23 May 2007.
24. *BS EN 1998-1-1:2024*; Eurocode 8. Design of Structures for Earthquake Resistance. General Rules and Seismic Action. BSI: London, UK, 2024.
25. *CSA S807-10*; Specification for Fibre-Reinforced Polymers. CSA Group: Toronto, ON, Canada, 2015; p. 44.
26. *ACI 440*; Guide for the Design and Construction of Externally Bonded FRP Systems for Strengthening Concrete Structures. ACI: Farmington Hills, MI, USA, 2002; p. 200.
27. *CSA S806-12*; Design and Construction of Building Structures with Fibre-Reinforced Polymer. Canadian Standards Association: Toronto, ON, Canada, 2012; p. 198.
28. *ACI 440.1R-15; CNR-DT 200 R1/2012*; Guide for the Design and Construction of Externally Guide for the Design and Construction of Externally Bonded FRP Systems for Strengthening Existing Structures Materials, RC and PC Structures, Masonry Structures. ACI: Farmington Hills, MI, USA, 2015.
29. Kišiček, T.; Renić, T.; Lazarević, D.; Hafner, I. Compressive Shear Strength of Reinforced Concrete Walls at High Ductility Levels. *Sustainability* **2020**, *12*, 4434. [CrossRef]
30. Badakhsh, A.; An, K.H.; Kim, B.J. Enhanced Surface Energetics of CNT-Grafted Carbon Fibers for Superior Electrical and Mechanical Properties in CFRPs. *Polymers* **2020**, *12*, 1432. [CrossRef] [PubMed]
31. Lu, Z.; Jiang, M.; Pan, Y.; Xian, G.; Yang, M. Durability of Basalt Fibers, Glass Fibers, and Their Reinforced Polymer Composites in Artificial Seawater. *Polym. Compos.* **2022**, *43*, 1961–1973. [CrossRef]

32. Maidi, M.; Shufrin, I. Evaluation of Existing Reinforced Concrete Buildings for Seismic Retrofit through External Stiffening: Limit Displacement Method. *Buildings* **2024**, *14*, 2781. [CrossRef]
33. Response-2000 Reinforced Concrete Sectional Analysis Using the Modifed Compression Field Theorty. 2016. Available online: http://www.hadrianworks.com (accessed on 9 January 2025).
34. Bazan, J.L.; Fernandez-Davila, V.I. Evaluation of the Experimental Curvature Ductility of RC Beams Externally Strengthened with CFRP Bands. *Structures* **2020**, *26*, 1010–1020. [CrossRef]
35. Ahmed, M.M.; Farghal, O.A.; Nagah, A.K.; Haridy, A.A. Effect of Confining Method on the Ductility of Over-Reinforced Concrete Beams. *J. Eng. Sci.* **2007**, *35*, 617–633. [CrossRef]
36. *ACI 318 ACI Committee 318;* Building Code Requirements for Structural Concrete (ACI 318–95) and Commentary (ACI 318R-95). American Concrete Institute: Detroit, MI, USA, 1995; Volume 163, p. 369.
37. *ACI 440R-96;* State-of-the-Art Report on Fiber Reinforced Plastic (FRP) Reinforcement for Concrete Structures. ACI: Farmington Hills, MI, USA, 2002; p. 157.
38. *EN 1992;* Design of Concrete Structures. European Committee for Standardization: Lausanne, Switzerland, 1992.
39. Lifshitz Sherzer, G.; Schlangen, E.; Ye, G.; Gal, A.E. Evaluating Compressive Mechanical LDPM Parameters Based on an Upscaled Multiscale Approach. *Constr. Build. Mater.* **2020**, *251*, 118912. [CrossRef]
40. Binici, B. An Analytical Model for Stress-Strain Behavior of Confined Concrete. *Eng. Struct.* **2005**, *27*, 1040–1051. [CrossRef]
41. Suriyati; Ridwan; Djauhari, Z.; Romey Sitompul, I. Seismic Performance of Building Reinforced with CFRP Bars. In *MATEC Web of Conferences;* EDP Sciences: Les Ulis, France, 2019; Volume 276, p. 1021.
42. Zhou, Y.; Zheng, Y.; Sui, L.; Hu, B.; Huang, X. Study on the Flexural Performance of Hybrid-Reinforced Concrete Beams with a New Cathodic Protection System Subjected to Corrosion. *Materials* **2020**, *13*, 234. [CrossRef]
43. Zhou, F.; Zhang, J.; Song, S.; Yang, D.; Wang, C. Effect of Temperature on Material Properties of Carbon Fiber Reinforced Polymer (CFRP) Tendons: Experiments and Model Assessment. *Materials* **2019**, *12*, 1024. [CrossRef]
44. Song, S.; Wang, G.; Min, X.; Duan, N.; Tu, Y. Experimental Study on Cyclic Response of Concrete Frames Reinforced by Steel-CFRP Hybrid Reinforcement. *J. Build. Eng.* **2021**, *34*, 101937. [CrossRef]

polymers

MDPI

Article

Enhancing Radiation Shielding Efficiency of *Nigella sativa* Eumelanin Polymer Through Heavy Metals Doping

Mohammad Marashdeh * and Nawal Madkhali

Department of Physics, College of Sciences, Imam Mohammad Ibn Saud Islamic University (IMSIU), Riyadh 13318, Saudi Arabia; namadkhali@imamu.edu.sa
* Correspondence: mwmarashdeh@imamu.edu.sa

Abstract: Gamma radiation shielding is necessary for many applications; nevertheless, lead creates environmental risks. Eumelanin, a natural polymer, is a viable alternative, although its effectiveness is limited to lower gamma-ray energy. This research looks at how doping the herbal eumelanin polymer (*Nigella sativa*) with heavy metals including iron (Fe), copper (Cu), and zinc (Zn) affects its gamma radiation shielding characteristics. The inclusion of these metals considerably increases the linear attenuation coefficient (μ) and mass attenuation coefficient (μ_m) of eumelanin, especially at lower photon energies where the photoelectric effect is prominent. The μ value of pure eumelanin is 0.193 cm^{-1} at 59.5 keV. It goes up to 0.309 cm^{-1}, 0.420 cm^{-1}, and 0.393 cm^{-1} when Fe, Cu, and Zn are added, in that order. Similarly, the mass attenuation coefficients increase from 0.153 cm^2/g for pure eumelanin to 0.230, 0.316, and 0.302 cm^2/g for the Fe-, Cu-, and Zn-doped samples. At intermediate and higher energies (661.7 keV-to-1332.5 keV), where Compton scattering is the main interaction, differences in attenuation coefficients between samples are not as noticeable, which means that metal additions have less of an effect. The mean free path (MFP) and radiation protection efficiency (RPE) also show these behaviors. For example, at 59.5 keV the MFP drops from 5.172 cm for pure eumelanin to 3.244 cm for Mel-Fe, 2.385 cm for Mel-Cu, and 2.540 cm for Mel-Zn. RPE values also go up a lot at low energies. For example, at 59.5 keV Cu-doped eumelanin has the highest RPE of 34.251%, while pure eumelanin only has an RPE of 17.581%. However, at higher energies the RPE values for all samples converge, suggesting a more consistent performance. These findings suggest that doping eumelanin with Fe, Cu, and Zn is particularly effective for enhancing gamma-ray shielding at low energies, with copper (Cu) providing the most significant improvement overall, making these composites suitable for applications requiring enhanced radiation protection at lower gamma-ray energies.

Keywords: eumelanin; mass attenuation coefficient; metal doping; mean free path; radiation protection efficiency

1. Introduction

Ionizing radiation is an omnipresent aspect of modern life, emanating from both natural and artificial sources. While it has beneficial applications in medicine, industry, and research, it poses significant threats to human health and the environment. Exposure to ionizing radiation can lead to serious health problems, including cancer, damage to the nervous system, and mutations in human DNA [1,2]. According to the World Health Organization (WHO), even low doses of radiation can increase the risk of adverse health effects, making it crucial to develop effective protective measures against such exposures [3].

This underscores the need for robust radiation shielding materials, particularly in environments where ionizing radiation is prevalent, such as hospitals, research laboratories, and nuclear facilities.

Ionizing radiation finds applications across various sectors, including healthcare, agriculture, military, and industrial processes. In healthcare, for instance, radiotherapy uses ionizing radiation to treat cancer, targeting malignant cells while minimizing damage to surrounding healthy tissue [4–9]. In agriculture, radiation can be employed for pest control and food preservation, enhancing crop yields and extending shelf life [6]. Military applications include the use of radiation detection equipment for national security purposes, while industrial radiography is essential for the non-destructive testing of materials and components [7,8]. Despite these benefits, the potential risks associated with ionizing radiation necessitate effective shielding to protect both personnel and the public.

Radiation shielding materials serve a critical function in mitigating the harmful effects of ionizing radiation. These materials are employed in various forms, including structural components of buildings, protective suits for personnel, containers for radioactive materials, and windows in medical imaging equipment [9,10]. For example, radiation shielding is vital in diagnostic imaging procedures such as X-rays and computed tomography (CT) scans to protect patients and healthcare providers from unnecessary exposure [11]. Furthermore, in nuclear power plants effective shielding is necessary not only for the protection of workers but also to ensure the safety of surrounding communities [10].

Historically, metallic lead and lead-based compounds have been the primary materials used for shielding against X-rays and gamma rays due to their high density and atomic number, which make them effective at attenuating radiation [11]. However, the use of lead comes with significant drawbacks. Lead is a toxic and carcinogenic substance, and its corrosive properties can lead to environmental contamination [12]. Moreover, lead is ineffective against neutron radiation, producing high-energy secondary ionizing radiation when exposed. These limitations have prompted researchers to explore alternative materials that are non-toxic and environmentally friendly [13,14]. One promising candidate is eumelanin, a complex biological polymer that plays a crucial role in determining the color of skin, hair, and eyes in various organisms [15,16]. Eumelanin is produced through the melanogenesis process, which involves the polymerization of the amino acid tyrosine. Melanin is categorized into categories based on the molecular precursor from which the pigment is derived. Eumelanin (black melanin) consists primarily of oligomers of 5,6-dihydroxyindole (DHI) and 5,6-dihydroxyindole-2-carboxylic acid (DHICA) [17].

Beyond its aesthetic functions, eumelanin possesses unique physicochemical properties that have attracted considerable scientific attention in recent years. Notably, eumelanin is capable of absorbing a wide range of electromagnetic radiation, including ultraviolet (UV), visible, and infrared (IR) light [17,18]. This broad absorption spectrum not only supports its protective functions against UV radiation but also opens new avenues for its use in materials science, electronics, and biomedical engineering [19,20]. Natural eumelanin is a complex biopolymer that plays a crucial role in protecting organisms from harmful UV radiation and oxidative stress. Additionally, scientists explore the effects of metal modification on eumelanin's optical properties after UV irradiation. This research shows that eumelanin, when doped with transition metals such as Fe, Co, and Zn, retains substantial absorption characteristics across a broad spectrum (200–500 nm). Density functional theory (DFT) analysis indicates that these metal modifications enhance eumelanin's structural properties and optical reactivity, potentially contributing to treatments for skin diseases associated with excessive eumelanin secretion [18,21,22]. Recent studies have highlighted the potential of eumelanin as a bioinspired material for advanced technological applications. For instance, researchers have investigated its role in developing novel materials for solar energy harvesting and photoprotection [17]. Eumelanin's ability to absorb and dissipate excess energy makes it an attractive candidate for various applications in nanotechnology and drug delivery systems [18]. Furthermore, its biocompatibility and biodegradability enhance its appeal in biomedical applications, where the safety and environmental impact of materials are of paramount importance [19]. The incorporation of metal elements into eumelanin has been shown to significantly enhance its physical and mechanical properties. Research indicates that metal-doped eumelanin exhibits improved mechanical strength, thermal stability, optical absorption, electrical conductivity, and radioprotective qualities [23]. For instance, the addition of metal ions such as silver, copper, and zinc has been shown to improve the radiation attenuation properties of eumelanin, making it a suitable candidate for use in radiation shielding applications. The enhanced characteristics of metal-doped eumelanin composites offer new opportunities for their integration into high-tech applications, including electronic devices, protective coatings, and specialized radiation shielding materials. The simulation results support the spatial arrangement hypothesis, which is that the organization of melanin in enclosure structure may give additional shielding. The arrangement of melanin on the cell membrane may limit the absorbed dose of highly energetic particles at the nucleus, but this impact is cumulative across the fungal colony. A lattice of melanin ghosts will provide consistent shielding for colony cells in an anisotropic radiation environment. The bulk Sepia melanin shielding effect towards the gamma radiation of 122–140 keV energies showed that 4 mm of melanin cut the dose by ~33% and the linear attenuation coefficient was calculated to be 1.01 cm^{-1} [24–28].

In the present study, we focus on synthesizing eumelanin composites using three metallic elements: iron (Fe), copper (Cu), and zinc (Zn). These metals were chosen due to their availability, cost-effectiveness, and potential for enhancing the radiation shielding properties of eumelanin. The synthesized eumelanin samples will undergo measurements of linear and mass attenuation coefficients within the photon energy range of 59.9 keV-to-1332 keV, which is critical for assessing their effectiveness against X-rays and gamma rays. We will compare our experimental results with those calculated using the XCOM program developed by Berger and Hubbell, which provides a reliable database for photon cross-sections over a wide energy range [29]. Additionally, this study will examine several key radiation shielding parameters, including the mean free path (MFP), radiation protection efficiency (RPE), half-value layer (HVL), and effective atomic number (Z_{eff}) for the fabricated eumelanin samples. These parameters are essential for evaluating the suitability of the synthesized composites for radiation shielding applications. By systematically exploring these aspects, this research aims to contribute valuable insights into the potential of metal-doped eumelanin as a viable alternative to conventional radiation shielding materials.

2. Materials and Methods

2.1. Sample Preparation

Natural eumelanin was extracted from *Nigella sativa*, as mentioned in Ref. [18]. To this, 15 g of *Nigella sativa* eumelanin was treated with 100 mL of sodium hydroxide (2 M) and sonicated using an ultrasound device for 3 h at 100 °C. About 50 mL of hydrochloric acid (2 M) was distilled with a burette while stirring the concentrated melanin mixture and monitoring to adjust the pH to 6.5.

For the preparation of iron-doped eumelanin (Fe-Mel), iron(II) chloride ($FeCl_3$) was added to the eumelanin solution in a 1:10 ratio. The resulting mixture was left in each sample for three weeks while the color changed from red to transparent in the melanin sample with iron chloride, red to transparent in the eumelanin sample with copper chloride, and white to transparent in the melanin sample with zinc oxide. Subsequently, the samples were centrifuged at 3000 rpm for 2 min, and then dried in an oven at 60 °C for 48 h. Similar procedures were followed to create copper-doped (Cu-Mel) and zinc-doped eumelanin (Zn-Mel) using $CuCl_2$ and $ZnCl_2$, respectively. All resulting compounds were all black in color. A PerkinElmer 580 B-IR spectrometer (industrialbid, Las Vegas, NV, USA) was used to obtain Fourier transform infrared (FTIR) spectra. The SERON-AIS 1800C scanning electron microscope (SEM), Seron Technologies Inc., Uiwang, Republic of Korea: was used to study the morphology and has a resolution of ∼20 µm. The APEX™ EDS software program was used for collecting and analyzing energy-dispersive X-ray spectroscopy (EDS/EDX). The synthesis process is illustrated in Figure 1. XRD measurements were conducted using a Discover diffractometer (Bruker (Billerica, MA, USA): D8 with Cu-Kα radiation, λ = 1.5406Å).

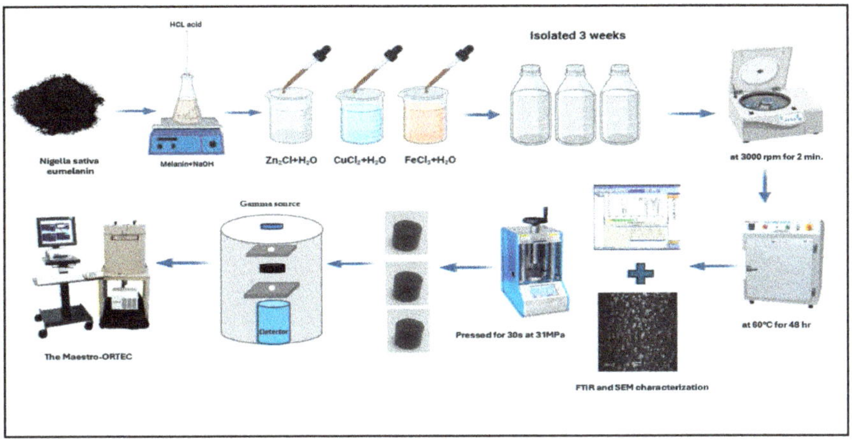

Figure 1. Schematic of the process followed for the synthesis of eumelanin doped with Zn, Fe, and Cu ions.

The samples were formed into discs after characterization by FTIR and SEM using a manual hydraulic press machine for 30 s at 31 MPa, as shown in Figure 1. The density of the samples was then determined by dividing their mass (in grams) by their volume (in cubic centimeters).

2.2. *Evaluations of Gamma-Ray Shielding Parameters*

Theory:

The primary parameter under investigation is the linear attenuation coefficient, μ, which can be calculated using Equation (1) [28].

$$I = I_0\, e^{-\mu x} \tag{1}$$

where x represents the thickness of the sample measured in (cm) and I and I_0 denote the beam intensity after traversing the thickness and the initial intensity, respectively. The linear attenuation coefficient (μ) was used to calculate the mass attenuation coefficient (μ_m) using the following formula:

$$\mu_m = \frac{\mu}{\rho} \tag{2}$$

This formula ensures an accurate evaluation of attenuation properties among materials with varying densities by establishing an individual relationship between μ and μ_m.

The error of the $\Delta\ \mu_m$ measurement is calculated using Equation (3).

$$\frac{\Delta\mu_m}{\mu_m} = \sqrt{\left(\frac{\Delta\rho x}{\rho x}\right)^2 + \left(\frac{\Delta I}{I}\right)^2 + \left(\frac{\Delta I_o}{I_o}\right)^2} \quad (3)$$

where $\Delta\ \rho x$, ΔI_o, and ΔI are the error of the mass density, unattenuated photon intensity, and attenuated photon intensity, respectively.

The half-value layer (HVL) is the thickness required to cut the amount of radiation in half and is calculated as shown in Equation (4) [30]:

$$\text{HVL} = \ln 2/\mu \quad (4)$$

The mean free path (MFP), defined as the average distance between two consecutive gamma photon interactions, is measured in centimeters (cm) [31]. The mean free path can be determined using Equation (5):

$$\text{MFP} = 1/\mu \quad (5)$$

where μ is the linear attenuation coefficient in inverse centimeters (cm^{-1}).

In addition, radiation protection efficiency (RPE) can be employed to evaluate the shielding effectiveness of composite samples based on the linear attenuation coefficient μ, as described by the following relation [32].

$$\text{RPE} = \left(1 - e^{-\mu}\right) \times 100 \quad (\%) \quad (6)$$

The mass attenuation coefficients are used to calculate the total atomic cross-section, which can be determined using Equation (7).

$$\sigma_a = \frac{(\mu_m)_{sample}}{N_A \sum_i^n (W_i / A_i)} \quad (7)$$

In this equation, N_A represents Avogadro's number, and A_i represents the atomic Weight of the constituent element in the sample. The total electronic cross-section for the element can be found using Equation (8) [33].

$$\sigma_l = \frac{1}{N_A} \sum_i^n \frac{f_i\ A_i}{Z_i} (\mu_m)_i \quad (8)$$

In this equation, fi represents the number of atoms of element I relative to the total number of atoms in the composite material, and Z_i represents the atomic number of the ith element in the sample. The selection of a suitable material for radiation dosimetry and detection relies heavily on the effective atomic number (Z_{eff}) and effective electron number (N_{eff}). To obtain the compound's effective atomic number (Z_{eff}), one may calculate the ratio between the total atomic cross-section and the total electronic cross-section using Equation (9) [23].

$$Z_{eff} = \frac{\sigma_a}{\sigma_l} \quad (9)$$

2.3. Experimental Setup

The fabricated eumelanin samples were irradiated using three typical point sources: ^{214}Am (59.5 keV), ^{137}Cs (661.7 keV), and ^{60}Co (1173.2 and 1332.5 keV). A sodium iodide (NaI (TI)) scintillation detector measuring 2″ × 2″ was employed to quantify the energy intensity. To protect the system from background radiation and scattering, a lead container was used. The Maestro-ORTEC (version 7.2) was utilized to analyze and identify gamma spectra. Each sample was exposed to 1800 s of radiation to provide reliable statistical results. Figure 2 illustrates the use of two collimators, each with a diameter of 0.3 cm, positioned in front of the source and detector. These collimators are made of lead and shaped like a disc with a thickness of 1.5 cm. The center of each collimator has a hole with a diameter of 0.3 cm. The distance between the point source and the sample was 10.3 cm, while the distance between the point source and the detector was 14.6 cm. Afterwards, the source was placed, and the initial gamma radiation, referred to as Io, was detected. Ultimately, the gamma counts for each material were determined at different thicknesses. To provide an unbiased evaluation, we calculated the relative counts (I/I_0) for different thickness values of the samples and subsequently evaluated them.

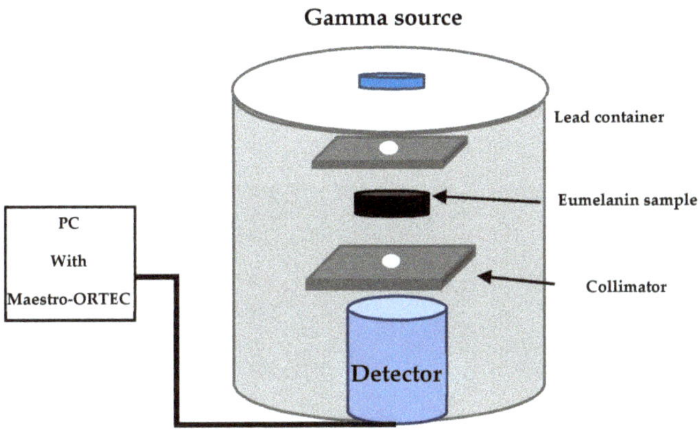

Figure 2. Schematic of the experimental setup.

3. Results and Discussion

Characterization

SEM examination can provide insight into the structural organization of eumelanin. Figure 3 shows that all samples lacked a typical structural structure. Eumelanin samples were shown as small compasses and aggregated likes hetero-polymer consisting of diverse units. That decreased after being doped with transition metals (Fe, Zn, and Cu). The EDX analysis was applied to analyze the elemental and chemical composition of the eumelanin. The EDX results indicated the amplest elements with the superlative compositions for carbon atoms with Weights from 90% to 62.78%. The EDX results showed the highest concentrations of elements, particularly carbon atoms, which ranged from 90% to 62.78%. In contrast, the percentage of oxygen decreased significantly from 22% after doping with transition metals, dropping to 9.8% in the zinc sample (Mel-Zn). Meanwhile, the copper and iron samples each exhibited oxygen levels of approximately 7%. This indicates that metal–oxide bonds are likely forming due to the doping process with transition metals. TEM images analyzed the nano-size and cluster -shape of eumelanin. The structure appears to be sheet-like, with a rather uniform texture. Small granules are observed on the surface, indicating nanoparticle dispersal. Doping with metal ions results in a more defined, organized structure than pure eumelanin.

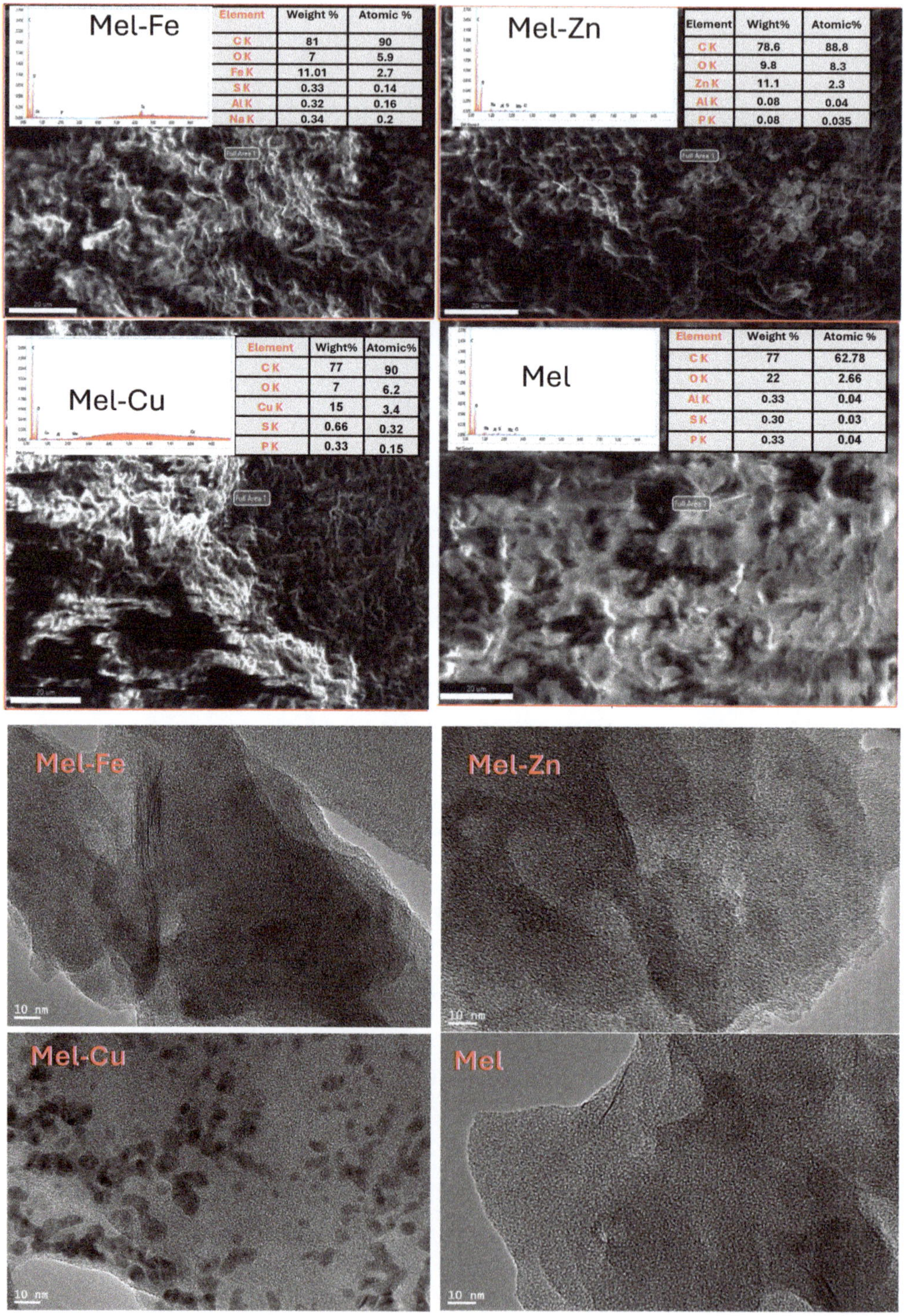

Figure 3. SEM micrographs and TEM images of eumelanin doped with Zn, Fe, and Cu ions.

As shown in Figure 4, The FTIR transmission spectra for pure eumelanin and metal-doped (Zn, Fe, and Cu) ions samples were analyzed in the range of 450-to-4000 cm^{-1}. The spectrum of eumelanin reveals the presence of the catechol (O-H) group, characterized by a broad band at 3400 cm^{-1}. Additionally, (C-H) stretching vibrations are indicated by bands at 2910 cm^{-1} and 2841 cm^{-1}. The band observed at 1638 cm^{-1} is attributed to the presence of carboxylate (COO^-) and/or aromatic (C=C) groups [24]. The signals in the range of 2800–3800 cm^{-1} arise from (O-H) and (N-H) stretching vibrations, reflecting the presence of both hydroxyl and amine groups. Furthermore, bands at 1042 cm^{-1} and 1530 cm^{-1} are associated with carboxylic acid and phenolic groups, while (C-H) vibrations contribute to the peak in the range of 671–698 cm^{-1}. Notably, changes occur in the FTIR spectrum upon bonding of Zn, Fe, and Cu ions to natural eumelanin. The (O-H) group shifts from 3404 cm^{-1} to a higher wavenumber, indicating that the metal ions are interacting with the oxygen atoms in the functional groups of eumelanin. The peak intensity at 1640 cm^{-1}, related to COO^-, decreases and shifts slightly to 1638 cm^{-1}, suggesting a bonding interaction between the metal ions and the ionized acid group. When Fe and Cu ions are present, the stretching modes associated with (C=C) or (COO^-) at 2912 cm^{-1} and 2840 cm^{-1} shift to 2926 cm^{-1} and 2856 cm^{-1}, respectively. In contrast, the FTIR spectra indicate that Zn ions do not significantly affect the structure of eumelanin. The XRD patterns show eumelanin before and after doping with Zn, Fe, and Cu ions. The XRD findings indicate that all of our samples have amorphous structures. The peak at 20° represents the production of eumelanin. Furthermore, there is a minor shift associated with a significant decrease in the strength of this peak in all co-doped samples, indicating that metal ions are connected to eumelanin. Furthermore, no additional Fe, Zn, or Cu peaks arise in the XRD patterns, indicating that the doped metal ions are well bound to the eumelanin molecule.

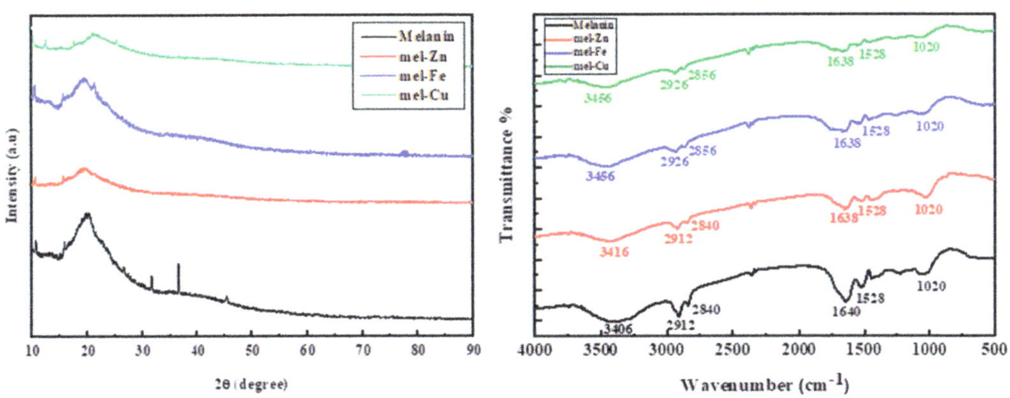

Figure 4. XRD patterns and FTIR spectra of eumelanin and Fe, Cu, and Zn ions.

The densities of Fe, Cu, and Zn samples with eumelanin were calculated to determine the physical properties of the compounds and to measure radiation properties. Table 1 presents the experimental values of the calculated densities, revealing that the density of the samples varies depending on the component compositions and the type of material added to the eumelanin compound. In general, the addition of Fe, Cu, and Zn compounds to eumelanin clearly increased the density of the samples in different proportions; the Mel-Fe sample showed the highest density, with percentage increases of 6.01%, 4.98%, and 3.16% compared to the Mel density of the Mel-Fe, Mel-Cu, and Mel-Zn samples, respectively.

Table 1. Composition and calculated density of fabricated eumelanin samples.

Samples	Element Composition (%)									Density (g/cm^3)
	C	O	Al	S	P	Fe	Cu	Na	Zn	
Mel	77.00	22.00	0.33	0.30	0.33	-	-	-	-	1.264
Mel-Fe	81.00	7.00	0.32	0.33	-	11.01		0.34	-	1.340
Mel-Cu	77.00	7.00	-	0.66	0.33	-	15.00	-	-	1.327
Mel-Zn	78.60	9.80	0.08	0.08	0.08	-	-	-	11.10	1.304

Energy-dispersive X-ray spectroscopy (EDX) is a key analytical technique for determining the elemental composition of samples, which is essential for various applications in radiation measurements. By identifying and measuring the elements present, EDX helps to theoretically calculate properties such as the mass attenuation coefficient. mass attenuation coefficient describes a material's ability to attenuate X-rays or gamma rays, and allows comparison with experimental results. This analysis is critical for developing materials with specific radiation attenuation properties, such as those used in radiation shielding, nuclear science, and radiology. EDX is often used in conjunction with techniques such as scanning electron microscopy (SEM) to provide comprehensive material characterization and microstructural analysis, allowing researchers to optimize materials for use in radiation environments. Table 1 displays the mass distribution ratios of the constituent materials in the fabricated eumelanin samples. The mass distribution ratios of 15% were used for all Fe, Cu, and Zn samples that were added to the eumelanin. However, the ratios of 11% were used for the Fe and Zn samples after they were made for theoretical and experimental calculations.

The addition of iron (Fe), copper (Cu), and zinc oxide (Zn) to eumelanin leads to notable changes in both the linear and mass attenuation coefficients, suggesting that these additives enhance eumelanin's ability to attenuate gamma radiation. The experimental results for four different gamma energies—59.5 keV (from ^{214}Am), 661.7 keV (from ^{137}Cs), 1173.2 keV, and 1332.5 keV (from ^{60}Co)—were obtained by averaging the results for each sample in each photon energy.

Figure 5 shows that the linear attenuation coefficient (μ) is higher for eumelanin samples with added Fe, Cu, and Zn compared to pure eumelanin (Mel), especially at lower gamma energies due to the higher atomic numbers of Fe, Cu, and Zn, which enhance photon absorption through the photoelectric effect. At 59.5 keV (^{214} Am), pure eumelanin (Mel) has a μ value of 0.193, while Mel-Fe increases significantly to 0.309, Mel-Cu shows an even higher value of 0.420, and Mel-Zn has an elevated μ value of 0.393. This substantial increase for Cu and Zn additives indicates their strong enhancement of photon absorption via the photoelectric effect, with Cu providing the most significant increase in attenuation at lower gamma energies. However, at higher gamma energies (661.7 keV from ^{137}Cs, 1173.2 keV from ^{60}Co, and 1332.5 keV from ^{60}Co), the differences in μ between the additives are less pronounced. For instance, at 661.7 keV the μ values for Mel-Cu, Mel-Fe, and Mel-Zn are 0.085, 0.087, and 0.088, respectively, which are close to each other. At these higher energies, all samples, even pure eumelanin, have μ values that are pretty close to each other. This means that Compton scattering, not the photoelectric effect, is the main way that the samples interact, with the addition of Fe, Cu, and Zn having almost no effect on the attenuation.

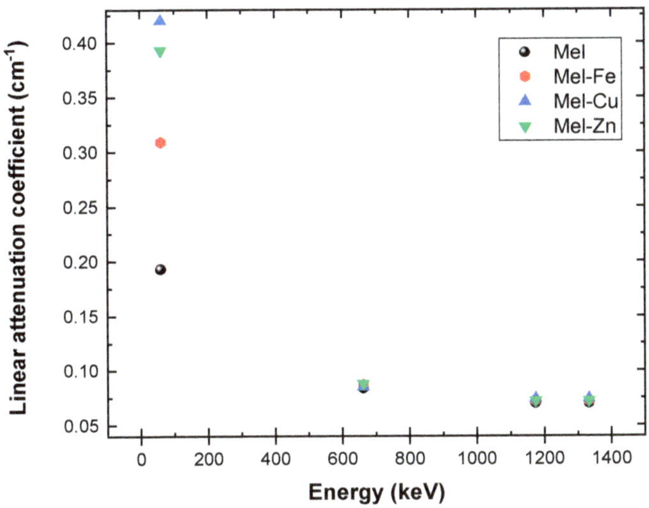

Figure 5. Measurement of the relationship between the linear attenuation coefficient and photon energy in fabricated eumelanin samples.

An essential measure for evaluating the gamma-ray shielding efficiency of materials, regardless of their density, is the mass attenuation coefficient (μ_m). As shown in Table 2, adding Fe, Cu, and Zn to eumelanin makes the mass attenuation coefficients much higher at lower gamma energies, like 59.5 keV, where the photoelectric effect is strongest. Eumelanin, in its pure form, has a μ_m value of 0.153. However, the inclusion of Fe, Cu, and Zn raises the μ_m values to 0.230, 0.316, and 0.302, respectively. This trend demonstrates that the addition of higher-atomic-number elements (Fe, Cu, and Zn) leads to an increase in photon absorption through the photoelectric effect. Among these additions, Cu and Zn exhibit the most notable increase. At gamma energies in the middle, like 661.7 keV, when Compton scattering is the main way that particles interact, the mass attenuation coefficients for all samples are very close to each other. They are Mel (0.067), Mel-Fe (0.065), Mel-Cu (0.064), and Mel-Zn (0.068). This suggests that the influence of the atomic number is less significant, and that electron density has a greater effect on attenuation in this range. At higher energy levels (1173.2 keV and 1332.5 keV from ^{60}Co), the mass attenuation coefficients are even more similar across all samples, with values that are very close to each other. This observation indicates that the impact of the additives is insignificant when Compton scattering is the dominating factor. Furthermore, the error values seen in Table 2 are indicative of the experimental difficulties encountered when measuring attenuation coefficients at various energies and for different materials. Higher error rates were seen at lower gamma energies (59.5 keV), ranging from 0.014 to 0.028. This could be because the method is more sensitive to changes in sample preparation, thickness, and photoelectric interactions, especially when high-Z additions like Cu and Zn are used. By comparison, the reduced errors seen at intermediate and high energies (661.7 keV, 1173.2 keV, and 1332.5 keV) indicate more dependable results. These data are characterized by a decreased sensitivity of the primary interaction mechanisms (Compton scattering) to these parameters, with error values ranging from 0.003 to 0.005.

Table 2. Linear and mass attenuation coefficients of fabricated eumelanin samples measured experimentally.

Samples	^{214}Am (59.5 keV)			^{137}Cs (661.7 keV)			^{60}Co (1173.2)			^{60}Co (1332.5 keV)		
	μ	μ_m	Error (±%)	μ	μ_m	Error (±%)	μ	μ_m	Error (±%)	μ	μ_m	Error (±%)
Mel	0.193	0.153	0.014	0.084	0.067	0.008	0.070	0.056	0.005	0.068	0.054	0.004
Mel-Fe	0.309	0.230	0.017	0.087	0.065	0.007	0.072	0.054	0.006	0.069	0.052	0.004
Mel-Cu	0.420	0.316	0.028	0.085	0.064	0.010	0.074	0.056	0.004	0.070	0.052	0.003
Mel-Zn	0.393	0.302	0.024	0.088	0.068	0.007	0.072	0.056	0.006	0.068	0.053	0.005

Figure 6 displays the quantitative mass attenuation coefficients obtained from experimental measurements in our investigation within the energy range of 59.5 keV–1332.5 keV, together with the theoretical values within the broader range of 10 keV–2000 keV. The observed trend is that the μ_m exhibits a rapid drop as the photon energy increases, while it increases with the Weight of the additive material (Fe, Cu, and Zn). It is clear that the μ_m is affected by both the filler mass and photon energy. This aligns with the findings reported in the study conducted by [34]. These findings indicate that the inclusion of Fe, Cu, and Zn is very efficient in reducing the intensity of low-energy gamma rays. This makes these composites appropriate for applications such as medical imaging or radiation shielding, where low-energy photons are common. Nevertheless, when dealing with higher-energy gamma rays, where Compton scattering is the primary factor, the impact of these additions on mass attenuation is insignificant. In such cases, other material characteristics, such as mechanical strength or cost, may become more crucial in the choice of materials for radiation shielding. This study is very important for making eumelanin-based composites that are specifically designed for certain radiation shielding tasks based on the gamma rays that are used.

The differences in the μ_m of the fabricated eumelanin samples, as presented in Table 3, can be attributed to many aspects associated with the samples' characteristics, the settings in which the experiments were conducted, and the constraints of the theoretical model. For example, the sample Mel exhibits a percentage deviation of 15.00% at 59.5 keV, which then drops to 1.82% at 1332.5 keV. The fact that the deviation decreases as the energy level rises suggests that lower photon energies are more affected by factors like sample composition, uneven density, and possible contaminants that can change attenuation measurements. Furthermore, Mel-Fe demonstrates a significant divergence of 21.23% at 59.5 keV and 5.45% at 1332.5 keV, suggesting a considerable influence of Fe inclusion on photon interactions, especially at lower energy levels when photoelectric absorption is the dominant factor. The different deviations seen in different samples, like Mel-Cu (19.59% at 59.5 keV and 3.70% at 1332.5 keV) and Mel-Zn (14.69% at 59.5 keV and 3.64% at 1332.5 keV), show how the different atomic numbers and cross-sections of the elements used have an effect. The greater disparities observed at lower energy levels may be attributed to the heightened probability of scattering and absorption phenomena, as well as the partial penetration of photons. Conversely, at higher energy levels a tighter correspondence between experimental and theoretical values suggests a diminished influence of these secondary effects. Furthermore, the observed values are crucially influenced by experimental parameters such as detector efficiency, source-to-sample distance, and beam alignment. These factors contribute to minor variations of WinXCOM's theoretical predictions, which assume optimal situations.

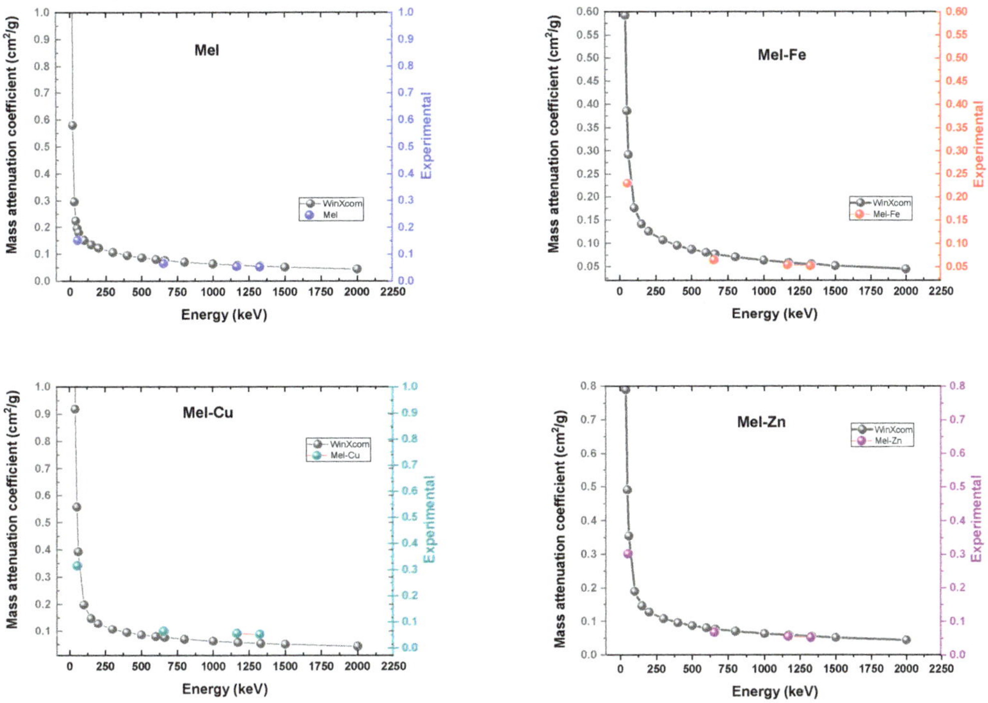

Figure 6. Theoretical (WinXCOM) and experimental mass attenuation coefficients for eumelanin samples.

Table 3. Comparison of experimental and WinXCOM results based on the mass attenuation coefficients of the fabricated eumelanin samples.

Samples	Energy (keV)	Mass Attenuation Coefficients (cm^2/g)		
		Experimental	Theoretical (WinXCOM)	Percentage Deviation%
Mel	59.5	0.153	0.180	15.00
	661.7	0.067	0.077	12.99
	1173.2	0.056	0.059	5.08
	1332.5	0.054	0.055	1.82
Mel-Fe	59.5	0.230	0.292	21.23
	661.7	0.065	0.077	15.58
	1173.2	0.054	0.058	6.90
	1332.5	0.052	0.055	5.45
Mel-Cu	59.5	0.316	0.393	19.59
	661.7	0.064	0.076	15.79
	1173.2	0.056	0.058	3.45
	1332.5	0.052	0.054	3.70
Mel-Zn	59.5	0.302	0.354	14.69
	661.7	0.068	0.077	11.69
	1173.2	0.056	0.058	3.45
	1332.5	0.053	0.055	3.64

Table 4 shows that comparing μ_m at 59.5 keV is crucial, as this low gamma energy is where the photoelectric effect dominates, making it highly relevant for evaluating the efficiency of shielding materials. With a μ_m of 0.153 cm^2/g (experimental) and 0.180 cm^2/g (theoretical), eumelanin showed a lot of promise as a radiation shield in this study. Compared to widely used polymeric shielding materials such as poly(methyl methacrylate) (μ_m = 0.109 cm^2/g) and polypropylene (0.126 cm^2/g), eumelanin achieves approximately 40%- and 21%-higher attenuation, respectively, highlighting its superior performance for use in radiation panels and medical dosimetry tools, where conventional materials often fall short in low-energy gamma shielding. Also, eumelanin works well against aromatic polymers that are known for having high attenuation properties. For example, phenol-formaldehyde resin = 0.173 cm^2/g and polycarbonate = 0.172 cm^2/g are only 11%- and 12%-different from eumelanin. These materials are commonly used in radiation detector housings and industrial shielding, but they lack the biocompatibility and eco-friendly properties that eumelanin offers. Unlike these conventional materials, eumelanin provides dual functionality, combining effective gamma-ray attenuation with free radical scavenging properties, making it highly valuable for biomedical applications, including wearable protective garments, medical implants, and device coatings, where both radiation protection and biological safety are critical.

Table 4. Mass attenuation coefficient values compared to other materials studied at 59.5 keV in the literature.

Study	Study Theme	Material	μ_m(cm^2/g)
Singh et al. [35]	Simulation (MCNP)	phenol-formaldehyde resin (ρ =1.36 g/cm^3)	0.173
	Simulation (MCNP)	Polycarbonate (ρ = 1.22 g/cm^3)	0.172
Kucuk et al. [36]	Experimental	Poly(methyl methacrylate) (ρ = 1.18 g/cm^3)	0.109
	Experimental	Polypropylene (ρ = 0.946 g/cm^3)	0.126
	Experimental	Polyethylene (ρ = 0.920 g/cm^3)	0.112
This study	Experimental	eumelanin (Mel)	0.153
	Theoretical (XCOM)	eumelanin (Mel)	0.180
	Experimental	Mel-Fe	0.230
	Experimental	Mel-Cu	0.316
	Experimental	Mel-Zn	0.302

The HVL values of eumelanin polymer samples with added Fe, Cu, and Zn were obtained at 59.5 keV, 661.7 keV, 1173.2 keV, and 1332.5 keV. At 59.5 keV, the half-value layer (HVL) measurements for Mel, Mel-Fe, Mel-Cu, and Mel-Zn were 3.591 cm, 2.243 cm, 1.650 cm, and 1.763 cm, respectively, as shown in Figure 7. Copper exhibited the greatest attenuation at 54.0%, followed by zinc at 50.9% and iron at 37.6%, signifying an increased photoelectric absorption attributable to their elevated atomic numbers. At 661.7 keV, the half-value layer (HVL) measurements were 8.250 cm for Mel, 7.966 cm for Mel-Fe, 8.153 cm for Mel-Cu, and 7.875 cm for Mel-Zn. Zinc had the most attenuation (4.5%) because it has the best atomic number and density for Compton scattering. At 1173.2 keV, the half-value layer (HVL) measurements were 9.900 cm (Mel), 9.625 cm (Mel-Fe), 9.365 cm (Mel-Cu), and 9.625 cm (Mel-Zn), with copper exhibiting the most significant drop at 5.4%. At 1332.5 keV, the half-value layer (HVL) measurements were 10.191 cm (Mel), 10.043 cm (Mel-Fe), 9.900 cm (Mel-Cu), and 10.191 cm (Mel-Zn), with copper exhibiting the lowest HVL value at 2.9%. Copper doping consistently showed better effectiveness across all energy levels, establishing it as the most effective dopant, but zinc and iron also enhanced shielding.

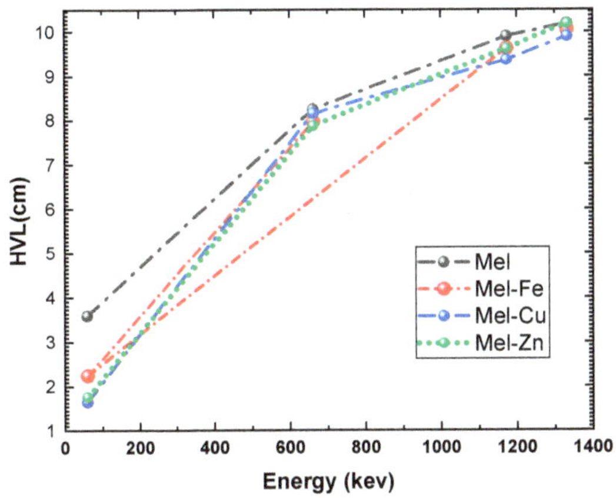

Figure 7. Half-value layer (HVL) of the fabricated eumelanin samples at photon energies of 59.5 keV, 661.7 keV, 1173.2 keV, and 1332.5 keV.

The mean free path (MFP) is an essential quantity in radiation shielding that quantifies the average distance covered by a photon within a material prior to its interaction. The mean free energy (MFP) values for eumelanin (Mel) and its composites with iron (Fe), copper (Cu), and zinc (Zn) at different photon energies (59.5 keV, 661.7 keV, 1173.2 keV, and 1302.5 keV) are shown in Figure 8. The results show that adding Fe, Cu, and Zn to eumelanin greatly lowers the MFP at lower energy levels, which suggests better attenuation properties. At 59.5 keV, pure eumelanin has an MFP of 5.172 cm. This value significantly reduces to 3.244 cm for Mel-Fe, 2.385 cm for Mel-Cu, and 2.540 cm for Mel-Zn. The observed decrease shows that some additions, especially copper, make it much easier for photons to be absorbed by the photoelectric effect. This results in fewer mean free pathways and better shielding at low energy levels. The experimental findings are in excellent agreement with the theoretical predictions (WinXCOM), with minor deviations likely caused by experimental circumstances and sample preparation.

The variations in the MFP among the samples become less significant at higher photon energies (661.7 keV, 1173.2 keV, and 1332.5 keV) when Compton scattering is the dominant factor. At an energy level of 661.7 keV, the MFP values exhibit considerable proximity—11.810 cm for Mel, 11.478 cm for Mel-Fe, 11.775 cm for Mel-Cu, and 11.280 cm for Mel-Zn—demonstrating little deviation. This behavior persists at 1173.2 keV and 1332.5 keV, when the MFP values for all samples approach a range of 13–15 cm. This suggests that the additives have a reduced influence on the attenuation characteristics at these higher energy levels. The findings indicate that the addition of Fe, Cu, and Zn to eumelanin greatly enhances its ability to protect against lower energy levels. Among these, Cu exhibits the highest level of efficacy, but its impact decreases at higher photon energies.

The radiation protection efficiency (RPE) values for eumelanin (Mel) and its composites with Fe, Cu, and Zn indicate significant variations in shielding effectiveness across different photon energies, as shown in Figure 9. At a low energy of 59.5 keV, pure eumelanin exhibits an RPE of 17.581%, which is notably increased by adding Fe, Cu, and Zn, resulting in RPE values of 26.529%, 34.251%, and 32.545%, respectively. This shows that Cu offers the highest improvement in radiation protection at this energy level. As the photon energy increases to 661.7 keV, the RPE values decrease, showing less differentiation between the samples—8.119% for Mel, 8.344% for Mel-Fe, 8.142% for Mel-Cu, and 8.483% for Mel-Zn—indicating a modest enhancement with these additives. At the even higher energies of 1173.2 keV and 1332.5 keV, the RPE values continue to converge, with Mel having 6.833%

and 6.597%, Mel-Fe having 6.982% and 6.733%, Mel-Cu having 7.162% and 6.668%, and Mel-Zn having 7.041% and 6.676%, respectively. These results suggest that while Fe, Cu, and Zn significantly improve the radiation protection efficiency of eumelanin at lower energies, their impact becomes more uniform and less pronounced at higher energies.

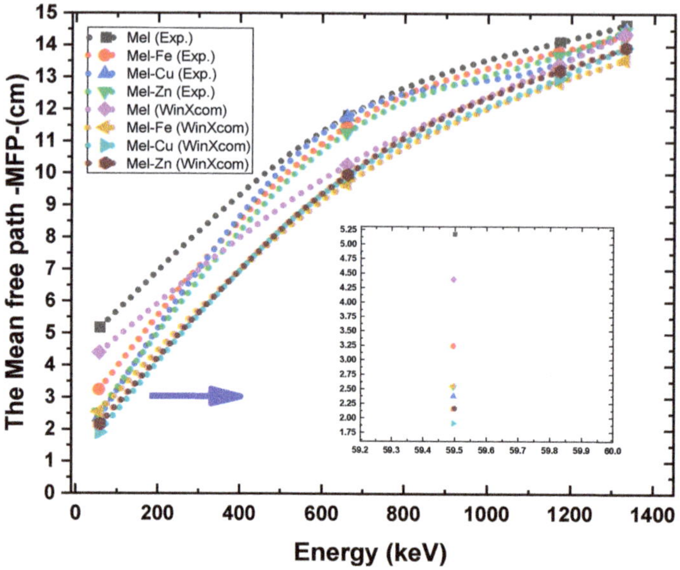

Figure 8. The mean free path (MFP) for the Mel, Mel-Fe, Mel-Cu, and Mel-Zn samples at photon energies of 59.5 keV, 661.7 keV, 1173.2 keV, and 1332.5 keV.

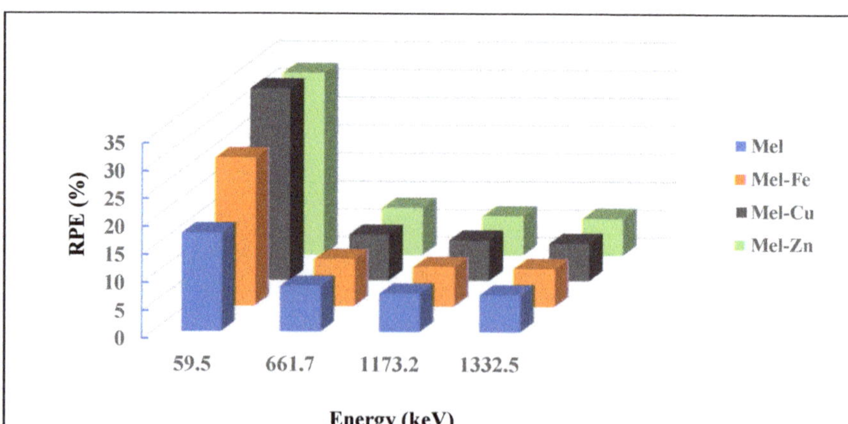

Figure 9. Relationship between RPE and energy in the Mel, Mel-Fe, Mel-Cu, and Mel-Zn samples.

As shown in Figure 10, the effective atomic number (Z_{eff}) results for pure eumelanin (Mel) and its composites with Fe, Cu, and Zn show a significant increase in radiation shielding effectiveness due to metal doping, particularly at lower photon energies. Figure 10 shows that at 59.9 keV the Z_{eff} increases from 5.478 for pure eumelanin to 7.514 for Mel-Fe, 9.655 for Mel-Cu, and 9.462 for Mel-Zn, demonstrating that Cu provides the highest enhancement. As the photon energy increases, the Z_{eff} values for all samples converge, indicating less variation between the doped and undoped samples. At 661.6 keV, the

Z_{eff} values are 5.544 for Mel, 5.645 for Mel-Fe, 5.831 for Mel-Cu, and 5.980 for Mel-Zn, while at 1332 keV they are 6.253, 6.342, 6.681, and 6.483, respectively. This convergence suggests that the impact of metal additives on Z_{eff} is more pronounced at lower energies, where radiation shielding improvements are more significant, highlighting the potential of Cu-doped eumelanin for optimized shielding applications.

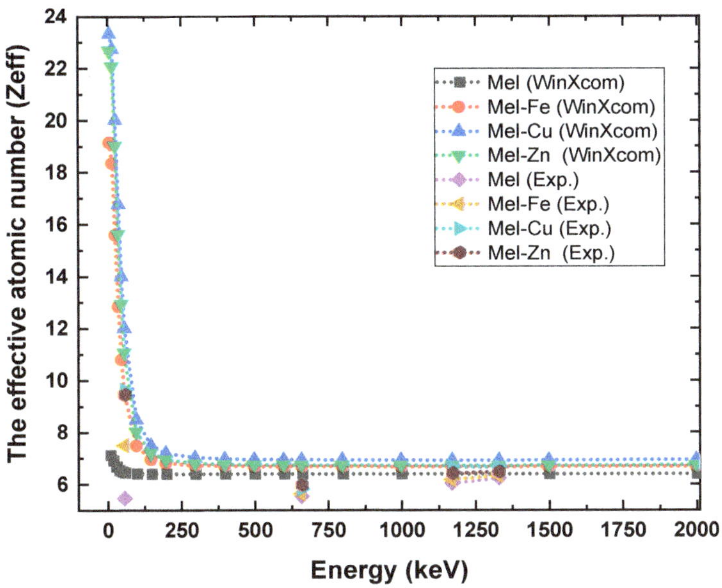

Figure 10. The effective atomic number (Z_{eff}) of the fabricated eumelanin samples.

4. Conclusions

This study demonstrates that doping eumelanin with metals like iron (Fe), copper (Cu), and zinc (Zn) significantly improves its gamma radiation shielding properties, particularly at lower photon energies. The results show that both the linear (μ) and mass attenuation coefficients (μ_m) are notably higher for doped samples compared to pure eumelanin. For instance, at a low energy of 59.5 keV pure eumelanin exhibits a μ value of 0.193, which increases to 0.309 for Mel-Fe, 0.420 for Mel-Cu, and 0.393 for Mel-Zn. Similarly, the mass attenuation coefficient increases from 0.153 cm2/g for pure eumelanin to 0.230, 0.316, and 0.302 cm^2/g for Mel-Fe, Mel-Cu, and Mel-Zn, respectively. These increases are attributed to the higher atomic numbers of Fe, Cu, and Zn, which enhance photon absorption via the photoelectric effect. However, at intermediate and higher energies (661.7 keV-to-1332.5 keV), where Compton scattering dominates, the differences in attenuation coefficients among the samples diminish, indicating a reduced impact of the metal additives.

Furthermore, the mean free path (MFP) and radiation protection efficiency (RPE) results further demonstrate the benefits of metal doping at a low photon energy. At 59.5 keV, the MFP is reduced from 5.172 cm for pure eumelanin to 3.244 cm for Mel-Fe, 2.385 cm for Mel-Cu, and 2.540 cm for Mel-Zn, indicating improved photon absorption and enhanced shielding effectiveness. The RPE values at the same energy level show a substantial increase, with pure eumelanin having an RPE of 17.581%, which rises to 26.529% for Mel-Fe, 34.251% for Mel-Cu, and 32.545% for Mel-Zn. According to these results, all three additives make eumelanin a better shielding material at low-energy gamma rays, but copper (Cu) does the best overall, showing the highest improvement in both attenuation coefficients and radiation protection efficiency.

Author Contributions: Conceptualization, M.M. and N.M.; Methodology, M.M. and N.M.; Validation, M.M; Formal analysis, M.M. and N.M.; Investigation, M.M. and N.M.; Resources, N.M.; Data curation, M.M. and N.M.; Writing—original draft, M.M.; Writing—review and editing, N.M.; Funding acquisition, M.M. All authors have read and agreed to the published version of the manuscript.

Funding: This work was supported and funded by the Deanship of Scientific Research at Imam Mohammad Ibn Saud Islamic University (IMSIU) (grant number IMSIU-DDRSP2502).

Institutional Review Board Statement: Not applicable.

Data Availability Statement: Data are contained within the article.

Conflicts of Interest: The authors declare no conflicts of interest.

References

1. Samarth, R.; Kumar, M.; Matsumoto, Y.; Manda, K. The Effects of Ionizing and Non-Ionizing Radiation on Health. *Recent Trends Adv. Environ. Health* **2020**, *179*, 180–203.
2. Thomas, G.A.; Symonds, P. Radiation exposure and health effects–is it time to reassess the real consequences? *Clin. Oncol.* **2016**, *28*, 231–236. [CrossRef] [PubMed]
3. World Health Organization. *Ionizing Radiation, Health Effects and Protective Measures*; World Health Organization: Geneva, Switzerland, 2016.
4. Harrison, C.; Weaver, S.; Bertelsen, C.; Burgett, E.; Hertel, N.; Grulke, E. Polyethylene/boron nitride composites for space radiation shielding. *J. Appl. Polym. Sci.* **2008**, *109*, 2529–2538. [CrossRef]
5. Hashim, A.; Agool, I.R.; Kadhim, K.J. Novel of (polymer blend-Fe_3O_4) magnetic nanocomposites: Preparation and characterization for thermal energy storage and release, gamma ray shielding, antibacterial activity and humidity sensors applications. *J. Mater. Sci. Mater. Electron.* **2018**, *29*, 10369–10394. [CrossRef]
6. Harish, V.; Nagaiah, N.; Prabhu, T.N.; Varughese, K.T. Preparation and characterization of lead monoxide filled unsaturated polyester based polymer composites for gamma radiation shielding applications. *J. Appl. Polym. Sci.* **2009**, *112*, 1503–1508. [CrossRef]
7. Marashdeh, M.; Abdulkarim, M. Determination of the attenuation coefficients of epoxy resin with carbopol polymer as a breast phantom material at low photon energy range. *Polymers* **2023**, *15*, 2645. [CrossRef] [PubMed]
8. Marashdeh, M.W.; Mahmoud, K.A.; Akhdar, H.; Tharwat, M. Structural, physical, optical, and gamma ray shielding properties of SnO^2-based boro-silicate glasses: The influence of substituting Na^2O by SnO^2. *Nucl. Eng. Technol.* **2024**, *56*, 3804–3811. [CrossRef]
9. Zohuri, B. *Heat Pipe Applications in Fission Driven Nuclear Power Plants*; Springer International Publishing: Berlin/Heidelberg, Germany, 2019.
10. Part, N.G. *Radiation Protection and Safety of Radiation Sources International Basic Safety Standards*, Interim ed.; International Atomic Energy Agency: Vienna, Austria, 2011.
11. AbuAlRoos, N.J.; Amin, N.A.B.; Zainon, R. Conventional and new lead-free radiation shielding materials for radiation protection in nuclear medicine: A review. *Radiat. Phys. Chem.* **2019**, *165*, 108439. [CrossRef]
12. Raj, K.; Das, A.P. Lead pollution: Impact on environment and human health and approach for a sustainable solution. *Environ. Chem. Ecotoxicol.* **2023**, *5*, 79–85. [CrossRef]
13. Safari, A.; Rafie, P.; Taeb, S.; Najafi, M.; Mortazavi, S.M.J. Development of Lead-Free Materials for Radiation Shielding in Medical Settings: A Review. *J. Biomed. Phys. Eng.* **2024**, *14*, 229–244. [PubMed]
14. Wozniak, A.I.; Ivanov, V.S.; Zhdanovich, O.A.; Nazarov, V.I.; Yegorov, A.S. Modern approaches to polymer materials protecting from ionizing radiation. *J. Port Sci. Res.* **2017**, *33*, 2148–2163. [CrossRef]
15. Karkoszka, M.; Rok, J.; Wrześniok, D. Eumelanin Biopolymers in Pharmacology and Medicine—Skin Pigmentation Disorders, Implications for Drug Action, Adverse Effects and Therapy. *Pharmaceuticals* **2024**, *17*, 521. [CrossRef] [PubMed]
16. Qiu, Z.; Wang, S.; Zhao, J.; Cui, L.; Wang, X.; Cai, N.; Li, H.; Ren, S.; Li, T.; Shu, L. Synthesis and structural characteristics analysis of eumelanin pigments induced by blue light in Morchella sextelata. *Front. Microbiol.* **2023**, *14*, 1276457. [CrossRef]
17. Madkhali, N.; Alqahtani, H.R.; Al-Terary, S.; Laref, A.; Hassib, A. Control of optical absorption and fluorescence spectroscopies of natural eumelanin at different solution concentrations. *Opt. Quantum Electron.* **2019**, *51*, 227. [CrossRef]
18. Madkhali, N.; Algessair, S. Exploring the optical properties of metal-modified eumelanin following ultraviolet irradiation: An experimental and theoretical study using density functional theory. *Heliyon* **2024**, *10*, e29287. [CrossRef]
19. Laref, A.; Madkhali, N.; Alqahtani, H.R.; Wu, X.; Laref, S. Electronic structures and optical spectroscopies of 3d-transition metals-doped eumelanin for spintronic devices application. *J. Magn. Magn. Mater.* **2019**, *491*, 165513. [CrossRef]
20. Xie, W.; Dhinojwala, A.; Gianneschi, N.C.; Shawkey, M.D. Interactions of Eumelanin with Electromagnetic Radiation: From Fundamentals to Applications. *Chem. Rev.* **2024**, *124*, 7165–7213. [CrossRef]

21. Madkhali, N.; Alqahtani, H.R.; Al-Terary, S.; Laref, A.; Haseeb, A. The doping effect of Fe, Cu and Zn ions on the structural and electrochemical properties and the thermostability of natural eumelanin extracted from *Nigella sativa* L. *J. Mol. Liq.* **2019**, *285*, 436–443. [CrossRef]
22. Madkhali, N. The optical and electronic properties of DHI monomer of eueumelanin doped with transition metals (TMs) based on LSDA approximation in DFT theory. *AIP Conf. Proc.* **2023**, *2872*, 020012.
23. Liu, H.; Yang, Y.; Liu, Y.; Pan, J.; Wang, J.; Man, F.; Zhang, W.; Liu, G. Eumelanin-like nanomaterials for advanced biomedical applications: A versatile platform with extraordinary promise. *Adv. Sci.* **2020**, *7*, 1903129. [CrossRef] [PubMed]
24. Araujo, M.; Xavier, J.R.; Nunes, C.D.; Vaz, P.D.; Humanes, M. Marine sponge melanin: A new source of an old biopolymer. *Struct. Chem.* **2012**, *23*, 115–122. [CrossRef]
25. Chen, L.-M.; Liu, Y.-N. Surface-enhanced Raman detection of eueumelaninon silver-nanoparticle-decorated silver/carbon nanospheres: Effect of metal ions. *ACS Appl. Mater. Interfaces* **2011**, *3*, 3091–3096. [CrossRef] [PubMed]
26. Wang, Y.; Su, J.; Li, T.; Ma, P.; Bai, H.; Xie, Y.; Chen, M.; Dong, W. A novel UV-shielding and transparent polymer film: When bioinspired dopamine–eumelanin hollow nanoparticles join polymers. *ACS Appl. Mater. Interfaces* **2017**, *9*, 36281–36289. [CrossRef] [PubMed]
27. Vasileiou, T.; Summerer, L. A biomimetic approach to shielding from ionizing radiation: The case of melanized fungi. *PLoS ONE* **2020**, *15*, e0229921. [CrossRef] [PubMed]
28. Dadachova, E.; Casadevall, A. Melanin Nanoshells for Protection Against Radiation and Electronic Pulses. U.S. Patent No. 8,586,090, 19 November 2013.
29. *NBSIR-87-3597*; XCOM: Photon Cross Sections on a Personal Computer. National Bureau of Standards: Washington, DC, USA, 1987.
30. Sharma, A.; Sayyed, M.; Agar, O.; Kaçal, M.; Polat, H.; Akman, F. Photon-shielding performance of bismuth oxychloride-filled polyester concretes. *Mater. Chem. Phys.* **2020**, *241*, 122330. [CrossRef]
31. Naseer, K.A.; Marimuthu, K.; Mahmoud, K.A.; Sayyed, M.I. Impact of Bi_2O_3 modifier concentration on barium–zincborate glasses: Physical, structural, elastic, and radiation-shielding properties. *Eur. Phys. J. Plus* **2021**, *136*, 116. [CrossRef]
32. Marashdeh, M.; Al-Hamarneh, I.F. Evaluation of Gamma Radiation Properties of Four Types of Surgical Stainless Steel in the Energy Range of 17.50–25.29 keV. *Materials* **2021**, *14*, 6873. [CrossRef] [PubMed]
33. Akhdar, H.; Marashdeh, M.W.; AlAqeel, M. Investigation of gamma radiation shielding properties of polyethylene glycol in the energy range from 8.67 to 23.19 keV. *Nucl. Eng. Technol.* **2022**, *54*, 701–708. [CrossRef]
34. Alshareef, R.; Marashdeh, M.; Almurayshid, M.; Alsuhybani, M. Study of radiation attenuation properties of HDPE/ZnO at energies between 47.5 and 266 keV. *Prog. Nucl. Energy* **2023**, *165*, 104909. [CrossRef]
35. Singh, V.P.; Shirmardi, S.P.; Medhat, M.E.; Badiger, N.M. Determination of mass attenuation coefficient for some polymers using Monte Carlo simulation. *Vacuum* **2015**, *119*, 284–288. [CrossRef]
36. Kucuk, N.; Cakir, M.; Isitman, N.A. Mass attenuation coefficients, effective atomic numbers and effective electron densities for some polymers. *Radiat. Prot. Dosim.* **2013**, *153*, 127–134. [CrossRef]

polymers

MDPI

Article

Investigations on Thermal Transitions in PDPP4T/PCPDTBT/AuNPs Composite Films Using Variable Temperature Ellipsometry

Paweł Jarka [1], Barbara Hajduk [2,*], Pallavi Kumari [2], Henryk Janeczek [2], Marcin Godzierz [2], Yao Mawuena Tsekpo [1] and Tomasz Tański [1]

[1] Department of Engineering Materials and Biomaterials, Silesian University of Technology, 18a Konarskiego Str., 41-100 Gliwice, Poland; pawel.jarka@polsl.pl (P.J.); yao.tsekpo@polsl.pl (Y.M.T.)

[2] Centre of Polymer and Carbon Materials, Polish Academy of Sciences, 34 Marie Curie-Skłodowska Str., 41-819 Zabrze, Poland; pkumari@cmpw-pan.pl (P.K.); hjaneczek@cmpw-pan.pl (H.J.); mgodzierz@cmpw-pan.pl (M.G.)

* Correspondence: bhajduk@cmpw-pan.pl

Abstract: Herein, we report a comprehensive investigation on the thermal transitions of thin films of poly [2,5-bis(2-octyldodecyl)pyrrolo[3,4-c]pyrrole-1,4(2H,5H)-dione -3,6-diyl)-alt-(2,2′;5′,2″;5″,2‴-quaterthiophen-5,5‴-diyl)]PDPP4T, poly[2,6-(4,4-bis-(2-ethy-lhexyl)-4H-cyclopenta [2,1-b;3,4-b′]dithiophene)-alt-4,7(2,1,3-benzothiadiazole)] PCPDTBT, 1:1 blend of PDPP4T and PCPDTBT, and their composites with gold nanoparticles (AuNPs). The thermal transitions of these materials were studied using variable temperature spectroscopic ellipsometry (VTSE), with differential scanning calorimetry (DSC) serving as the reference method. Based on obtained VTSE results, for the first time, we have determined the phase diagrams of PDPP4T/PCPDTBT and their AuNPs composites. The VTSE measurements revealed distinct thermal transitions in the thin films, including characteristic temperatures corresponding to the pure phases of PDPP4T and PCPDTBT within their blends. These transitions were markedly different in the AuNPs composites compared to the neat materials, highlighting the unique interactions between the polymer matrix and AuNPs. Additionally, we explored the optical properties, surface morphology, and crystallinity of the materials. We hypothesize that the observed variations in thermal transitions, as well as the improvement in optical properties and crystallinity, are likely influenced by localized surface plasmon resonance (LSPR) and passivation phenomena induced by the AuNPs in the composite films. These findings could have important implications for the design and optimization of materials for optoelectronic applications.

Keywords: variable temperature ellipsometry; thermal transitions; polymer/nanoparticle composites; thin films

1. Introduction

The widespread integration of semiconducting organic polymers into the organic electronics industry has yet to be fully realized. However, due to their low cost and high efficiency, these materials have been proven to be a promising candidate for future renewable energy sources (RESs). One of the most critical challenges currently being addressed by numerous scientific research groups is the thermal stability of these materials. The study of thermal transitions in organic semiconductor materials is one of the essential factor for optimizing the performance and stability of organic electronic devices. Addressing this issue requires comprehensive studies of their morphology (both surface and internal),

https://doi.org/10.3390/polym17050704

structural properties, and the impact of fabrication parameters [1,2], which influence the chemical and physical structure of active layers in devices, such as organic light-emitting diodes (OLEDs) [3,4] and organic field-effect transistors (OFETs) [5,6]. These studies typically focus on factors such as polymer molecular weight, thin-film deposition conditions, solvent selection, and variations in the composition of donor-acceptor polymer blends, all of which significantly affect charge transport and the overall performance of electronic devices [7–11]. A major research direction involves exploring polymer with inorganic metal oxides and metal nanoparticle (NP) composites for enhancing the electronic and optical properties of polymers [12,13], which are widely utilized across multiple industries [14–18]. In the field of organic electronics, the most commonly used inorganic metal nanoparticles are gold and silver nanoparticles [19,20]. These metal nanoparticles enhance light absorption primarily through the effect of their surface plasmon resonance (SPR). When exposed to light, the nanoparticles support the collective oscillations of conduction electrons at specific frequencies that resonate with the incident light. This phenomenon significantly amplifies the light absorption, with the degree of enhancement depending on the size and morphology of the nanoparticles. For instance, Pillai et al. demonstrated a 33% increase in photocurrent for silicon solar cells based on thin films by incorporating metallic nanoparticles. This improvement results from the enhanced absorption or scattering of light, driven by the size-dependent properties of the particles, which enable more efficient light harvesting [21]. Gold nanoparticles (AuNPs), in particular, are a highly effective tool in manipulating light–matter interactions within organic solar cell (OSC) architectures. AuNPs can be used in both single and mixed morphologies, providing plasmonic enhancement and light trapping properties. The unique SPR characteristics of AuNPs allow for the manipulation and control of light at the nanoscale, optimizing the interaction between light and the material [22].

To improve the thermal and electrical characteristics of organic electronics, one potential research area in the accounts is the study of composite films that combine semiconducting organic polymers with inorganic nanoparticles, namely gold nanoparticles (AuNPs). One such promising composite system is based on two semiconducting polymers, PDPP4T and PCPDTBT, which are known for their small optical band gaps and high charge carrier mobility, making them ideal candidates for organic solar cells [23,24]. In PDPP4T, the diketopyrrolopyrrole (DPP) unit enhances electron affinity, charge transport, and π-π stacking, while the thiophene segments improve backbone conjugation and charge mobility. Similarly, PCPDTBT incorporates cyclopenta[2,1-b;3,4-b′]dithiophene (CPDT) for high charge mobility and strong π-π interactions, while benzothiadiazole (BT) boosts electron affinity and charge separation. The alkyl side chains of both polymers improve solubility and processability making it well-suited for solution-based fabrication [25]. Incorporating gold nanoparticles (AuNPs) into blends of these polymers offers a unique opportunity to study how nanoparticles influence thermal transitions in these materials, ultimately affecting the material properties and device performance.

In this study, we utilize variable temperature ellipsometry to investigate the thermal transitions in PDPP4T/PCPDTBT/AuNPs composite films. This powerful technique enables the precise measurement of the film's optical properties as a function of temperature, providing valuable insights into the behavior of the composite films. Key thermal transitions, such as the glass transition temperature (T_g), cold crystallization temperature (T_{cc}), and melting temperature (T_m), are critical for understanding the stability and operational limits of organic materials in electronic applications [26–30]. By studying these transitions, we aim to uncover how the incorporation of AuNPs affects the thermal properties of the polymers and polymer composites and explore potential pathways for improving the performance and longevity of organic electronic devices like organic solar cells and light-emitting diodes.

A significant innovation of this work is the development of a phase diagram for the PDPP4T/PCPDTBT blend, created for the first time using variable temperature spectroscopic ellipsometry, with differential scanning calorimetry (DSC) serving as the reference technique [26,27]. This approach builds upon previous studies where thermal transitions were determined using raw ellipsometric data (temperature-induced changes in ellipsometric angles Ψ and Δ at a selected wavelength). Along the thermal transition, we also studied the optical properties and surface morphology. Our results indicate that incorporating AuNPs significantly impacts the thermal stability and crystallization behavior of the material, which is crucial for applications in organic field-effect transistors (OFETs). This work presents a novel contribution to the field, shedding light on the detail investigation on thermal transition of PDPP4T/PCPDTBT/AuNPs composites and their potential for use in organic electronics.

2. Experimental

2.1. Materials

We have used two commercial, semiconducting polymers, poly[2,6-(4,4-bis-(2-ethylhexyl)-4H-cyclopenta [2,1-b;3,4-b′]dithiophene)-alt-4,7(2,1,3-benzothiadiazole)]M2096A3—PCPDTBT (molar mass M_w = 66.499 kDa) [31–33] and poly[2,5-bis(2-octyldodecyl)pyrrolo[3,4-c]pyrrole-1,4(2H,5H)-dione -3,6-diyl)-alt-(2,2′;5′,2″;5″,2‴-qua-terthiophen-5,5‴-diyl)] M333—PDPP4T (molar mass M_w = 84 kDa) [25,34,35], which were supplied by Ossila (Ossila Ltd., Sheffield, UK). The chemical structures of these polymers are presented in Figure 1a,b. The gold nanoparticles—AuNPs with a 20 nm diameter were supplied by Sigma-Aldrich (Merck KGaA, Darmstadt, Germany).

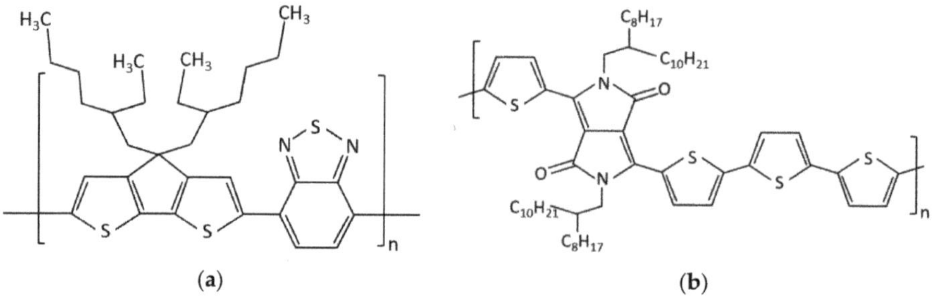

Figure 1. Chemical structures of PCPDTBT (**a**) and PDPP4T (**b**).

2.2. Samples Preparation

The polymers PCPDTBT and PDPP4T were dissolved in chloroform at a concentration of 10 mg/mL. Solutions containing AuNPs were prepared with a 10% weight concentration of nanoparticles, while maintaining a constant overall concentration of 10 mg/mL. The specific weight concentrations are provided in Tables 1 and 2.

Table 1. The individual concentrations of pure materials and their blends.

Sample No	1	2	3	4	5
PCPDTBT [%]	100	75	50	25	0
PDPP4T [%]	0	25	50	75	100

Table 2. The individual concentrations of the composites of polymers and their blends.

Sample No	6	7	8	9	10
PCPDTBT [%]	90	70	45	20	0
PDPP4T [%]	0	20	45	70	90
Au [%]	10	10	10	10	10

All solutions were placed in an ultrasonic bath at 40 °C for approximately 30 min. The films were prepared using spin-coating and casting techniques onto silicon substrates (covered with 90 nm of SiO_2) and microscope slides (0.1 mm thick). The casting technique was used for the very thick samples intended for DSC measurements, where the material was scratched from the surface. The thickness of the prepared films was determined ellipsometrically and is presented in Table 3, below.

Table 3. The individual thickness values of prepared samples using spin coating, deposited onto silicon and microscopic glass substrates (number of samples are the same with Tables 1 and 2).

		1	2	3	4	5	6	7	8	9	10
d [nm]	SiO_2	111	113	114	93	110	90	100	116	111	97
	glass	416	511	480	473	490	503	489	470	501	490

The substrates were cleaned following a standard procedure, which included rinsing with deionized water, ultrasonic cleaning with organic solvents (acetone and methanol), and drying in a centrifuge. The prepared films were annealed on a hot plate for 10 min at 120 °C to remove any residual solvent, then stored at room temperature in a dry laboratory box. The films were kept in a rubber-sealed storage container, partially filled with a hygroscopic gel, and stored in a nitrogen glove box. The films deposited onto silicon substrates were used for ellipsometric. In case of films deposited onto microscopic glass substrates, the prepared samples were used for XRD and microscopic analysis.

2.3. Methods

Ellipsometric measurements were conducted using a SENTECH SE 850E ellipsometer (Sentech, Krailling, Germany), operated with SpectraRay 3 software, covering a spectral range of 240–2500 nm. Temperature control during measurements was achieved using a temperature chamber under low vacuum conditions (10^{-1} Tr). The temperature was regulated by an INSTEC mK 1000 controller (Instec, Boulder, CO, USA), utilizing an electrical heater and a liquid nitrogen circuit. The optical windows of the vacuum chamber allowed for measurements at a 70° incidence angle. A standard temperature protocol was applied, where each sample was heated to 300 °C for 2 min and then rapidly cooled to −100 °C. Temperature curves were recorded during the heating cycles at a rate of 2 °C/min.

Differential scanning calorimetry (DSC) measurements were performed using a DSC Q2000 (TA Instruments, Newcastle, DE, USA) with aluminum sample pans. Thermal characteristics were assessed under a nitrogen atmosphere with a gas flow rate of 50 mL/min. The instrument was calibrated using high-purity indium standards, and the heating and cooling rates were set to 20 °C/min. DSC measurements were taken for the powders of PDPP4T and PCPDTBT, as well as for their 50% blend, which were scratched from the glass substrate.

X-ray diffraction (XRD) scans were carried out on polymer films deposited on Si/SiO_2 substrates using a D8 Advance diffractometer (Bruker, Karlsruhe, Germany) with a Cu-Kα

cathode (λ = 1.54 Å) in coupled Two-Theta/Theta mode. The scan rate was 1.2°/min with a step size of 0.02° in the 2θ range of 2° to 60° (dwell time 1 s). Background subtraction, accounting for air scattering, was performed using the DIFFRAC.EVA software V5.1.

The surface morphology of the prepared thin films, including pure materials, their blends, and composites, was analyzed using an Park Systems XE 100 atomic force microscope, operated by XEI computer program (Park Systems, Suwon, Republic of Korea), operating in contact mode in air with a constant force regime.

3. Results and Discussion

3.1. Optical Analysis

Spectroscopic ellipsometry is a reflection optical technique that examines the change in polarization parameters of light reflected from the surface of a sample. Light incident on the sample is linearly polarized, while after reflection, it is elliptically polarized. The two main parameters measured by the ellipsometer are called ellipsometric angles—Ψ and Δ. The first parameter, Ψ, is the change of the polarization amplitude, while Δ is the phase shift between the electrical vectors −p and −s of the electromagnetic beam before and after reflection from the sample. The relation that connects these parameters is called the main equation of ellipsometry, describing the complex reflectance coefficient ρ.

$$\rho = e^{i\Delta} tan\Psi \quad (1)$$

Measurements of ellipsometric angles Ψ and Δ were performed in the variable angle mode (VASE) of the ellipsometer, in the range of angles 40–70°, every 10°, at room temperature, in the wavelength range 240–930 nm. The absorption spectra were derived using an ellipsometric model consisting of four layers, as depicted in Scheme 1. The layers representing air, polymer, SiO_2, and silicon were treated as film-type layers, where the modulation was achieved by fitting the refractive index (n) and extinction coefficient (k) at each wavelength (λ). It is important to note that these parameters were fitted as pairs, with each value of n and k corresponding to the same λ. The silicon oxide layer was modelled using a Cauchy layer, with the corresponding relationship detailed in our previous works [26,27]. The mean square error (MSE) for the fitting results was approximately MSE ≈ 0.7.

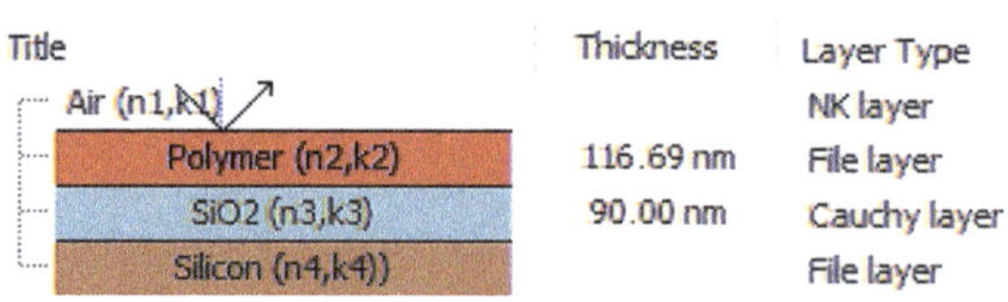

Scheme 1. The ellipsometric model used for the measurement fittings.

Figure 2a,b displays the optical absorption spectra and energy band gaps for pure PCPDTBT, PDPP4T, their 1:1 blend, and the corresponding AuNPs (10%) composites for clarity. The remaining absorption spectra are provided in the Supplementary Materials, as shown in Figure S1.

The spectra of pure materials are denoted in dark blue, while the spectra of their composites are shown in dark yellow. In all cases, the first strong absorption band, occurring between 1.25 and 2.25 eV, is associated with π-π^* electronic transitions, while the absorption band in the range of 2.25 to 3.5 eV corresponds to n-π^* transitions. In the subsequent step, we calculated the absorption coefficient to determine the energy gaps. The absorption coefficient α was obtained using the following relation:

$$\alpha = \frac{(2.302 \cdot A)}{d} \quad (2)$$

Next, using the relationship $(\alpha E)^{(1/2)}$ as a function of energy, the energy gap values for each layer were determined through a graphical method by fitting a tangent to the edge of the first absorption band. The results for all samples are presented in Table 4. The increased absorption in the PCPDTBT/Au composite compared to the PCPDTBT spectrum suggests the presence of a plasmonic effect in the material. This is consistent with the observed decrease in the energy gap value. It is well known that the plasmonic effect can simultaneously lead to both a shift in the absorption band and an increase in absorption intensity [36]. This effect arises from changes in the local oscillations of electrons within gold nanoparticles, which influence electronic transition processes in the polymer. AuNPs enhance absorption within specific wavelength ranges, contributing to the overall increase in absorption. In the case of the PDPP4T/Au composite, we observed a slightly lower absorption intensity compared to pure PCPDTBT, along with a narrowing of the absorption band. Additionally, a slight decrease in the energy gap (E_g) value was noted. We attribute this effect to the local passivation of gold nanoparticles [37,38]. We believe this mechanism is due to the interaction of the polymer material with the surface charge of the nanoparticles. The DPP group contains oxygen atoms that bind to AuNPs via dipole attraction. Additionally, this polymer contains long side chains (attached to the DPP group right next to the oxygen atoms) that are electrostatically attracted also to the nanoparticles surface. In this case, long, branched chains present in PDPP4T can physically surround the gold nanoparticles and create a steric barrier that prevents their aggregation and can stabilize them electrostatically.

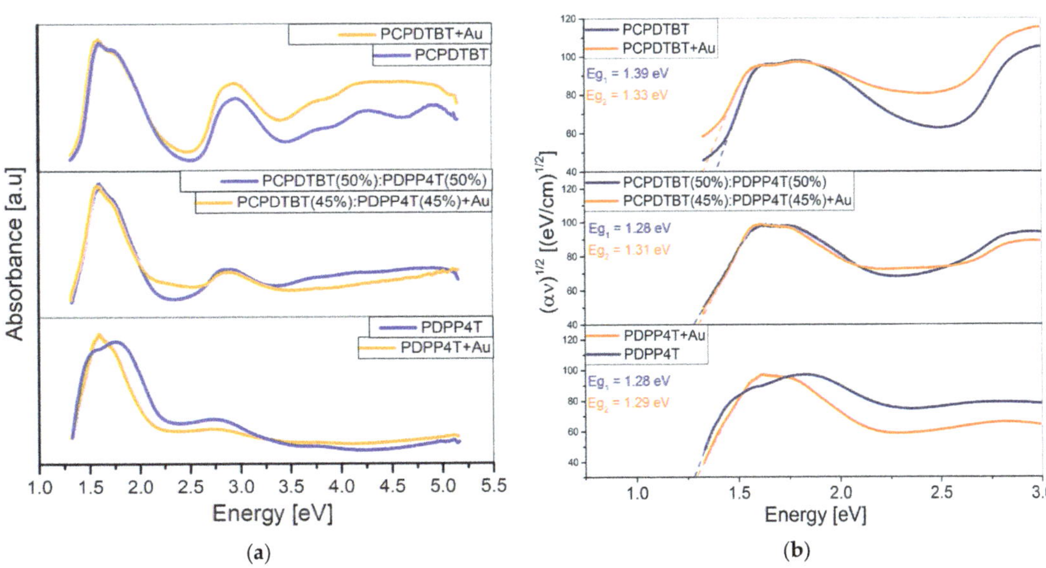

Figure 2. Absorption spectra, determined using Spectra Ray 3 (**a**), and the energy gaps, determined with the Tauc graphical method (**b**).

Table 4. Band gap values for PCPDTBT, PDPP4T, their blends, and nanocomposites with Au nanoparticles.

PCPDTBT [%]	PDPP4T [%]	Eg [eV]	Eg [eV] (+Au)
100	0	1.39	1.33
75	25	1.32	1.38
50	50	1.28	1.31
25	75	1.22	1.29
0	100	1.28	1.29

The minimal changes in the UV-Vis spectrum of the PCPDTBT/PDPP4T/AuNPs composite, compared to the spectrum of the blend, suggest that both the plasmonic effect and passivation occur locally within the prepared samples. We believe that the passivation effect of gold nanoparticles, resulting from the interaction of side chains and oxygen atoms contained in the PDPP4T polymer, quite strongly suppresses the plasmonic effect that appears in the PCPDTBT: AuNPs composite. This is probably caused by the relatively strong screening of nanoparticles by the PDPP4T polymer side chains. However, we believe that the plasmonic effect is still present in the PCPDTBT: PDPP4T: AuNPs composite mixture. This is indicated by the slightly increased optical absorption intensity (Figure 2b) in the region of 2.1–2.45 eV, in the spectrum of the composite compared to the spectrum of the mixture without nanoparticles. Plasmonic bands, in the UV-Vis spectrum, usually appear at values of 520–580 nm wavelength, i.e., 2.1–2.3 eV) [39,40].

3.2. Thermal Analysis

Thermal transition analysis was performed using two methods, variable temperature spectroscopic ellipsometry (VTSE) and differential scanning calorimetry (DSC). In previous studies [26,27], we presented ellipsometric angles (Ψ or Δ) as a function of temperature for a selected wavelength (λ). For this study, we used λ = 930 nm, which exhibited the lowest dispersion of points. Ellipsometric scans were conducted during heating cycles, with measurements taken every 10 s. The VTSE temperature scans for pure PCPDTBT, PDPP4T, their 1:1 blend, and their AuNPs composites (with a nanoparticle concentration of 10%) are shown in Figure 3a–f. The corresponding DSC curves (Figure 4a–c) were obtained for pure materials and their 1:1 blend. Notably, DSC measurements were not conducted for the AuNPs composites, primarily due to potential nanoparticle aggregation and their size. The small sample volume used for the measurements may not ensure uniform nanoparticle distribution, and the nanoparticles could induce localized temperature variations, disrupting the uniform heat flow within the sample.

Based on the obtained results, it can be concluded that the primary thermal transitions detectable in both the individual materials and their mixtures are characteristic of this type of polymer. These transitions include the glass transition temperature (T_g), cold crystallization temperature (T_{cc}), and melting temperature (T_m). For pure materials, the temperatures detected by the DSC and VTSE methods are generally consistent, with any observed differences attributed to several factors. These include the state of the material being tested—where DSC analyzes the powder, while the ellipsometric method examines the prepared film—and differences in the heating rates. In the VTSE method, the heating rate during the temperature cycle is 2 °C/min, while in DSC, it ranges from 10 to 20 °C/min. This manuscript presents results for the most significant materials, i.e., pure polymers, their 1:1 blends, and their composites. Additional thermal data can be found in the Supplementary Materials, specifically in Figure S4a–j. The thermal transitions measured using the VTSE method for pure PCPDTBT and PDPP4T are presented in Figure 3a and 3c, respectively. Based on the DSC results, we have identified several glass transitions in both materials.

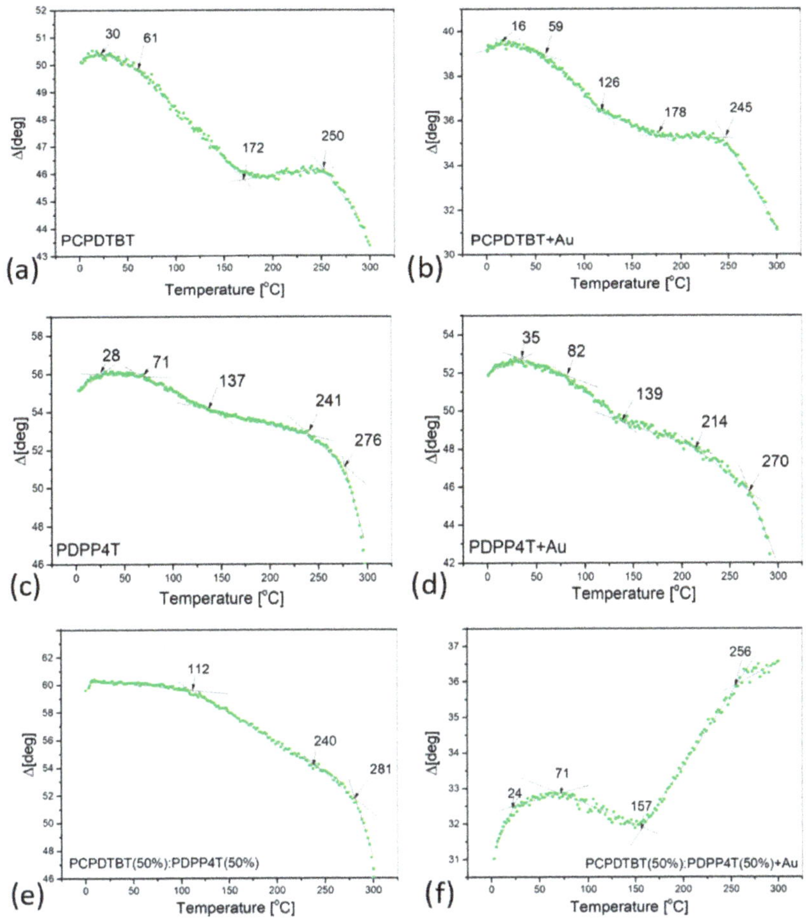

Figure 3. Δ at 930 nm as a function of temperature for PCPDTBT (**a**), PDPP4T (**c**), their 50% blend (**e**), and their AuNPs composites with 10% Au (**b**,**d**,**f**).

For PCPDTBT, the glass transitions were observed around 30 °C, 61 °C, and 172 °C. The corresponding DSC curve revealed T_g values at approximately 72 °C, 167 °C, and 197 °C. Additionally, the cold crystallization temperature (T_{cc}) observed during the heating cycle at around 250 °C was not detected by DSC. This finding aligns with the literature, which suggests that polymers with a similar chemical structure typically exhibit multiple glass transitions [41]. The first glass transition, occurring around 30 °C, is likely attributed to the DPP group, which is commonly observed around 23 °C in DSC measurements. The two remaining T_g values are likely related to the collective motion of the aromatic rings within the π-conjugated fragments in the polymer's main chain and the aliphatic side chains. The glass transition at approximately 61 °C is consistent with our earlier findings [26,27,41].

For pure PDPP4T, the glass transition temperatures (T_g) were detected around 28 °C, 71 °C, and 137 °C, with two cold crystallization temperatures (T_{cc}) observed at approximately 241 °C and 276 °C. The corresponding DSC results showed T_g at around 65 °C and

290 °C. Additionally, the melting point (T_m) was observed around 330 °C. These results are consistent with those from our previous work [41], where we presented the phase diagram for PDPP4T: PDBPyBT for the first time, confirming that the temperatures for pure PDPP4T match those in the phase diagram.

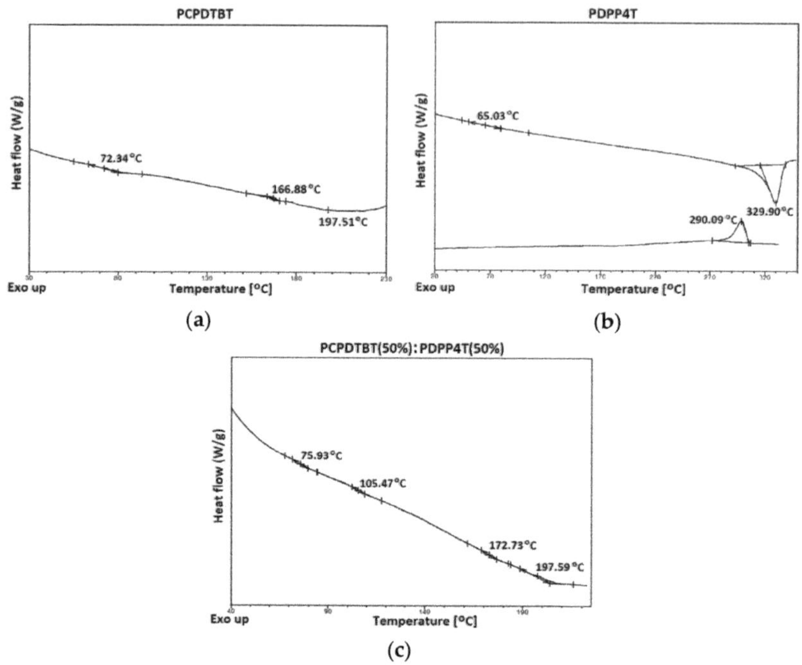

Figure 4. DSC plots, with a heating rate of 20 °C/min, for pure PCPDTBT (**a**) and PDPP4T (**b**) and for PCPDTBT: PDPP4T (1:1) blend (**c**).

In the 1:1 PCPDTBT: PDPP4T blend, the glass transition was detected around 112 °C, with cold crystallization occurring at around 240 °C and 281 °C. The DSC results revealed additional glass transitions at approximately 75 °C, 105 °C, 172 °C, and 197 °C. This suggests that the internal morphology of the blend film is more uniform compared to the sample prepared for DSC measurement. Notably, the glass transitions observed in the pure materials are not present in the VTSE graph of the mixture.

In the case of PCPDTBT with gold nanoparticles, the glass transitions appeared at about 16, 59, 126, 178 °C. One additional temperature (126 °C) probably originates from local interaction of Au nanoparticles with polymer. As we noted earlier, the presence of gold nanoparticles influences the thermal properties of the material. We believe that gold nanoparticles can weaken interchain interactions in amorphous regions. This can result in higher flexibility of the polymer matrix in the composite compared to the pure material, which results in the occurrence of thermal transitions at lower temperatures. It should be emphasized, however, that these differences are not high. An additional T_g at 126 °C indicates the existence of a second phase in the composite, which could result from the separation of domains with varying gold contents, for instance. This could be the result of areas with varying chain mobility, forming as a result of nanoparticles interacting with particular polymer segments.

In the PDPP4T composite with gold nanoparticles, thermal transitions occur at temperatures of 35, 82, 139, 214, and 270 °C. The last two temperatures are the T_{cc} temperatures, and the rest are the T_g temperatures. In the case of pure PDPP4T, these temperatures are pronounced at about 28, 71, 137, 241, and 276 °C. Higher T_g values in composite in comparison to pure PDPP4T may indicate that the composite has slightly better crystalline order than the pure material. The lower T_{cc} value in the case of composite can result from fact that the energy necessary to form the crystalline phase may be decreased if nanoparticles alter the packing of the polymer chains, and the polymer matrix may have a greater ability to form well-developed crystalline regions. In the case of the 1:1 PCPDTBT: PDPP4T mixture, thermal transitions were noticed at about 112, 240, and 281 °C. In the case of the composite of the aforementioned mixture with 10% gold nanoparticles, thermal transitions were present at 24, 71, 157, and 256 °C. We assume that introducing the AuNPs can reduce the mutual compatibility of PCPDTBT and PDPP4T, leading to the formation of separate polymer phases (a higher number of T_g than in the pure material), each with its own glass transition. The separated phases also favor the occurrence of LSPR and the passivation of AuNPs at the same time. The reduction in the glass transition temperature of the polymer matrix in the composite compared to the pure mixture suggests increased crystallinity of the composite.

Figure 5 represents the phase diagram, which was constructed using data from DSC and temperature ellipsometry. Thermal transitions are indicated, as follows: light green circles for VTSE measurements, dark green open circles for DSC transitions in pure materials and their blends, and red open circles for VTSE transitions in material/AuNPs composites (10%). The colored regions represent the following specific thermal characteristics: glass transition temperature (T_g) in pink, cold crystallization temperature (T_{cc}) in blue, and melting temperature (T_m) in orange.

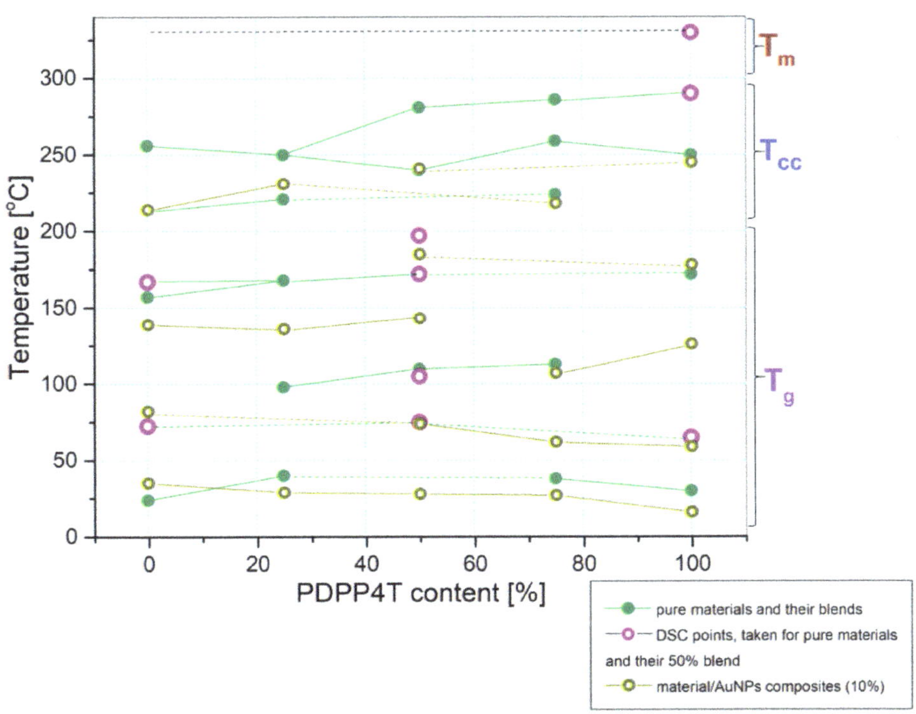

Figure 5. Phase diagram of PCPDTBT, PDPP4T, their PCPDTBT: PDPP4T blend films and their AuNPs composites.

The phase diagram clearly illustrates that thermal transitions differ between pure materials, their blends, and their composites. As previously noted, some transition temperatures are either elevated or reduced. Notably, in the polymer blends, glass transitions occur at approximately 98, 110, and 113 °C for 25%, 50%, and 75% PDPP4T content, respectively. However, these transitions are absent in the composites, suggesting strong interactions between AuNPs and polymer chains.

The most plausible explanation for this phenomenon is a combination of the local plasmonic effect, localized surface plasmon resonance (LSPR) [25,36], and the passivation of AuNPs [37,38]. In the case of the plasmonic effect, nanoparticles influence the glass transition process, rendering the polymer matrix in the composite more fluid over a wider temperature range, thereby diminishing the visibility of the blend's T_g. Conversely, if passivation dominates, gold nanoparticles stabilize the polymer structure, preventing the distinct glass transition observed in the blends. These findings align well with the optical results, further supporting the proposed mechanisms.

3.3. XRD Analysis

The XRD patterns of pure PCPDTBT and PDPP4T, their 1:1 blend, and the corresponding AuNPs composites are presented in Figure 6a–c; the remaining XRD patterns have been added in the Supplementary Materials in Figure S3. In the case of pure PDPP4T (Figure 6a), a broad amorphous background is observed along with several low-intensity peaks. While the peaks at approximately (100), (400), and (020) exhibit weak intensity in the pure material, their intensity increases in the composite, with additional peaks appearing around (001) and (200). These results indicate that pure PDPP4T consists of a mixed amorphous and crystalline structure. However, in the composite pattern, the presence of additional peaks at (001), (200), and (400), alongside those originating from the pure polymer, suggests that the nanoparticles significantly influence the crystallization process. Moreover, the nanoparticles may form their own crystalline structures, contributing to the diffraction pattern. This suggests the formation of new crystalline domains or strong interactions between the nanoparticles and the polymer matrix. From (Figure 6b), in the spectrum of pure PCPDTBT, two peaks corresponding to crystallization planes appear at approximately (100) and (300). In the PCPDTBT/AuNPs composite spectrum, an additional peak emerges around (200), while the amorphous background is notably reduced. The amorphous background in the pure PCPDTBT pattern suggests that its structure is not highly crystalline. The presence of (100) and (300) peaks indicate a mixed-phase structure, comprising both amorphous and crystalline regions. Comparing this to the composite, the increased intensity and the appearance of the (200) peak suggest a higher degree of crystallinity, likely due to AuNPs acting as nucleation sites around which polymer chains become more ordered [42,43].

This type of pattern indicates an edge-on crystalline orientation, where the crystallization planes are arranged parallel to each other, with each plane positioned perpendicularly to the sample surface.

The XRD patterns of the 1:1 PCPDTBT/PDPP4T blend and its corresponding AuNPs composite are compared in Figure 6c. Both spectra exhibit a broad amorphous background. In the pure blend, weak intensity peaks appear at approximately (100), (200), (400), and (020). However, in the composite, the (100) peak becomes more intense, the (200) peak disappears, the (300) peak emerges, and the intensity of the (400) and (020) peaks remains unchanged. These results indicate that both polymers retain a certain degree of crystallinity. The low-intensity peaks and broad background suggest a predominantly amorphous structure with small regions of ordered crystallization. In the composite, the appearance of the (300) peak suggests the formation of new crystalline structures and interactions between AuNPs and the polymer matrix.

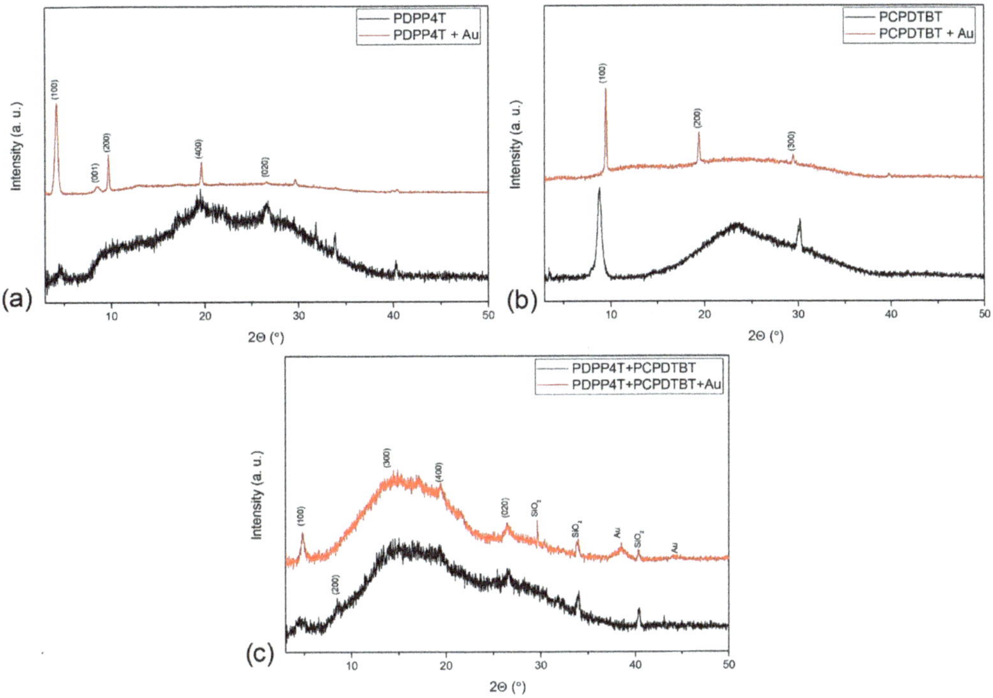

Figure 6. The XRD patterns of PCPDTBT (**a**), PDPP4T (**b**), 1:1 blend (**c**), and their AuNPs composites.

Using Scherrer's law for the (100) peak, the crystallite sizes were calculated (Table 5). The results show that the crystallite size in PCPDTBT with gold nanoparticles is three times larger than in the pure polymer. In contrast, the crystallization of pure PDPP4T is less affected by the introduction of AuNPs compared to PCPDTBT. For the PCPDTBT/PDPP4T blend, the increase in crystallite size after AuNPs incorporation follows a trend similar to that of pure PCPDTBT, with the crystallite size being three times larger.

Table 5. The size of crystallites calculated for pure materials and their composites.

Sample	Crystal Size (nm)
PCPDTBT	16.1
+Au	47.3
PDPP4T	18.0
+Au	31.6
BLEND	21.5
+Au	60.3

3.4. Microscopic Analysis

Topographic analysis was conducted on layers of pure polymer materials, their blends, and nanocomposite materials containing AuNPs additives. The resulting topographic images are presented in Figure 7. The root mean square roughness (R_q) was used to characterize the surface of all the tested samples.

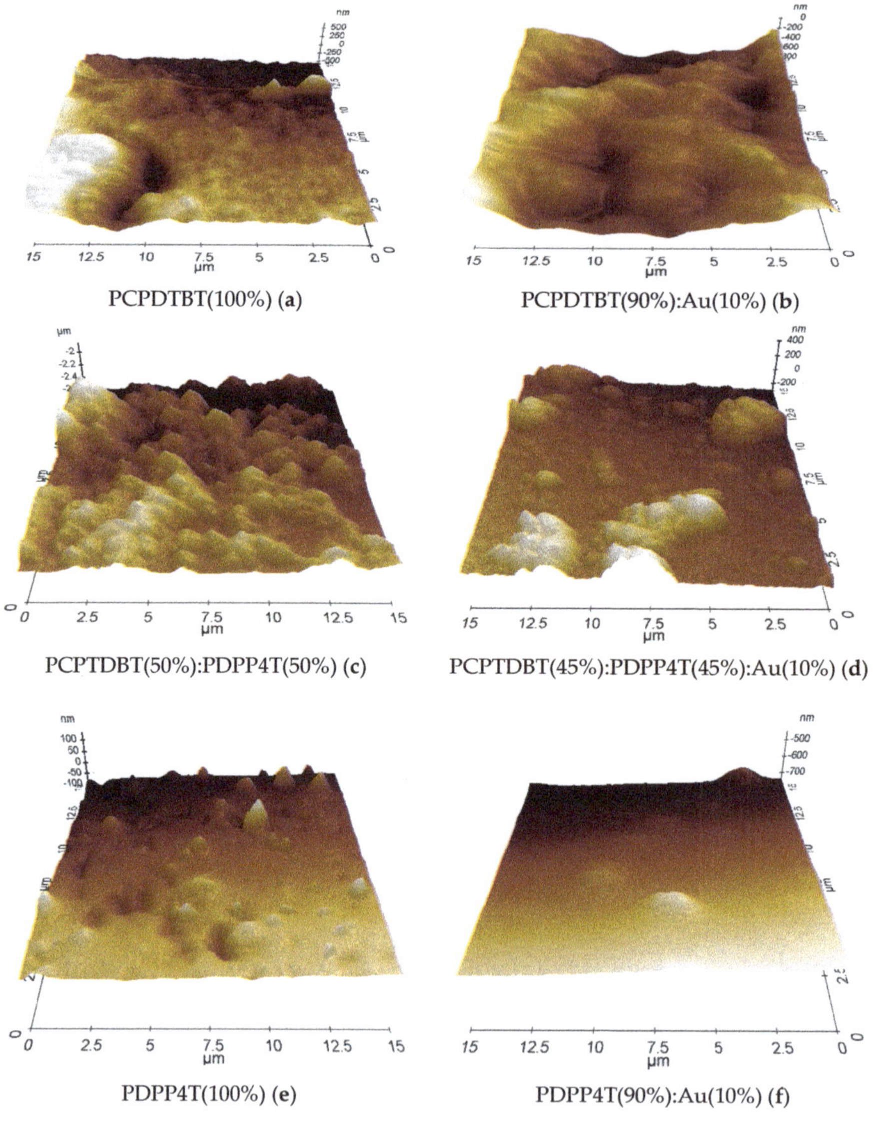

Figure 7. 15 × 15 μm 3D topographic surface images of PCPDTBT (**a**), PDPP4T (**c**), their 1:1 blend (**e**), and their corresponding AuNPs composites (**b**,**d**,**f**) obtained using AFM microscope.

It is defined with Equation (3) as in [44]:

$$R_q = \sqrt{\frac{1}{m}\sum_{i=1}^{m}\left(Z_i - \overline{Z}\right)^2} \quad (3)$$

where m is the number of sampled points, Z_i is the height of each point, and $\overline{Z}$ is the mean height value [44]. R_q was determined for three surface sizes, 1 × 1, 2 × 2 and 10 × 10 μm, and is shown in Figure 8.

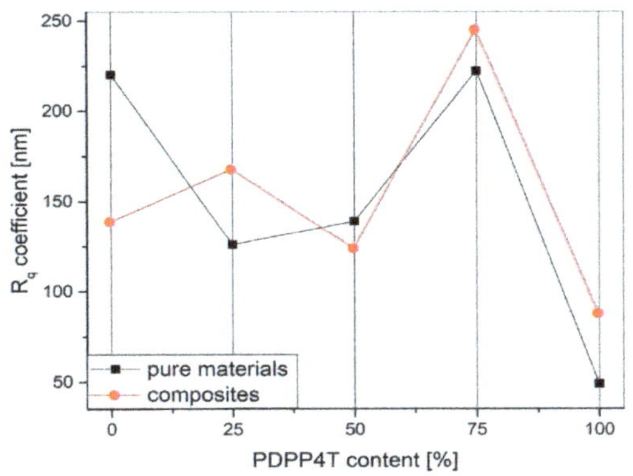

Figure 8. The mean square root of the roughness of PCPDTBT, PDPP4T, their PCPDTBT: PDPP4T blend films and their AuNPs composites.

For the clarity of this manuscript, here, we present the morphology of PCPDTBT, PDPP4T, their 1:1 blend PCPDTBT/PDPP4T, and their corresponding composites. The rest of the AFM morphology pictures are presented in the Supplementary Materials in Figure S2a–d).

The results of the study suggest that the pure PCPDTBT polymer thin film (Figure 7a) has a relatively homogeneous surface, with clearly visible loss and agglomeration. However, water defects on the surface of the thin film are most likely caused by the agglomeration of the polymer material caused by the volatility of solvent and interactions of polymer solvent pair [45]. The remaining surface of the thin film remains smooth, which is characteristic of PCPDTBT layers resulting from solid state ordering induced by racemic side chains [46]. The addition of AuNPs caused the development of the film surface, which, however, remained continuous and relatively smooth. The results shown in (Figure 7b) indicate that the addition of Au nanoparticles induced microstructural changes, likely due to agglomeration occurring during the mixing of different components in the film [47].

Thin films of blended materials (Figure 7c) exhibit a homogeneous surface with distinct crystalline structures. Small agglomerated polymer formations are visible; however, the absence of larger agglomerates (above 2 μm) suggests that blending promotes the ordering of polymer chains. However, considering the findings of other researchers, such as Wang et al. [48], it is expected that pure polymer semiconductor layers generally have lower surface roughness compared to blended films. For the blended film (Figure 7d), the addition of gold nanoparticles led to a reduction in surface roughness. This effect may be attributed to the influence of the nanoadditive on the ordering of polymer chains or a phenomenon observed by Nathanael et al. [49]. In their study, an increase in TiO_2 nanoparticle concentration enhanced the uniformity of the HAp/TiO_2 film, while reducing its roughness. Similarly, research by Predoi et al. [50] demonstrated that decreased composite layer roughness resulted from the formation of uniformly distributed nanoconglomerates on the surface.

The pure PDPP4T films (Figure 7e) exhibit a more homogeneous surface compared to the pure PCPDTBT layers, with no material agglomerates larger than 2 μm. Crystalline fractions are visible on the film's surface, a characteristic feature of this polymer due to the crystallization tendency of its alkyl chains attached to the DPP unit. These chains act as high-solubilizing groups, promoting tight molecular packing in the film [51]. Similar to PCPDTBT and blend thin films, the incorporation of AuNPs results in a smoother surface. In the case of PDPP4T/AuNPs layers (Figure 7f), this smoothing effect may be attributed to the formation of Au nanoclusters within the layers or the development of crystalline or semi-crystalline structures.

The surface topography of the produced layers was quantitatively analyzed by determining the surface roughness coefficient (R_q). The addition of Au nanoadditives to both pure PCPDTBT and its blend resulted in a reduction in surface roughness. This effect was more pronounced in pure PCPDTBT, likely due to the presence of larger agglomerates in the pristine material. Interestingly, for PDPP4T films, the R_q coefficient increased after incorporating the nanoadditive, despite the qualitatively smoother surface observed in microscopic analysis. However, given that the initial roughness of pure PDPP4T is significantly lower than that of PCPDTBT, it can be inferred that the formation of Au nanoclusters contributed to the increase in roughness. Furthermore, the results suggest that the R_q coefficient of the blended film indicates the effective mixing of the materials, supporting the hypothesis of improved uniformity in the blend structure.

3.5. General Discussion

The results obtained for pure materials and their composites indicate that several phenomena occur simultaneously in the tested samples. These include the plasmonic effect, the passivation of gold nanoparticles, and the role of AuNPs as nucleation centers influencing polymer crystallization. The plasmonic effect is most pronounced in the PCPDTBT/Au composite. A comparison of its optical absorption spectrum with that of the pure material reveals that the gold nanoparticles significantly alter the electronic structure. This corresponds to a localized plasmonic effect, where surface plasmons are excited directly on zero-dimensional AuNPs. These plasmons resonate around the nanoparticles, as illustrated in Scheme 2. It is important to note that the plasmonic effect is primarily associated with AuNPs that are well-isolated within the polymer matrix [52,53].

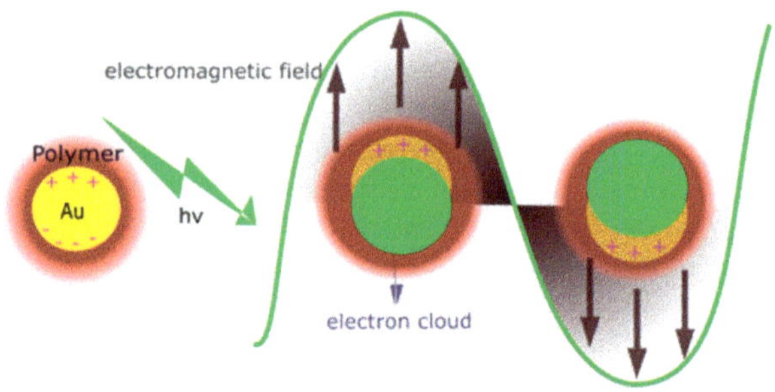

Scheme 2. Representation of the localized surface plasmonic effect in the polymer/AuNPs composite.

We noticed that the plasmonic effect also indirectly influences the thermal transitions of the polymer. The most evident example is the lower cold crystallization temperature (T_{cc}) in the PCPDTBT/Au composite, where T_{cc} decreases to 214 °C compared to 256 °C in pure PCPDTBT. These changes in thermal transitions arise from the ability of AuNPs to absorb light, generating localized heating through localized surface plasmon resonance (LSPR). This phenomenon enhances polymer chain mobility, accelerating the crystallization process. This effect is clearly reflected in the XRD spectra, where an increase in the crystalline peak intensity is observed.

For PDPP4T, the addition of gold nanoparticles leads to a noticeable reconstruction of the polymer's crystal structure and improved structural stability. The results suggest that AuNPs promote the formation of lamellae, which align in parallel planes perpendicular to the sample substrate. As previously mentioned, this corresponds to an edge-on structure, which facilitates charge transport via π-stacking interactions. In the case of the PCPDTBT/PDPP4T blend, the internal structure remains predominantly amorphous, with localized crystalline phases. X-ray diffraction results confirm this, showing a strong amorphous background alongside weak peaks corresponding to crystalline planes. Additionally, thermal and microscopic analyses indicate a high degree of miscibility between the two polymers. The characteristic thermal transitions of pure PCPDTBT and PDPP4T phases are nearly absent in the 1:1 blend, suggesting effective mixing at the molecular level. For the corresponding PCPDTBT/PDPP4T/Au composite, the emergence of additional crystalline peaks in the XRD pattern suggests the formation of new crystalline domains or strong interactions between the nanoparticles and the polymer matrix. This indicates that both plasmonic and passivation effects are present in the composite, but to a lesser extent than in the Au nanocomposites of the individual polymers.

The presence of this crystalline arrangement indicates the potential application of PDPP4T/Au composites as active materials in organic field-effect transistors (OFETs).

4. Conclusions

In this study, we investigated the thermal and physical properties of a novel polymer blend—PCPDTBT: PDPP4T—as well as its composite with gold nanoparticles (AuNPs). Our findings indicate that the local plasmonic effect was not uniformly present in all samples but was most pronounced in the PCPDTBT/Au composite. This was evidenced by a significant reconstruction of the electronic band structure observed in the UV-Vis spectra.

AuNPs had a pronounced influence on the thermal transitions within the polymer matrix. Notable effects included a significant decrease in the cold crystallization temperature (T_{cc}) of PCPDTBT and the disappearance of the glass transition temperature associated with the PCPDTBT/PDPP4T blend phase at different PDPP4T concentrations in the composites. In the case of the PCPDTBT: PDPP4T (1:1) mixture layer and its composite with AuNPs, the obtained thermal test results indicate the high miscibility of the tested polymers. Changes in thermal transitions, visible in the prepared phase diagram, suggest that, in the case of the composite, the separated phases promote the occurrence of passivation of gold nanoparticles and the plasmonic effect at the same time, which is strongly reflected in the optical absorption spectra also.

Furthermore, the incorporation of AuNPs had a notable impact on the internal structure of the tested layers, particularly their crystallinity. The most significant structural changes were observed in the PDPP4T/Au composite, where gold nanoparticles acted as nucleation centers, what is pronounced with the bigger, average size of appeared crystallites. Through interactions with the polymer chains, the AuNPs promoted a higher degree of molecular ordering compared to the pure polymer, leading to the formation of an edge-on crystalline orientation. For the PCPDTBT/PDPP4T composite, we identified the simultaneous occurrence of two key phenomena: a plasmonic effect—though weaker than that observed in the PCPDTBT/Au composite—and enhanced structural ordering. This was confirmed by the presence of additional peaks in the XRD spectra, indicating the formation of new crystalline domains.

In conclusion, the observed surface plasmon resonance and passivation effects induced by the presence of AuNPs in these materials presents exciting prospects for future research. Further investigations should focus on determining whether these phenomena can enhance the quantum efficiency of the investigated materials, opening up new possibilities for their application in optoelectronic devices.

Supplementary Materials: The following supporting information can be downloaded at: https://www.mdpi.com/article/10.3390/polym17050704/s1, Figure S1: Absorption spectra of pure PCPDTBT, PDPP4T, their blends, AuNPs composites (a) and their energy band gaps (b), Figure S2: 15 × 15 μm 3D topographic surface images of PCPTDBT(75%):PDPP4T(25%) (a) PCPTDBT(70%):PDPP4T(20%):Au(10%) (b) PCPTDBT(25%):PDPP4T(75%) (c) and PCPTDBT(20%):PDPP4T(70%):Au(10%) (d), Figure S3: XRD patterns of PCPDTBT, PDPP4T, their blends and AuNPs composites, Figure S4: Ellipsometric Ψ i Δ temperature scans at wavelength λ = 930 nm: (a) PCPDTBT(100%), (b) PCPDTBT(90%) + Au(10%), (c) PCPDTBT(75%):PDPP4T(25%), (d) PCPDTBT(70%):PDPP4T(20%) + Au(10%), (e) PCPDTBT(50%):PDPP4T(50%), (f) PCPDTBT(45%):PDPP4T(45%) + Au(10%), (g) PCPDTBT(25%):PDPP4T (75%), (h) PCPDTBT(20%):PDPP4T(70%) + Au(10%), (i) PDPP4T(100%), (j) PDPP4T(90%) + Au(10%).

Author Contributions: Conceptualization, B.H. and P.J.; methodology, B.H. and P.J.; validation, B.H. and P.J.; formal analysis, B.H. and P.J.; investigation, B.H., P.J., P.K., M.G., H.J. and Y.M.T.; data curation, B.H. and P.J.; writing—original draft preparation, B.H. and P.J.; writing—review and editing, B.H., P.J. and P.K.; visualization, B.H., P.J. and P.K.; supervision, B.H. and T.T. All authors have read and agreed to the published version of the manuscript.

Funding: The research was co-financed by the Silesian University of Technology under the Initiative program Excellence—Research University.

Institutional Review Board Statement: Not applicable.

Data Availability Statement: Data are contained within the article and Supplementary Materials.

Conflicts of Interest: The authors declare no conflict of interest.

References

1. Sulaiman, A.K.; Yap, C.C.; Touati, F. Binary blend based dye sensitized photo sensor using PCPDTBT and MEH-PPV composite as a light sensitizer. *Synth. Met.* **2015**, *210*, 392–397. [CrossRef]
2. Li, Y.; Wei, Y.; Feng, K.; Hao, Y.; Pei, J.; Zhang, Y.; Sun, B. Introduction of PCPDTBT in P3HT:Spiro-OMeTAD blending system for solid-state hybrid solar cells with dendritic TiO_2/Sb_2S_3 nanorods composite film. *J. Sol. St. Chem.* **2019**, *276*, 278–284. [CrossRef]
3. Banerjee, S.; Singh, P.; Purkayastha, P.; Ghosh, S.K. Evolution of Organic Light Emitting Diode (OLED) Materials and their Impact on Display Technology. *Chem. Asian J.* **2024**, *20*, e202401291. [CrossRef] [PubMed]
4. Nayak, D.; Choudhary, R.B. A survey of the structure, fabrication, and characterization of advanced organic light emitting diodes. *Microelectron. Reliab.* **2023**, *144*, 114959. [CrossRef]
5. Luo, L.; Liu, Z. Recent progress in organic field-effect transistor-based chem/bio-sensors. *View* **2022**, *3*, 20200115. [CrossRef]
6. Zhao, Y.; Wang, W.; He, Z.; Peng, B.; Di, C.-A.; Li, H. High-performance and multifunctional organic field-effect transistors. *Chin. Chem. Lett.* **2023**, *34*, 108094. [CrossRef]
7. Kim, O.; Kwon, J.; Kim, S.; Xu, B.; Seo, K.; Park, C.; Do, W.; Bae, J.; Kang, S. Efect of PVP-capped ZnO nanoparticles with enhanced charge transport on the performance of P3HT/PCBM polymer solar cells. *Polymers* **2019**, *11*, 1818. [CrossRef]
8. Liu, C.; Zhang, D.; Li, Z.; Zhang, X.; Guo, W.; Zhang, L.; Ruan, S.; Long, Y. Decreased charge transport barrier and recombination of organic solar cells by constructing interfacial nanojunction with annealing-free ZnO and Al flms. *ACS Appl. Mater. Interfaces* **2017**, *9*, 22068–22075. [CrossRef]
9. Mohtaram, F.; Borhani, S.; Ahmadpour, M.; Fojan, P.; Behjat, A.; Rubahn, H.G.; Madsen, M. Electrospun ZnO nanofber interlayers for enhanced performance of organic photovoltaic devices. *Sol. Energy* **2020**, *197*, 311–316. [CrossRef]
10. Swart, H.C.; Ntwaeaborwa, O.M.; Mbule, P.S.; Dhlamini, M.S.; Mothudi, B.B. P3HT: PCBM based solar cells: A short review focusing on ZnO nanoparticles bufer layer, post-fabrication annealing and an inverted geometry. *J. Mater. Sci. Eng. B* **2015**, *5*, 12–35. [CrossRef]
11. Wanninayake, A.P.; Church, B.C.; Abu-Zahra, N. Efect of ZnO nanoparticles on the power conversion efciency of organic photovoltaic devices synthesized with CuO nanoparticles. *AIMS Mater. Sci.* **2016**, *3*, 927–937. [CrossRef]
12. Dong, J.; Yan, C.; Chen, Y.; Zhou, W.; Peng, Y.; Zhang, Y.; Wang, L.-N.; Huang, Z.-H. Organic semiconductor nanostructures: Optoelectronic properties, modification strategies, and photocatalytic applications. *J. Mater. Sci. Technol.* **2022**, *113*, 175–198. [CrossRef]
13. Cheng, Z.; Javed, N.; O'Carroll, D.M. Optical and Electrical Properties of Organic Semiconductor Thin Films on Aperiodic Plasmonic Metasurfaces. *ACS Appl. Mater. Interfaces* **2020**, *12*, 35579–35587. [CrossRef] [PubMed]
14. Balakrishnan, T.S.; Sultan, M.T.H.; Shahar, F.S.; Basri, A.A.; Shah, A.U.M.; Sebaey, T.A.; Łukaszewicz, A.; Józwik, J.; Grzejda, R. Fatigue and impact properties of kenaf/glass-reinforced hybrid pultruded composites for structural applications. *Materials* **2024**, *17*, 302. [CrossRef]
15. Liu, S.; Yang, J.; Yu, Y.; Liang, D.; Li, Y.; Si, X.; Song, S.; Meng, M.; Zhang, J.; Zhang, Y. Multifunctional flexible carbon fiber felt@nickel composite films with core–shell heterostructure: Exceptional Joule heating capability, thermal management, and electromagnetic interference shielding. *Chem. Eng. J.* **2024**, *494*, 153221. [CrossRef]
16. Huang, Y.; Sultan, M.T.H.; Shahar, F.S.; Grzejda, R.; Łukaszewicz, A. Hybrid fiber-reinforced biocomposites for marine applications: A review. *J. Compos. Sci.* **2024**, *8*, 430. [CrossRef]

17. Shahar, F.S.; Sultan, M.T.H.; Grzejda, R.; Łukaszewicz, A.; Oksiuta, Z.; Krishnamoorthy, R.R. Harnessing the potential of natural composites in biomedical 3D printing. *Materials* **2024**, *17*, 6045. [CrossRef]
18. Ibrahim, N.I.; Sultan, M.T.H.; Łukaszewicz, A.; Shah, A.U.M.; Shahar, F.S.; Józwik, J.; Najeeb, M.I.; Grzejda, R. Characterization and isolation method of Gigantochloa scortechinii (*Buluh Semantan*) cellulose nanocrystals. *Int. J. Biol. Macromol.* **2024**, *272*, 132847. [CrossRef]
19. Wang, T.C.; Su, Y.H.; Hung, Y.K.; Yeh, C.S.; Huang, L.W.; Gomulya, W.; Lai, L.H.; Loi, M.A.; Yang, J.S.; Wu, J.J. Charge collection enhancement by incorporation of gold–silica core–shell nanoparticles into P3HT:PCBM/ZnO nanorod array hybrid solar cells. *Phys. Chem. Chem. Phys.* **2015**, *17*, 19854–19861. [CrossRef]
20. Gao, H.; Meng, J.; Sun, J.; Deng, J. Enhanced performance of polymer solar cells based on P3HT:PCBM via incorporating Au nanoparticles prepared by the micellar method. *J. Mater. Sci. Mat. Electr.* **2020**, *31*, 10760–10767. [CrossRef]
21. Pillai, S.; Catchpole, K.R.; Trupke, T.; Green, M.A. Surface plasmon enhanced silicon solar cells. *J. Appl. Phys.* **2007**, *101*, 093105. [CrossRef]
22. Sharifi, M.; Attar, F.; Saboury, A.A.; Akhtari, K.; Hooshmand, N.; Hasan, A.; El-Sayed, M.A.; Falahati, M. Plasmonic gold nanoparticles: Optical manipulation, imaging, drug delivery and therapy. *J. Control. Release* **2019**, *311–312*, 170–189. [CrossRef] [PubMed]
23. Ghosh, S.; Shankar, S.; Philips, D.S.; Ajayaghosh, A. Diketopyrrolopyrrole-based functional supramolecular polymers: Next-generation materials for optoelectronic applications. *Mater. Today Chem.* **2020**, *16*, 100242. [CrossRef]
24. Kotowicz, S.; Małecki, J.G.; Pająk, A.K.; Siwy, M.; Łuszczyńska, B.; Selerowicz, A. Optical, electrochemical and thermal investigations with DFT calculations of new 1,8-naphthal-phthalimide compounds. *J. Mol. Struct.* **2025**, *1332*, 141711. [CrossRef]
25. Yi, Z.; Sun, X.; Zhao, Y.; Guo, Y.; Chen, X.; Qin, J.; Yu, G.; Liu, Y. Diketopyrrolopyrrole-Based π Conjugated Copolymer Containing β Unsubstituted Quintetthiophene Unit: A Promising Material Exhibiting High Hole-Mobility for Organic Thin-Film Transistors. *Chem. Mater.* **2012**, *24*, 4350–4356. [CrossRef]
26. Hajduk, B.; Jarka, P.; Bednarski, H.; Janeczek, H.; Kumari, P.; Farcas, A. Thermal Transitions and Structural Characteristics of Poly(3,4-ethylenedioxythiophene/cucurbit[7]uril) Polypseudorotaxane and Polyrotaxane Thin Films. *Materials* **2024**, *17*, 1318. [CrossRef]
27. Hajduk, B.; Bednarski, H.; Jarząbek, B.; Nitschke, P.; Janeczek, H. Phase diagram of P3HT:PC70BM thin films based on variable-temperature spectroscopic ellipsometry. *Polym. Test.* **2020**, *84*, 106383. [CrossRef]
28. Wang, T.; Pearson, A.J.; Dunbar, A.D.F.; Staniec, P.A.; Watters, D.C.; Coles, D.; Yi, H.; Iraqi, A.; Lidzey, D.G.; Jones, R.A.L. Competition between substrate-mediated π-π stacking and surface-mediated Tg depression in ultrathin conjugated polymer films. *Eur. Phys. J. E* **2012**, *35*, 129. [CrossRef]
29. Kim, J.H.; Jang, J.; Zin, W.-C. Estimation of the thickness dependence of the glass transition temperature in various thin polymer films. *Langmuir* **2000**, *16*, 4064–4067. [CrossRef]
30. Keddie, J.L.; Jones, R.A.; Cory, R.A. Size-dependent depression of the glass transition temperature in polymer films. *Europhys. Eur. Lett.* **1994**, *27*, 59–64. [CrossRef]
31. Pratyusha, T.; Sivakumar, G.; Yella, A.; Gupta, D. Novel Ternary Blend of PCDTBT, PCPDTBT and PC70BM for the Fabrication of Bulk Heterojunction Organic Solar Cells. *Mater. Today Proc.* **2017**, *4*, 5067–5073. [CrossRef]
32. Fischer, F.S.U.; Trefz, D.; Back, J.; Kayunkid, N.; Tornow, B.; Albrecht, S.; Yager, K.G.; Singh, G.; Karim, A.; Neher, D.; et al. Ludwigs, Highly Crystalline Films of PCPDTBT with Branched Side Chains by Solvent Vapor Crystallization: Influence on Opto-Electronic Properties. *Adv. Mater.* **2015**, *27*, 1223–1228. [CrossRef] [PubMed]
33. Ahmad, Z.; Touati, F.; Muhammad, F.F.; Najeeb, M.A.; Shakoor, R.A. Shakoor Effect of ambient temperature on the efficiency of the PCPDTBT:PC71BM BHJ solar cells. *Appl. Phys. A* **2017**, *123*, 486. [CrossRef]
34. Liu, Y.; Liu, F.; Wang, H.W.; Nordlund, D.; Sun, Z.; Ferdous, S.; Russell, T.P. Sequential Deposition: Optimization of Solvent Swelling for High-Performance Polymer Solar Cells. *ACS Appl. Mater. Interfaces* **2015**, *7*, 653–661. [CrossRef] [PubMed]
35. Bijleveld, J.C.; Verstrijden, R.A.M.; Wienka, M.M.; Janssen, R.A.J. Copolymers of diketopyrrolopyrrole and thienothiophene for photovoltaic cells. *J. Mater. Chem.* **2011**, *21*, 9224–9231. [CrossRef]
36. Notarianni, M.; Vernon, K.; Chou, A.; Aljada, M.; Liu, J.; Motta, N. Plasmonic effect of gold nanoparticles in organic solar cells. *Sol. Energy* **2014**, *106*, 23–37. [CrossRef]
37. Vienna, H.A.; Klimpovuz, C.R.; Turchetti, D.A.; Rossi, A.L.; Oliveira, M.M.; Marletta, A.; Akcelrud, L.C. Polyfluorene-based conjugated nanocomposites with in-situ gold nanoparticles: Synthesis via rational chemical passivation and characterization of supramolecular fibrillar structures. *J. Mol. Struct.* **2025**, *1322*, 140431. [CrossRef]
38. Li, Z.; Liu, C.; Guo, J.; Zhou, Y.; Shen, L.; Guo, W. Passivation effect of composite organic interlayer on polymer solar cells. *Org. Electron.* **2018**, *63*, 129–136. [CrossRef]
39. Guo, C.; Hall, G.N.; Addison, J.B.; Yarger, J.L. Gold nanoparticle-doped silk film asbiocompatible, SERS substrate. *RSC Adv.* **2015**, *5*, 1937. [CrossRef]

40. Long, N.V.; Ohtaki, M.; Yuasa, M.; Yoshida, S.; Kuragaki, T.; Thi, C.M.; Nogami, M. Synthesis and Self-Assembly of Gold Nanoparticles by Chemically Modified Polyol, Methods under Experimental Control. *J. Nanomater.* **2013**, *2013*, 93125. [CrossRef]
41. Hajduk, B.; Jarka, P.; Tański, T.; Bednarski, H.; Janeczek, H.; Gnida, P.; Fijalkowski, M. An Investigation of the Thermal Transitions and Physical Properties of Semiconducting PDPP4T:PDBPyBT Blend Films. *Materials* **2022**, *15*, 8392. [CrossRef] [PubMed]
42. Elashmawi, I.S.; Abdelghany, A.M.; Meikhail, M.S.; Aldhabi, A.A. Effect of gold and silver mixed nanofillers and gamma irradiation on the structure and physical properties of PVA/PVP nanocomposite. *Radiat. Phys. Chem.* **2025**, *230*, 112558. [CrossRef]
43. Masirek, R.; Szkudlarek, E.; Piorkowska, E.; Slouf, M.; Kratochvil, J.; Baldrian, J. Nucleation of isotactic polypropylene crystallization by gold nanoparticles. *J. Polym. Sci. B Polym. Phys.* **2010**, *48*, 469–478. [CrossRef]
44. Borgesi, A.; Tallarida, G.; Amore, G.; Cazzaniga, F.; Queirolo, F.; Alessandri, M.; Sassela, A. Influence of roughness and grain dimension on the optical functions of polycrystalline silicon films. *Thin Solid Film.* **1998**, *313–314*, 243–247. [CrossRef]
45. Spangler, L.L.; Torkelson, J.M.; Royal, J.S. Influence of solvent and molecular weight on thickness and surface topography of spin-coated polymer films. *Polym. Eng. Sci.* **1990**, *30*, 644–653. [CrossRef]
46. Fronk, S.L.; Wang, M.; Ford, M.; Coughlin, J.; Mai, C.-K. Effect of chiral 2-ethylhexyl side chains on chiroptical properties of the narrow bandgap conjugated polymers PCPDTBT and PCDTPT. *Chem. Sci.* **2016**, *7*, 5313–5321. [CrossRef]
47. Wang, Y.; An, T.; Xue, J. Realizing high detectivity organic photodetectors in visible wavelength by doping highly ordered polymer PCPDTBT. *Org. Electron.* **2020**, *82*, 105700. [CrossRef]
48. Wang, L.; Hu, M.; Zhang, Y.; Yuan, Z.; Hu, Y.; Zhao, X.; Chen, Y. High molecular weight polymeric acceptors based on semi-perfluoroalkylated perylene diimides for pseudo-planar heterojunction all-polymer organic solar cells. *Polymer* **2022**, *255*, 125114. [CrossRef]
49. Nathanael, A.J.; Mangalaraj, D.; Ponpandian, N. Controlled growth and investigations on the morphology and mechanical properties of hydroxyapatite/titania nanocomposite thin films. *Compos. Sci. Technol.* **2010**, *70*, 1645–1651. [CrossRef]
50. Predoi, D.; Talu, S.; Ciobanu, S.C.; Iconaru, S.L.; Matos, R.S.; da Fonseca Filho, H.D. Exploring the physicochemical traits, antifungal capabilities, and 3D spatial complexity of hydroxyapatite with Ag^+–Mg^{2+} substitution in the biocomposite thin films. *Micron* **2024**, *184*, 103661. [CrossRef]
51. Ha, J.S.; Kim, K.H.; Choi, D.H. 5-Bis(2-octyldodecyl)pyrrolo[3,4-c]pyrrole-1,4-(2H,5H)-dione-Based DonorAcceptor Alternating Copolymer Bearing 5,5′-Di (thiophen-2-yl)-2,20 -biselenophene Exhibiting 1.5 cm2V-1s-1Hole Mobility in Thin-Film Transistors. *J. Am. Chem. Soc.* **2011**, *133*, 10364–10367. [CrossRef] [PubMed]
52. Hamad-Schifferli, K. Applications of Gold Nanoparticles in Plasmonic and Nanophotonic Biosensing. *Adv. Biochem. Eng. Biotechnol.* **2024**, *187*, 185–221. [CrossRef] [PubMed]
53. Xiong, Y.; Chikkaraddy, R.; Readman, C.; Hu, S.; Xiong, K.; Peng, J.; Lin, Q.; Baumberg, J.J. Metal to insulator transition for conducting polymers in plasmonic nanogaps. *Light Sci. Appl.* **2024**, *13*, 3. [CrossRef] [PubMed]

polymers

MDPI

Article

Microcanonical Analysis of Semiflexible Homopolymers with Variable-Width Bending Potential

Matthew J. Williams * and Michael C. Gray

School of Engineering, Murray State University, Murray, KY 42071, USA
* Correspondence: mwilliams72@murraystate.edu

Abstract: Understanding the structural dynamics of semiflexible polymers in an implicit solvent under varying conditions provides valuable insights into their behavior in diverse environments. In this work, we systematically investigate the effect of the angular width of the bending potential on structural state behavior and conformational variability using microcanonical analysis. A range of angular widths is explored, with the widest value corresponding directly to the classic semiflexible polymer model, which exhibits a diverse set of structural states, including Two-Strand, Three-Strand, Four-Strand, Ring, Random Coil, and Globule configurations. As the angular width narrows, structural variability within states decreases, overlap between structural states is reduced, and conformations become more stable, leading to an expansion of the parameter space dominated by individual structures. By examining microcanonical entropy and its derivatives, we identify transitions analogous to first-, second-, and third-order thermodynamic transitions, providing a deeper understanding of the configurational landscape of semiflexible polymers.

Keywords: semiflexible polymers; microcanonical analysis; replica-exchange Monte Carlo; structural phases

1. Introduction

Biopolymers, such as proteins, nucleic acids, and polysaccharides, serve crucial functions within physiological environments, with their roles inherently linked to their structural configurations [1–4]. Understanding the structural dynamics of these polymer systems under various conditions is crucial in numerous fields, including drug delivery, tissue engineering, and nanomedicine [5,6]. Computer simulations have become essential tools for probing the behaviors of macromolecules. However, while all-atom simulations provide detailed molecular descriptions, they often face computational limitations for large systems and yield case-specific insights, limiting their generality. To address these challenges, coarse-grained effective potential models simplify interactions while preserving essential physical behavior, enabling broader applicability to mesoscopic polymer systems [7,8].

Semiflexible polymer models, like the worm-like chain model [9] and coarse-grained bead-spring models [10], are popular due to their simplicity and ability to approximate biological systems while capturing essential features like bending stiffness. These models enable the representation of a broad range of structural states by adjusting model parameters, energy scales, and incorporating additional effective potentials [11]. Analytical models such as Flory mean-field theory [12] and self-avoiding walk models [13] provide insights into polymer phase transitions in the thermodynamic limit, where the chain length approaches infinity. However, the present study focuses on finite mesoscopic polymer systems, where structural transitions are analyzed within systems of limited size. Unlike models designed for extrapolation to bulk thermodynamic behavior, our approach examines how bending flexibility and angular constraints influence polymer configurations at experimentally relevant mesoscopic scales. Applications include simulations of helical structures [14,15], adsorption on surfaces [16,17], multiple-chain aggregation [18,19], dense packing [20], liquid crystalline ordering in dense polymer solutions [21], crystallization in semiflexible polymer melts [22], and behavior within confined environments [23].

In this study, we introduce a modified semiflexible polymer model that incorporates an angular width parameter, w, into a cosine-based bending potential. This parameter influences the structural stability and diversity of the polymers by scaling the range of angular offsets where bending energy is felt. For $w = 1.0$, the model reproduces standard bead-spring semiflexible polymer behavior, while smaller values (e.g., $w = 0.5$) narrow the potential well, leading to enhanced structural stability and expanded dominance of certain structural states. The angular width parameter w is introduced specifically for use in coarse-grained polymer models and is not intended for chemically realistic, atomistic-level modeling. While a similar width factor could be introduced for a harmonic bending potential, a change in width would be equivalent to a change in well depth, whereas in the cosine potential, there are two degrees of freedom that can be tuned independently. This versatility offers new opportunities for exploring structural transitions under diverse conditions.

The modified model is simulated using a two-dimensional replica-exchange Monte Carlo approach, which efficiently samples structural configurations for systems exhibiting complex structural transitions [24–26]. The histogram reweighting of canonical ensembles allows us to calculate microcanonical entropy, providing thermodynamic markers of structural transitions, including first-, second-, and third-order transitions [27,28].

Our findings highlight the dual role of the angular width parameter in promoting structural stability and diversity, making the variable-width semiflexible polymer model a interesting tool for studying the structural transitions and dynamics of semiflexible polymers.

2. Materials and Methods

2.1. Model Description

This study employs a coarse-grained semiflexible homopolymer model in which the bending potential includes an angular width parameter which scales the range of angular offsets over which the cosine bending restraint is felt. The energy, E, of a polymer chain with N monomers includes three potentials: a bonded interaction between adjacent monomers along the polymer chain, a non-bonded interaction between monomers in physical proximity but not directly bonded by the FENE potential, and a bending potential associated with bond angles. In this study, we present data for polymer chains of the lengths $N = 30$ and $N = 40$.

Bonded monomers interact according to the finitely extensible nonlinear elastic (FENE) potential [29]. The FENE potential, presented in Equation (1), depends solely on the distance between the two bonded monomers, r. A minimum bond energy is achieved when $r = r_0 \equiv 1$ and the maximum deviation from this value is $R \equiv 3/7r_0$.

Non-bonded monomers separated by a distance less than $r_c \equiv 2.5\sigma$ interact according to the Lennard–Jones (LJ) potential, which is provided in Equation (2) [30]. The Lennard–Jones potential has a minimum when monomers are separated by a distance r_0. To achieve this, we use a parameter $\sigma \equiv 2^{-1/6}r_0$. In order to avoid a discontinuity in energy at r_c, we shift the potential by $v_c = 4[(\sigma/r_c)^{12} - (\sigma/r_c)^6]$. The bonded and non-bonded interactions are given by

$$v_{\rm FENE}(r) = -\log\{1 - [(r - r_0)/R]^2\} \tag{1}$$

$$v_{\rm LJ}(r) = 4[(\sigma/r)^{12} - (\sigma/r)^6] - v_c \tag{2}$$

The widely used cosine-based semiflexible homopolymer model is parameterized as $v(\theta) = -\cos(\theta - \theta_0)$ [31,32]. We use a reference bending angle of $\theta_0 = 0$. With this choice, there is an energy penalty for any deviation from the straight chain configuration. In this paper, we introduce a generalization of this formulation that incorporates an angular width parameter, w which controls the range of angular offsets over which the bending potential is applied. This modification allows for tunable flexibility in the polymer backbone, affecting both structural stability and conformational variability. The modified bending potential is given by

$$v_\theta(\theta) = \begin{cases} -w\cos(1/w(\theta-\theta_0)) & \text{if } |\theta-\theta_0| < \pi w, \\ w & \text{otherwise.} \end{cases} \tag{3}$$

The inclusion of the angular width factor, w, determines the width of the potential well, as shown in Figure 1. When $w = 1$, this model reduces to the standard semiflexible model. Choosing a smaller value of w narrows the width of the cosine, causing energetic penalties to take effect for smaller deviations from θ_0. When comparing systems with different values of w, we find it useful to scale v_θ by w to maintain an equivalent maximum gradient of the potential. In exploratory simulations without this scaling, structural comparisons across different w values were less direct. We empirically observed that including the scaling factor facilitated a more direct practical comparison of the structures generated under varying angular widths. Introducing the angular width parameter allows for the simulation of a wider range of systems exhibiting qualitatively distinct structural behavior.

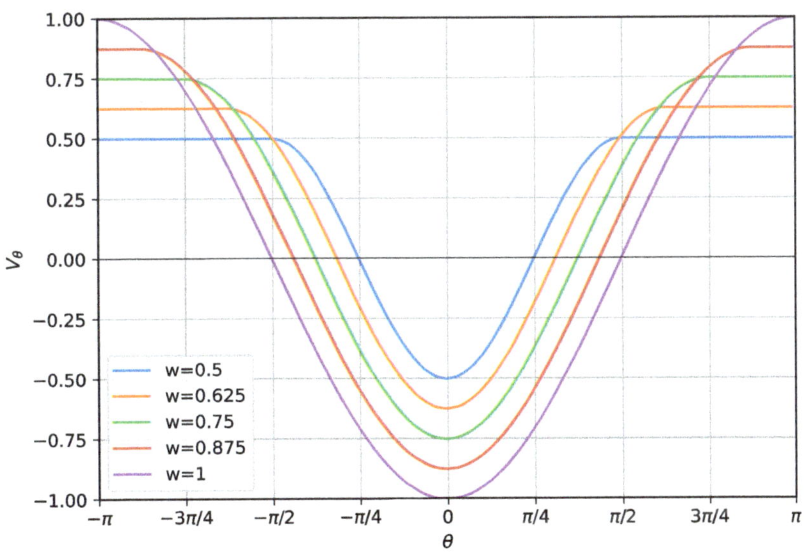

Figure 1. Plot of v_θ as a function of θ for $\kappa = 1$. Curves are shown for values of w ranging between 0.5 and 1. Note that the maximum gradient of v_θ is consistent across for all values of w due to the scaling of the potential.

The total potential energy of a particular polymer configuration $\mathbf{X}$ can be calculated using the Hamiltonian provided in Equation (4). The Hamiltonian consists of all three potentials applied across the entire polymer, with each term being weighted by an appropriate energy scale. We use $S_{\text{FENE}} = \frac{1}{2}KR^2$ with $K = (98/5)r_0^2$ and $S_{\text{LJ}} = 1$ for the FENE and LJ potentials. In this study, the values of S_θ vary linearly between a minimum value of $S_{\theta,\text{min}} = 8$ and a maximum value of $S_{\theta,\text{max}} = 19$. This provides a sufficient range for observing distinct structural transitions. These values were selected to span a broad range of structural behaviors, based on exploratory simulations.

$$H(\mathbf{X}) = S_{\text{LJ}} \sum_{i>j} v_{\text{LJ}}(r_{ij}) + S_{\text{FENE}} \sum_{i} v_{\text{FENE}}(r_{i\,i+1}) + S_\theta \sum_{k} v_\theta(\theta_k) \tag{4}$$

2.2. *Parallel Tempering*

All simulations in this paper consist of an array of simulation threads, with each thread generating a canonical ensemble at temperature T and obeying Hamiltonian H. Threads independently proceed with Monte Carlo updates. Each update changes the energy of the structure by ΔE. Any update that separates two bonded monomers by more than $r_0 + R$ or brings them closer than $r_0 - R$ is immediately rejected. Otherwise, updates are accepted with probability $P_{\textbf{acc}} = \min(1, e^{-\beta \Delta E})$, where $\beta = 1/k_B T$. We use units in which $k_B \equiv 1$.

Most Monte Carlo updates consist of single monomer displacements in which a random monomer is chosen and then displaced to a random location within a box surrounding its original location. To enhance sampling efficiency, additional update moves include the collective displacement of a randomly chosen group of monomers and the rotation of a polymer segment around an axis perpendicular to adjacent bonds, modifying a single bend angle.

This study implements a two-dimensional replica-exchange Monte Carlo technique, also known as two-dimensional parallel tempering. Each thread attempts temperature exchanges with its neighboring threads at higher and lower temperatures, as well as Hamiltonian exchanges with neighboring threads with larger or smaller values of S_θ. In this way, replicas perform a two-dimensional random walk in the S_θ-T space, allowing for the efficient sampling of threads across both temperature and bending energy scales.

Exchanges are attempted every 500 Monte Carlo steps. In an exchange, two neighboring threads attempt to swap configurations, where thread i proposes sending structure X_i to thread j while receiving structure X_j in return. The exchange is accepted with probability

$$P_{\textbf{exch}} = \min\left(1, \frac{e^{\beta_i H_i(X_i)} e^{\beta_j H_j(X_j)}}{e^{\beta_i H_i(X_j)} e^{\beta_j H_j(X_i)}}\right). \tag{5}$$

In the case of a temperature exchange, $H_i = H_j \equiv H$ and Equation (5) becomes

$$P_{\textbf{exch},T} = \min\left(1, e^{(\beta_i - \beta_j)(H(X_i) - H(X_j))}\right). \tag{6}$$

For the Hamiltonian exchange, $\beta_i = \beta_j \equiv \beta$ and Equation (5) becomes

$$P_{\textbf{exch},H} = \min\left(1, e^{\beta\left(H_i(X_i) + H_j(X_j) - H_i(X_j) - H_j(X_i)\right)}\right). \tag{7}$$

In this study, variations in the Hamiltonian H correspond to changes in the bending energy scale factor, S_θ. The temperature values T are logarithmically spaced from $T_{\text{min}} = 0.2$ to $T_{\text{max}} = 1.6$, facilitating efficient sampling across a broad thermal spectrum. For each set of S_θ and T, simulations are performed for values of w ranging from 0.5 to 1.0, where $w = 1.0$ recovers the standard semiflexible polymer model.

3. Results

3.1. Structure Types

A variety of distinct structural conformations are produced in each simulation. Representative configurations of key structural types are presented in Figures 2 and 3. Example configurations for Two-Strand, Three-Strand, Four-Strand, and Five-Strand hairpin structures are shown in Figure 2, with one representative structure from the $w = 1.0$ simulation and one from the $w = 0.5$ simulation placed side by side for comparison. Figure 3 includes Six-Strand structures and Ring structures from the $w = 1.0$ simulation. Ring structures were not observed for $w = 0.5$, as discussed previously.

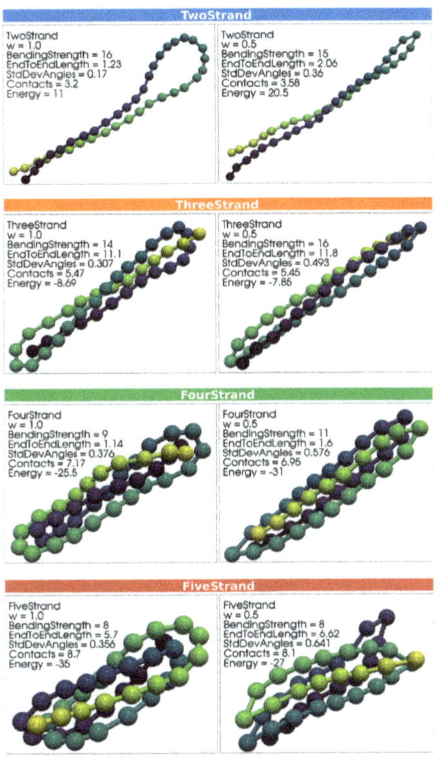

Figure 2. Representative structures for Two-Strand, Three-Strand, Four-Strand, and Five-Strand hairpin structures are shown. For each structure type, examples from both the $w = 1.0$ and $w = 0.5$ simulations are displayed. Within each panel, S_θ, L_{e2e}, C, and E are indicated. While these examples illustrate general structural characteristics, the structural variability within each type is broader than what is captured in this graphic. Rectangles labeling structural representations are colored according to structure typs, consistent with the point-color scheme in Figures 5–7. Monomers within each polymer structure are colored according to a gradient solely to visually distinguish overlapping strands. Colors do not indicate any additional quantitative property.

Structural variability extends beyond what is depicted in these figures. Example structures were selected to be as close as possible to the center of their respective structural clusters, based on σ_θ, C, L_{e2e}, and E.

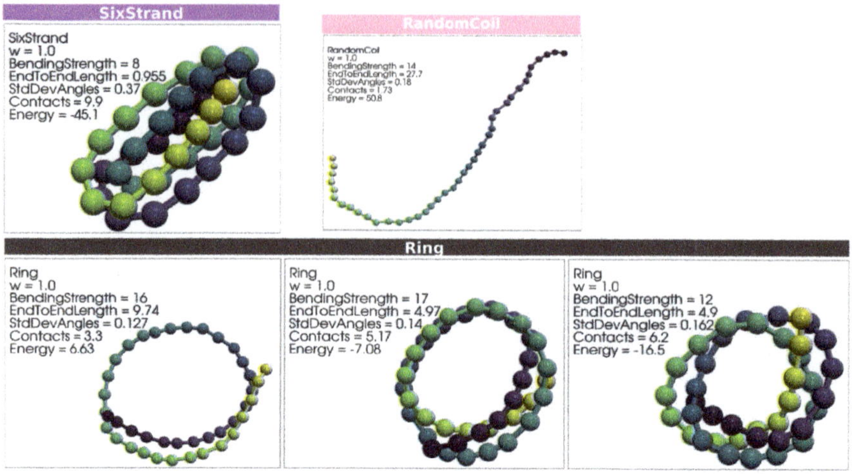

Figure 3. A representative Six-Strand hairpin structure from the $w = 1.0$ simulation is shown. No Six-Strand structures were produced in the $w = 0.5$ simulations. A representative Random Coil structure from the $w = 1.0$ simulation is shown. Without systematic variation with w, we did not feel it necessary to include a $w = 0.5$ example. Ring structures from each of the distinct Ring clusters identified in Figure 5b are displayed in the second row. Within each panel, the bending strength, End-to-End Length, Average Number of Contacts, and energy are indicated. While these examples illustrate the general structural characteristics, the structural variability within each type extends beyond what is captured in this figure. Rectangles labeling structural representations are colored according to structure types, consistent with the point-color scheme in Figures 5–7. Monomers within each polymer structure are colored according to a gradient solely to visually distinguish overlapping strands. Colors do not indicate any additional quantitative property.

The top row of Figure 2 presents two representative Two-Strand structures. The first example is from the $w = 1.0$ simulation, and the second is from the $w = 0.5$ simulation. The $w = 1.0$ example structure features a broad, arcing joint that minimizes individual bending angle deviations from 0 degrees. The vertical extension of the Two-Strand structure cluster in Figure 5b reflects variability in joint types, ranging from the broad arcing joint shown here to a narrower joint more closely resembling the $w = 0.5$ example structure. This variation results from a trade-off between maximizing the number of contacts and minimizing the total bending energy penalty. In contrast, all Two-Strand structures in the $w = 0.5$ simulation exhibit narrow joints, where a single bending angle bears most of the bending penalty, allowing other angles to remain near 0 degrees.

Hairpin structures with more strands exhibit similar features to the Two-Strand case. As in the Two-Strand structures, joints in the $w = 1.0$ simulations range from broad, arcing forms to narrow configurations, while the $w = 0.5$ simulations exclusively produce structures with narrow joints. This variability in joint type is reflected in the vertical extension of clusters within Figure 5b, highlighting a continuum of bending energy trade-offs similar to those observed in the Two-Strand case.

One representative Six-Strand structure is shown in the top row of Figure 3. This structure was generated within the $w = 1.0$ simulation, and no such structures were observed in the $w = 0.5$ simulation. Multiple clusters of Ring structures were produced in the $w = 1.0$ simulation. A single representative structure is selected from the center of each cluster in Figure 5b. As the number of wraps in a Ring structure increases, so does the number of contacts, with successive clusters in Figure 5b representing increasingly wrapped conformations along the horizontal axis.

3.2. Microcanonical Analysis

To identify structural transitions from the simulation results, we locate inflection points in the microcanonical entropy and its derivatives. The microcanonical entropy, defined as $S(E) = k_B \log(g(E))$, is calculated by reweighting the histograms produced by individual simulation threads. Each simulation thread i, operating at temperature T, generates a histogram $h_i(E)$, which estimates the density of states as $\overline{g}_i(E) = h_i(E)e^{\beta E}$. The statistical accuracy of this estimate at a given energy value E depends on the number of histogram counts in that energy bin, with more counts leading to improved reliability.

If histograms from threads at different temperatures overlap sufficiently in terms of energy space, they can be weighted to produce a density of states, $g(E)$, across the full range of E [33]. To achieve this, we initialize the partition function with $Z = 1$ at all temperatures. An estimate of $\hat{g}(E)$ is then computed using Equation (8), where M_i represents the total number of energy samples (histogram counts) from thread i. This estimate normalizes the histogram data across threads, enabling an approximate and consistent density of the states. The partition function estimate is then iteratively refined using Equation (9) until convergence.

$$\hat{g}(E) = \frac{\sum_i h_i(E)}{\sum_i M_i Z_i^{-1} e^{-\beta_i E}}. \tag{8}$$

$$Z_i = \sum_E \hat{g}(E) e^{-\beta_i E}. \tag{9}$$

From the density of states, the microcanonical entropy can be calculated. Bezier smoothing is applied to reduce statistical noise in S and improve numerical stability when computing its derivatives while preserving underlying trends [34]. The derivatives of $S(E)$, denoted as $\beta(E) \equiv dS/dE$, $\gamma(E) \equiv d^2S/dE^2$, and $\delta(E) \equiv d^3S/dE^3$, are used to analyze structural transitions.

An nth-order structural transition in a finite system appears as a region of reduced variation in the $(n-1)^{\text{th}}$ derivative of the microcanonical entropy, leading to a zero crossing in the $(n+1)^{\text{th}}$ derivative of the entropy. These zero crossings correspond to an inflection point that is two orders lower, which serves as an indicator of a structural transition [35].

Figure 4 provides example plots of the entropy and its first four derivatives for a bending strength of $S_\theta = 10$. Most of the structural transitions identified using this method align well with the qualitative changes in structure type, as described in Section 3.3.

In the $w = 0.5$ case, four distinct structural transitions of orders 1–3 are identified. A first-order structural transition occurs at approximately $E = 30$, indicated by an inflection point in the entropy curve, a peak in the first derivative, and a zero crossing in the second derivative. Three third-order structural transitions appear as regions of reduced variation in the second derivative, dips in the third derivative, and zero crossings in the fourth derivative, occurring at approximately $E = -19, 9$, and 23. While further derivatives could be used to identify fourth-order transitions, distinguishing qualitatively distinct structures at such higher orders would require additional structural analysis.

For the $w = 1.0$, five structural transitions of orders 1–3 are identified. A first-order structural transition occurs at $E = 12$, marked by an inflection point in the entropy curve, a slight peak in the first derivative, and a zero crossing in the second derivative. Three second-order structural transitions, indicated by regions of reduced variation in the first derivative and zero crossings in the third derivative, occur at $E = -33, -21$, and 29. A single third-order transition is observed at $E = 0$, characterized by a region of reduced variation in the second derivative and a zero crossing in the fourth derivative.

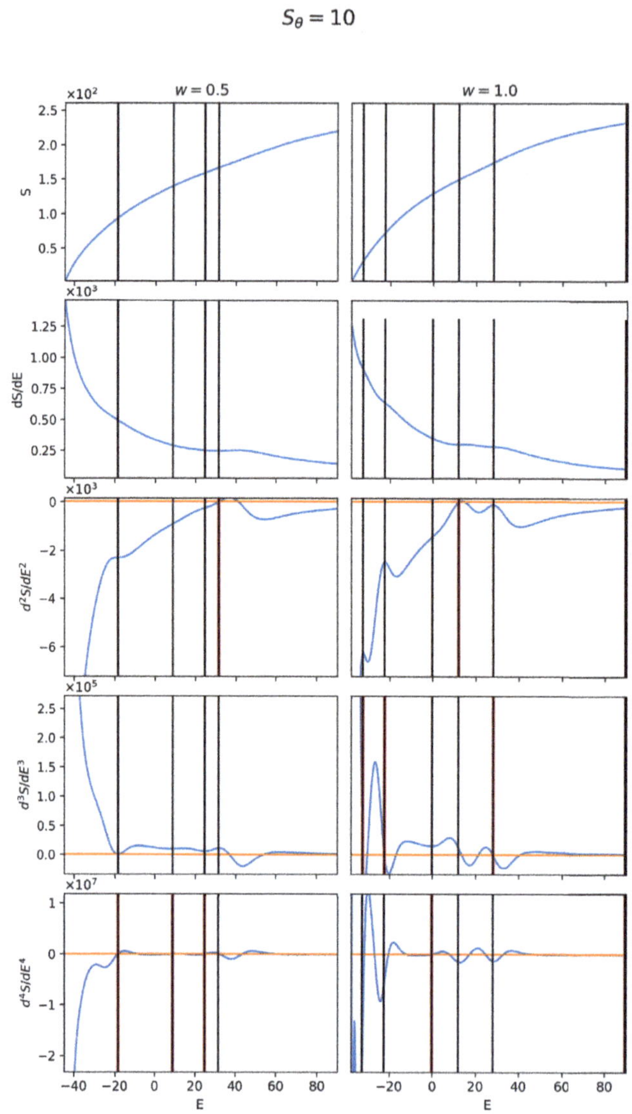

Figure 4. Example plots of the first four derivatives of S for bending strength of $S_\theta = 10$. The left column corresponds to $w = 0.5$, and the right $w = 1.0$. Zero lines are shown in orange, and vertical red lines indicate zero crossings in the derivatives that correspond to structural transitions. Vertical black lines mark the transition energies, aligning these crossings across all derivative orders.

3.3. *Structural Classification and Diagram Comparison*

Structures are classified by analyzing the clustering of example structures across an array of carefully chosen structural parameters. The primary parameters used to distinguish structures are the standard deviation of bending angles (σ_θ) and the per-monomer number of contacts (C). Scatter plots illustrating the structures produced in selected simulation runs are presented in Figure 5. Each panel represents a sampling of structures from every thread of a single two-dimensional parallel tempering simulation. Structure clusters are identifiable through visual inspection. Three-dimensional models of representative structures within each cluster are examined, and all structures are labeled according to their respective clusters. While some clusters overlap in the two-dimensional representations shown here, they can be distinguished by considering their positions in additional structural parameters, such as energy (E) and end-to-end length (L_{e2e}). This multi-parameter classification helps ensure that structural distinctions are robust. Three-dimensional representations of example structures from each cluster are presented in Section 3.1.

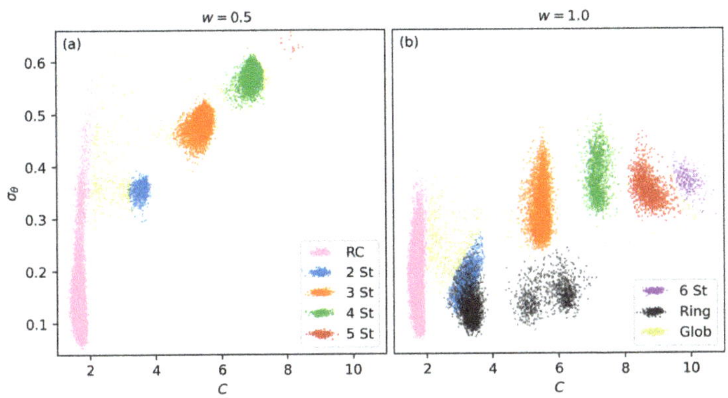

Figure 5. Structural classification based on clustering within the σ_θ-C structural parameter space is illustrated, alongside the structure representation across temperature and energy, for polymers of length 40 and $w = 1.0$. (**a**,**b**) display scatter plots of structural clusters for $w = 0.5$ and $w = 1.0$, respectively. The color key for structure types used throughout this paper is shown in these two panels.

Once the structures are classified according to their clustering within structural parameter space, they are plotted as a function of S_θ and T. Figure 6 illustrates this representation. To mitigate overlapping points, we apply Gaussian-distributed jitter, which spreads the points and provides a clearer depiction of each structure type's distribution. This approach effectively creates a canonical structural transition diagram. Similarly, a microcanonical representation is generated, as shown in Figure 7, by plotting structures according to S_θ and E. Microcanonical structural transitions are also included in this representation.

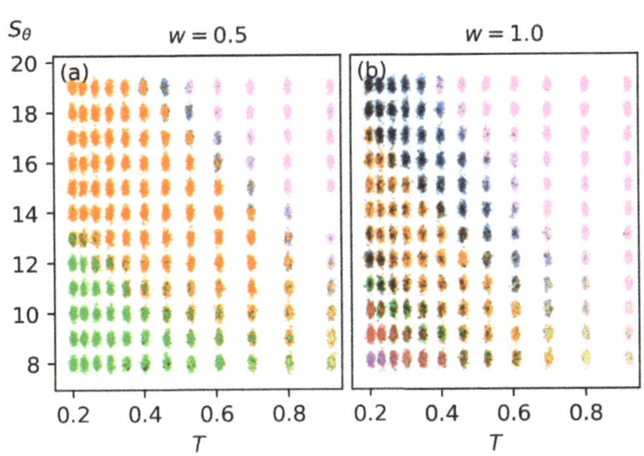

Figure 6. Structural transition diagram showing structure distribution across bending strength (S_θ) and temperature (T). Colors representing each structure type are the same as in Figure 5. (**a**,**b**) correspond to $w = 0.5$ and $w = 1.0$, respectively.

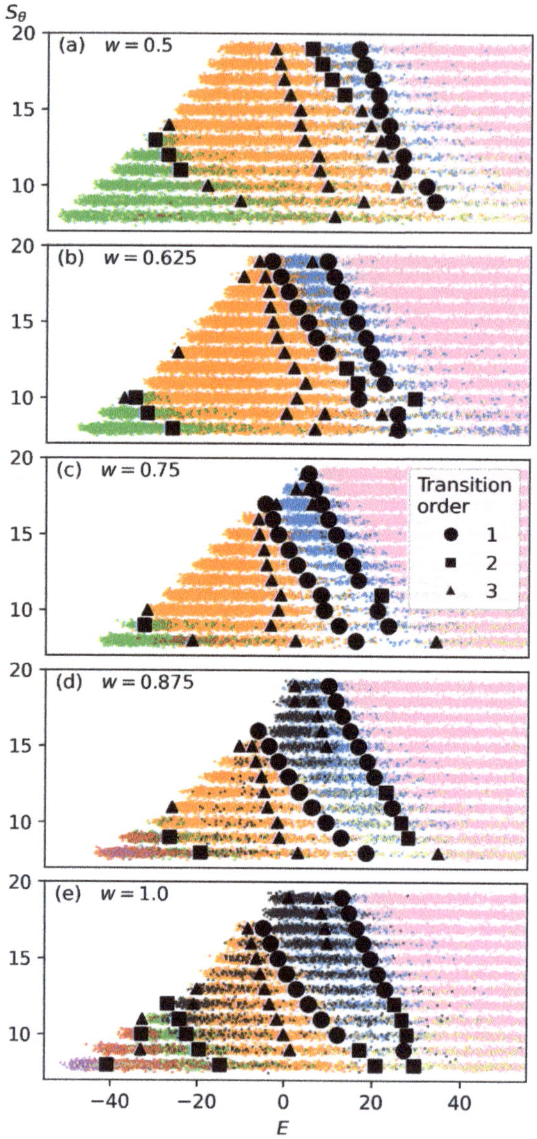

Figure 7. Microcanonical structural transition diagram, plotting structures as a function of bending strength (S_θ) and energy (E), Transition points are marked by circles for first-order transitions, squares for second-order transitions, and triangles for third-order transitions. The colors representing each structure type are the same as in Figure 5.

3.3.1. $w = 1.0$ Simulation

In the $w = 1.0$ simulation, eight distinct structure types are represented. Random Coil (RC) structures are flexible, single-strand linear polymers that do not exhibit a persistent structural organization. Hairpin structures, characterized by multiple relatively straight strands connected by bends or joints, are labeled Two-Strand through Six-Strand (2 St – 6 St). The bends connecting the strands can vary in sharpness, influencing overall polymer shape. Ring structures, where the polymer wraps around itself with overlapping ends, are divided into several distinct clusters based on the degree of overlap. For the purposes of this paper, we do not distinguish between rings with differing degrees of overlap. Finally, Globules (Glob) exhibit a dense, disordered arrangement with significant contacts between non-neighboring monomers but lack a persistent secondary structure or long-range order.

In both the canonical and microcanonical representations, significant structural coexistence can be observed. This occurs among various Hairpin structures, where the probability of one structure type gradually decreases as another becomes more dominant. Ring structures are present across a broad range of structural space but typically appear as a minority within most structural regimes.

First-, second-, and third-order structural transitions from the microcanonical analysis are identified for the $w = 1.0$ simulations. Strong first-order transitions, characterized by sharp entropy changes, are consistently observed between the Random Coil state and the Two-Strand state, as well as between the Two-Strand and Three-Strand states. For some S_θ values, particularly at weaker bending strengths, these transitions are classified as second-order instead of first-order. A consistently second-order transition is observed between the Three-Strand and Four-Strand states, as well as between the Five-Strand and Six-Strand states. A mixture of second- and third-order transitions occurs between the Four-Strand and Five-Strand states. At a $S_\theta = 8$, Four-Strand structures are not sufficiently dominant, and no clear transition to Five-Strand structures is observed.

Additionally, third-order structural transitions are identified in several regions. One occurs between the region dominated by Two-Strand structures and the low-energy, high-bending-strength region where Ring structures dominate. Another consistent third-order transition is found on the high-energy side of the Three-Strand structural state, though no clear structural change has been identified at this transition, suggesting the need for further analysis. Finally, in the low-energy, high-bending-strength region of the Three-Strand structural state, a third-order transition to Ring structures is observed.

3.3.2. $w = 0.5$ Simulation

The structure types observed in the $w = 0.5$ results are largely similar to those in the $w = 1.0$ simulation, with the notable exception that Ring structures are entirely absent, as the narrowed bending potential suppresses their formation. In Figure 5, more distinct clustering of structure types is observed, with tighter clusters and fewer ambiguous structures bridging distinct structural states. Additionally, there is significantly more structural separation within both the canonical and microcanonical representations. Structural coexistence between various Hairpin structures is reduced and Ring structures do not appear in this regime.

Two-Strand structures dominate a significantly smaller region of the T vs. S_θ structural space compared to the $w = 1.0$ simulation. Extending the $w = 0.5$ simulation to higher bending strengths may further expand the structural space where Two-Strand structures are dominant, warranting further investigation.

Due to the additional structural stability provided by the narrowed bending restraint in the $w = 0.5$ case, Four-Strand structures occupy a significantly larger region of structural space compared to the $w = 1.0$ case. This expansion entirely displaces the Five-Strand and Six-Strand structures from dominating any region of structural space for $w = 0.5$. In contrast, for $w = 1.0$, there is a low-temperature, low-bending-strength region where Five-Strand structures dominate. In the $w = 0.5$ simulation, Five-Strand structures were extremely rare and did not form a dominant structural region. However, with an expanded simulation space that includes lower bending strengths and temperatures, it is possible that a distinct Five-Strand structural region could develop.

Similar to the $w = 1.0$ case, we observe a consistent first-order transition between the Random Coil structural region and the Two-Strand structural region, extending even to areas where Two-Strand structures are not dominant. The transition between the dominant Two-Strand and Three-Strand structural regions is second-order. As bending strength decreases, this transition shifts to third-order. We again observe a third-order transition within the middle of the Three-Strand structural region, though no apparent structural change is detected at this transition, suggesting the need for further analysis. Between Three-Strand and Four-Strand structures, we see either second- or third-order transitions, depending on bending strength. While we can observe a sparse representation of Five-Strand structures, we have not explored sufficiently low bending strengths to identify a region where Five-Strand structures dominate.

3.3.3. Intermediate Values of w

Microcanonical structural space plots for intermediate values of w are shown in Figure 7. As w increases, the Three-Strand structural region progressively shrinks, leading to a broader distribution of structures across both high and low bending-strength values. At high bending strengths, the Two-Strand structural region expands, while at lower bending strengths, Five-Strand and Six-Strand structures become more prominent.

With increasing w, there is also a progressive increase in structural coexistence. This is evident both between neighboring structural regions and in the increased presence of Ring structures across multiple structural regions, particularly in the $w = 0.875$ and $w = 1.0$ cases. This increase in structural coexistence corresponds to the decreased structural stability for all low-temperature structures under the conditions of a wider bending restraint. The broader distribution of structures underscores the role of w in modulating structural diversity and stability across temperature and energy spaces.

4. Discussion

This study demonstrates the diverse array of semiflexible polymer behaviors that emerge from introducing an angular width parameter (w) into the bending potential. We find that this parameter strongly influences the range of structural states observed, the degree of structural coexistence, and the structural variability of semiflexible polymers. Through a combination of canonical and microcanonical analyses, we identified a range of structural states, including Two-Strand, Ring, and Globule structures, and their dependence on bending strength and temperature. Narrowing the bending potential ($w = 0.5$) increases structural stability while reducing structural coexistence and variability. The increased structural stability leads to the Four-Strand structural state dominating over other structure types, such as the Five-Strand, Six-Strand, and Ring structures observed in the $w = 1.0$ simulation. Additionally, the expansion of the Three-Strand structural state significantly reduces the structural space representation of Two-Strand structures.

The first four derivatives of the microcanonical entropy reveal clear markers of first-, second-, and third-order structural transitions, offering deeper insight into the thermodynamic properties of these systems. These findings have important implications for understanding the self-assembly and structural behavior of polymers under physical constraints, with potential applications in nanomedicine, materials science, and soft matter physics.

Future work could explore several promising avenues to extend and deepen the current analysis. First, simulations could cover a broader range of bending strengths, to more clearly capture multistrand bundle formation, especially in cases of narrower bending potentials. Additionally, extending the simulations over a wider temperature range would enable the clearer identification of structural transitions and their thermodynamic characteristics. Exploring smaller values for the angular width parameter could provide valuable insights as the bending potential approaches the fully flexible (freely jointed) polymer limit. Furthermore, performing canonical analyses to investigate structural transitions based on temperature-dependent properties, such as the specific heat, would complement the microcanonical findings presented here. Finally, comparing our coarse-grained results to more applied polymer systems could strengthen the relevance of this new potential for practical polymer modeling.

Author Contributions: Conceptualization, M.J.W.; methodology, M.J.W. and M.C.G.; software, M.J.W. and M.C.G.; validation, M.J.W. and M.C.G.; formal analysis, M.J.W. and M.C.G.; investigation, M.J.W. and M.C.G.; resources, M.J.W.; data curation, M.J.W.; writing—original draft preparation, M.J.W. and M.C.G.; writing—review and editing, M.J.W.; visualization, M.J.W. and M.C.G.; supervision, M.J.W.; project administration, M.J.W. All authors have read and agreed to the published version of the manuscript.

Funding: This research received no external funding.

Institutional Review Board Statement: Not applicable.

Informed Consent Statement: Not applicable.

Data Availability Statement: The data that support the findings of this study are available from the corresponding author upon reasonable request.

Conflicts of Interest: The authors declare no conflicts of interest.

References

1. Kmiecik, S.; Gront, D.; Kolinski, M.; Wieteska, L.; Dawid, A.E.; Kolinski, A. Coarse-grained protein models and their applications. *Chem. Rev.* **2016**, *116*, 7898. [CrossRef] [PubMed]
2. Yuan, C.; Chen, H.; Lou, X.W.; Archer, L.A. DNA bending stiffness on small length scales. *Phys. Rev. Lett.* **2008**, *100*, 018102. [PubMed]
3. Sluysmans, D.; Willet, N.; Thevenot, J.; Lecommandoux, S.; Duwez, A.-S. Single-molecule mechanical unfolding experiments reveal a critical length for the formation of α-helices in peptides. *Nanoscale Horiz.* **2020**, *5*, 671.
4. Wu, J.; Cheng, C.; Liu, G.; Zhang, P.; Chen, T. The folding pathways and thermodynamics of semiflexible polymers. *J. Chem. Phys.* **2018**, *148*, 184901.
5. Gao, P.; Nicolas, J.; Ha-Duong, T. Supramolecular organization of polymer prodrug nanoparticles revealed by coarse-grained simulations. *J. Am. Chem. Soc.* **2021**, *143*, 17412.
6. Ferreira, L.G.; Santos, R.N.D.; Oliva, G.; Andricopulo, A.D. Molecular docking and structure-based drug design strategies. *Molecules* **2015**, *20*, 13384. [CrossRef]
7. Katzgraber, H.G.; Trebst, S.; Huse, D.A.; Troyer, M. Feedback-optimized parallel tempering Monte Carlo. *J. Stat. Mech.* **2006**, P03018.
8. Singh, N.; Li, W. Recent advances in coarse-grained models for biomolecules and their applications. *Int. J. Mol. Sci.* **2019**, *20*, 3774. [CrossRef]
9. Bustamante, C.; Marko, J.F.; Siggia, E.D.; Smith, S. Entropic elasticity of λ-phage DNA. *Science* **1994**, *265*, 1599.
10. Kremer, K.; Grest, G.S. Dynamics of entangled linear polymer melts: A molecular-dynamics simulation. *J. Chem. Phys.* **1990**, *92*, 5057.
11. Aierken, D.; Bachmann, M. Impact of bending stiffness on ground-state conformations for semiflexible polymers. *J. Chem. Phys.* **2023**, *158*, 214905. [CrossRef] [PubMed]
12. Lifshitz, I.M.; Grosberg, A.Y.; Khokhlov, A.R. Some problems of the statistical physics of polymer chains with volume interaction. *Rev. Mod. Phys.* **1978**, *50*, 683. [CrossRef]
13. Grosberg, A.Y.; Khokhlov, A.R.; Stanley, H.E.; Mallinckrodt, A.J.; McKay, S. Statistical physics of macromolecules. *Comput. Phys.* **1995**, *9*, 171. [CrossRef]
14. Williams, M.J.; Bachmann, M. Significance of bending restraints for the stability of helical polymer conformations. *Phys. Rev. E* **2016**, *93*, 062501. [CrossRef]
15. Williams, M.J. Microcanonical Analysis of Helical Homopolymers: Exploring the Density of States and Structural Characteristics. *Polymers* **2023**, *15*, 3870. [CrossRef]
16. Zierenberg, J.; Marenz, M.; Janke, W. Dilute semiflexible polymers with attraction: Collapse, folding and aggregation. *Polymers* **2016**, *8*, 333. [CrossRef]

17. Junghans, C.; Bachmann, M.; Janke, W. Statistical mechanics of aggregation and crystallization for semiflexible polymers. *Europhys. Lett.* **2009**, *87*, 40002. [CrossRef]
18. Milchev, A.; Binder, K. Semiflexible polymers interacting with planar surfaces: Weak versus strong adsorption. *Polymers* **2020**, *12*, 255. [CrossRef]
19. Sintes, T.; Sumithra, K.; Straube, E. Adsorption of semiflexible polymers on flat, homogeneous surfaces. *Macromolecules* **2001**, *34*, 1352. [CrossRef]
20. Martínez-Fernández, D.; Herranz, M.; Foteinopoulou, K.; Karayiannis, N.C.; Laso, M. Local and global order in dense packings of semi-flexible polymers of hard spheres. *Polymers* **2023**, *15*, 551. [CrossRef]
21. Milchev, A.; Egorov, S.A.; Binder, K.; Nikoubashman, A. Nematic order in solutions of semiflexible polymers: Hairpins, elastic constants, and the nematic-smectic transition. *J. Chem. Phys.* **2018**, *149*, 174909. [CrossRef] [PubMed]
22. Shakirov, T.; Paul, W. Crystallization in melts of short, semiflexible hard polymer chains: An interplay of entropies and dimensions. *Phys. Rev. E* **2018**, *97*, 042501. [CrossRef]
23. Milchev, A.; Egorov, S.A.; Vega, D.A.; Binder, K.; Nikoubashman, A. Densely packed semiflexible macromolecules in a rigid spherical capsule. *Macromolecules* **2018**, *51*, 2002. [CrossRef]
24. Kimura, K.; Higuchi, S. Extension of the constant exchange probability method to multi-dimensional replica exchange Monte Carlo applied to the tri-critical spin-1 Blume–Capel model. *J. Stat. Mech.* **2016**, *2016*, 123207. [CrossRef]
25. Mitsutake, A.; Okamoto, Y. From multidimensional replica-exchange method to multidimensional multicanonical algorithm and simulated tempering. *Phys. Rev. E* **2009**, *79*, 047701. [CrossRef]
26. Fukunishi, H.; Watanabe, O.; Takada, S. On the Hamiltonian replica exchange method for efficient sampling of biomolecular systems: Application to protein structure prediction. *J. Chem. Phys.* **2002**, *116*, 9058. [CrossRef]
27. Schnabel, S.; Seaton, D.T.; Landau, D.P.; Bachmann, M. Microcanonical entropy inflection points: Key to systematic understanding of transitions in finite systems. *Phys. Rev. E* **2011**, *84*, 011127. [CrossRef]
28. Qi, K.; Bachmann, M. Classification of phase transitions by microcanonical inflection-point analysis. *Phys. Rev. Lett.* **2018**, *120*, 180601. [CrossRef]
29. Bird, R.B.; Armstrong, R.C.; Hassager, O.; Bird, R.B. *Dynamics of Polymeric Liquids. 2: Kinetic Theory*, 2nd ed.; Wiley: New York, NY, USA, 1987.
30. Qi, K.; Liewehr, B.; Koci, T.; Pattanasiri, B.; Williams, M.J.; Bachmann, M. Influence of bonded interactions on structural phases of flexible polymers. *J. Chem. Phys.* **2019**, *150*, 054904. [CrossRef]
31. Marenz, M.; Janke, W. Knots as a topological order parameter for semiflexible polymers. *Phys. Rev. Lett.* **2016**, *116*, 128301. [CrossRef]
32. Aierken, D.; Bachmann, M. Secondary-structure phase formation for semiflexible polymers by bifurcation in hyperphase space. *Phys. Chem. Chem. Phys.* **2023**, *25*, 30246. [CrossRef] [PubMed]
33. Conrad, P.B.; de Pablo, J.J. Comparison of histogram reweighting techniques for a flexible water model. *Fluid Ph. Equilib.* **1998**, *150–151*, 51. [CrossRef]
34. Bachmann, M. *Thermodynamics and Statistical Mechanics of Macromolecular Systems;* Cambridge University Press: Cambridge, UK, 2014.
35. Sitarachu, K.; Bachmann, M. Evidence for additional third-order transitions in the two-dimensional Ising model. *Phys. Rev. E* **2022**, *106*, 014134. [CrossRef] [PubMed]

polymers

MDPI

Article

Deformation Characterization of Glass Fiber and Carbon Fiber-Reinforced 3D Printing Filaments Using Digital Image Correlation

Vivien Nemes [1,2], Szabolcs Szalai [1,2], Brigitta Fruzsina Szívós [1,2], Mykola Sysyn [3], Dmytro Kurhan [4] and Szabolcs Fischer [1,2,*]

[1] Central Campus Győr, Széchenyi István University, H-9026 Győr, Hungary; nemes.vivien@sze.hu (V.N.); szalaisz@sze.hu (S.S.); szivos.brigitta.fruzsina@sze.hu (B.F.S.)
[2] Vehicle Industry Research Center, Széchenyi István University, H-9026 Győr, Hungary
[3] Department of Planning and Design of Railway Infrastructure, Technical University Dresden, D-01069 Dresden, Germany; mykola.sysyn@tu-dresden.de
[4] Department of Transport Infrastructure, Ukrainian State University of Science and Technologies, UA-49005 Dnipro, Ukraine; d.m.kurhan@ust.edu.ua
* Correspondence: fischersz@sze.hu; Tel.: +36-(96)-503-400

Abstract: The paper offers an in-depth deformation study of glass fiber-reinforced and carbon composite filaments of 3D printers. During the certification, the authors used DIC (Digital Image Correlation) as a full-field strain measurement technique to explore key material traits as a non-contact optical measurement method. The insights captured through the DIC technology enabled to better understand the localized strain distributions during the loading of these reinforced filaments. The paper analyzes the glass fiber and carbon fiber filaments used in 3D printing that are reinforced with these materials and are subjected to bending and compressive loading. The segment presents how loading affects the performance of reinforced filaments when varying such factors as the deposition patterns, layer orientation, and other process parameters. Different types and combinations of reinforcements and printing variables were tested, and the resulting dependencies of mechanical parameters and failure modes were established for each case. Key conclusions demonstrate that the mechanical behavior of both carbon- and glass fiber-reinforced filaments is strongly affected by the 3D printing parameters, particularly infill density, pattern, and build orientation. The application of Digital Image Correlation (DIC) allowed for a precise, full-field analysis of strain distribution and deformation behavior, offering new insights into the structural performance of fiber-reinforced 3D printed composites. The findings from the study provide guidance for the proper choice of filling material and the optimal parameters for the 3D printing process of models with high-performance indexes and seamless applications in the automotive and industrial manufacturing sectors.

Keywords: carbon composite; glass fiber-reinforced; deformation analysis; material properties; DIC; 3D printing; mechanical testing

1. Introduction

The rapid development of 3D printing has enabled the more efficient production of parts with more complex geometries, which can be used in various industrial applications [1,2]. FDM (Fused Deposition Modeling) technology is one of the most widely used 3D printing processes, which builds objects from thermoplastic polymer filaments such as PLA and PETG in layers [1–3]. Fiber-strengthened composites such as carbon and glass fibers are often used to optimize mechanical properties and improve printed structures' strength, stiffness, and coefficient of thermal expansion [4].

As reviewed by Kohutiar et al. [5], the effectiveness of these reinforcements strongly depends on the manufacturing technology used. Layer orientation, print parameters, and the type of polymer matrix significantly influence the mechanical performance of the final composite [5].

The increasing demand for lightweight and high-performance materials has led to the widespread application of fiber-reinforced composites, where both natural and synthetic fibers—such as glass and carbon—play a crucial role in improving mechanical behavior under complex loading conditions [6]. Previous studies have shown that glass fiber composites exhibit excellent mechanical properties, particularly in terms of fatigue resistance and impact strength [7].

In materials science, the study of the mechanical behavior of materials and the effects of different loads is widely used, with particular emphasis on exploring the relationships between mechanical and geometrical compositional properties [8,9].

The importance of understanding the interplay between microstructure and mechanical performance in composite systems has been highlighted even in food-grade materials, such as restructured pimiento alginate–guar gels, where internal structuring significantly influenced texture and mechanical integrity [10].

Digital Image Correlation (DIC) is an effective solution for accurately assessing the mechanical performance of objects and can be used in a wide range of applications, whether in engineering, materials science, or electronics [11,12]. It provides a non-contact measurement procedure that allows a detailed analysis of the deformation and stress distribution of materials, giving a more accurate picture of the behavior of the structures under test [12,13].

Increasing industrial competitiveness, raw material efficiency, sustainability considerations, and the potential to develop complex products have made additive manufacturing (AM) technologies even more promising in recent years [1,2]. Additive manufacturing is a technology in which materials are built up layer by layer, allowing objects to be created from 3D model data [1,2]. This process is particularly advantageous for complex geometric shapes, although its drawbacks include limited surface quality and geometric accuracy [1,2].

3D printing is one part of additive manufacturing. There are many different 3D printing methods, the most common of which include Stereolithography (SLA), Selective Laser Sintering (SLS), Lamination Object Manufacturing (LOM), and Fused Deposition Modeling (FDM) [1,2].

FDM (Fused Deposition Modeling) is the most widely known 3D printing technology for industrial applications, mainly due to the low cost of printers and the wide range of fiber materials available at affordable prices. FDM 3D printers are often used in various industries such as automotive, aerospace, construction, and medical, especially in the field of rapid prototyping [1–3]. The printing material is a thermoplastic polymer, such as polylactic acid (PLA), acrylonitrile butadiene styrene (ABS), or polyamide (PA) filament [1–3].

The use of ABS filament, one of the most common materials in the world of 3D printing, is declining due to its disadvantages in printability.

The 8th PLA World Congress was held in Munich in 2024, during which it was said that PLA is a very advanced material, but it has functional and mechanical limitations. For this reason, the aim is to design and develop special PLA composites with increased heat resistance, printability, tensile strength, and impact strength compared to basic PLA materials with pseudo reinforcement [14].

The increase in the popularity of PLA is also reflected in the growing number of PLA 3D printing-related papers in research databases (ScienceDirect and Scopus), which overtake ABS, as illustrated in Figure 1.

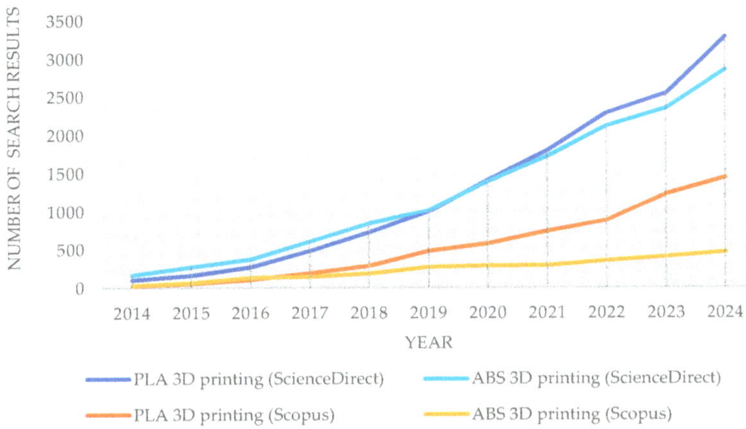

Figure 1. Number of search results for PLA and ABS 3D printing on ScienceDirect and Scopus databases.

From this point of view, it was considered important to investigate these dynamically evolving base materials, e.g., composite or carbon fiber-reinforced PLA matrix materials, to see what properties can be achieved with different reinforcements.

Conventional PETG was considered important to include in the materials to be tested because of its good impact resistance and toughness, which are superior to basic PLA, and its printability properties are better than ABS.

Due to the microstructural anisotropy and the layer-by-layer architecture of FDM technology, the resulting parts' mechanical performance and manufacturing quality are generally inferior to those made using conventional manufacturing processes [15]. During FDM printing, the individual layers do not always adhere perfectly to each other, which can affect the strength and structural integrity of the final product, especially in the bonding planes between the layers [15].

Bembenek et al.'s [16] research shows that the printing angle directly affects Young's modulus [16]. The results suggest that the printing direction plays a significant role in the final mechanical properties, and selecting the adequate angle can be a key factor in achieving the mechanical performance appropriate for the application [16].

However, pure polymers are unsuitable for printing structures where the printed object must have electrical conductivity or significant mechanical properties [1,4]. The solution is to add different reinforcements and fillers to the polymer matrixes, thereby increasing their structural stability and adding functional properties to the reinforcements [1,4].

Carbon, glass, and aradmite fibers are the most commonly used reinforcing fibers. These fibers can be continuous or discontinuous. Tests show that composites reinforced with continuous fibers have better mechanical properties than those reinforced with short fibers [4].

In addition to glass and carbon fibers, nanocellulose, as a natural reinforcing material, also presents a promising opportunity to enhance the mechanical properties of 3D printed composites [17].

The fiber-reinforced polymer filaments used for FDM 3D printing can be either short fiber-reinforced thermoplastic (SFRT) composites or continuous fiber-reinforced thermoplastic (CFRT) composites. In the case of continuous fiber embedding, embedding can occur before the printing process or directly in the print head [2].

For example, acrylonitrile butadiene styrene polymer (ABS) or polylactic acid (PLA) can be the basis for the matrix of FDM-printed continuous fiber composites (CF composites) [18]. Carbon fiber reinforcement is often used in PLA matrix, which increases mechanical strength by enabling excellent mechanical performance, and is lightweight and can be used in aerospace applications, including aircraft, spacecraft, and various engineering applications [1,3,4].

The application of composite materials in additive manufacturing is gaining increasing attention in both biomedical and industrial applications, as they optimize mechanical properties and provide a sustainable alternative to traditional polymers [19].

The results of Maqsood et al. [4] showed that the average flexural stress value of continuous carbon fiber-reinforced PLA (PLA-CCF) specimens is 103% higher than that of PLA specimens, whereas the average flexural stress value of short carbon fiber-reinforced PLA (PLA-SCF) specimens is 91.6% of that of PLA. In addition, it was found that PLA-CCF has the highest flexural modulus, but the flexural modulus of PLA-SCF also exceeds that of PLA [4].

In the research by Heidari–Rarani et al. [2], an experiment was conducted on a self-made continuous fiber-reinforced PLA composite, and the results showed that its maximum flexural strength increased by 109% and its flexural modulus by 367.6% compared to pure PLA [2].

Durga Prasada Rao et al. [20] used short carbon fiber-reinforced PLA filament for FDM 3D printing. By investigating two filling patterns, cubic and quarter cubic, the results showed that specimens fabricated with cubic patterns have high tensile strength values [20].

Based on the research of Giani et al. [18], it is known that polymer matrixes reinforced with carbon fiber reduce the coefficient of thermal expansion of the material and increase its thermal conductivity [18]. This results in printed objects with less warping and increased dimensional accuracy [18]. In addition, it has been found that PLA material with 10 wt% recycled carbon fiber, when 3D printed at 0° to the applied stress orientation, will exhibit approximately twice the elastic modulus and maximum fe-stress compared to the 90° orientation [18].

In addition to carbon fiber, glass fiber is a fiber reinforcement that can be used to increase the strength of PLA [21].

Glass fiber reinforcements dominate the composites industry, accounting for up to a significant percentage of all fiber reinforcements worldwide [21]. The thin surface coating applied during fiberglass production, known as sizing, significantly affects composite materials' mechanical properties, durability, and manufacturability, reducing fiber breakage [21]. Film-forming materials such as polyvinyl acetates, polyurethanes, and epoxy resins provide protection and facilitate processing, while coupling agents such as APTES, GPTMS, and MPTMS improve fiber–matrix adhesion [21].

Wang et al. [22] investigated the extent to which glass fiber reinforcement can improve the inherently limited mechanical performance of PLA using silane-modified glass fibers (m-GF) blended with PLA at different wt% (5, 10, 15, and 20 wt%) [22]. It has been shown that with the addition of glass fiber, the tensile strength and stiffness increased almost linearly with increasing glass fiber content, and when 20 wt% GF was used, these values increased almost twofold compared to pure PLA [22]. The impact strength showed an even more visible increase: compared to 30.9 J/m for the base PLA, the impact resistance increased to 102.8 J/m for the composite containing 20 wt% GF, an improvement of more than three times [22]. In addition, the glass fibers were shown to limit thermal deformation but did not in themselves increase the thermal stability of PLA, yet positively affected its processability and foamability [22].

Begum et al. [23] investigated GF/PLA filaments printed in 0°, 45°, and 90° orientations at 40%, 50%, and 60% fill densities [23]. In the flexural test, the specimen with 60% fill density and 90° orientation showed the highest strength, which was 25 MPa, 60% higher than the specimen with 0° orientation [23]. At 90° orientation, the material can resist bending forces more effectively, distributing stress and improving the adhesion and structural integrity of the layer [23]. The material's microstructure at 90° orientation allows for better deformation and energy dissipation under impact [23].

Liu et al. [24] investigated the mechanical properties of wood-, ceramic-, copper-, and carbon fiber PLA-based composites [24]. By FDM printing the specimens with different orientations and screen angles, they found that the specimens with edge-matched +45°/−45° screen angles exhibited the highest mechanical strength [24]. Due to the weak bonding between the layers, the vertically printed composite test specimens showed the weakest mechanical strength and modulus [24].

Comparing different fiber reinforcements, wood and carbon fiber-reinforced objects showed the lowest mechanical properties [24].

Ismail et al. [25] prepared composites using several methods. The methods involved the integration of composite fibers into thermoplastic matrixes; the methods differed in the techniques of integrating the fibers into the matrix (before printing, during extrusion, and during the printing process) [25]. Subsequently, different polylactic acid (PLA) and three different glass fiber-reinforced polylactic acid (GFPLA) composites were 3D printed with 1.02%, 2.39%, and 4.98% glass fiber content, respectively, and subjected to tensile tests [25]. The tensile strength of the composites increased from GFPLA-1 (1.02% glass fiber content) to GFPLA-2.4 (2.39% glass fiber content) but decreased dramatically at GFPLA-5 (4.98% glass fiber content), although it was still higher than that of pure PLA [25]. Finally, SEM studies showed that the porosity area decreased with increasing glass fiber content [25].

Research has shown that these properties improve proportionally with increasing fiber content, although fragility may also increase in individual cases, so it is important to find the optimum ratio [25].

The literature shows that adding carbon and glass fiber in the additive manufacturing of thermoplastics improves the mechanical properties of the product [26].

Goh et al. [27] tested objects made of carbon- and glass fiber-reinforced thermoplastic materials. From the results of the indentation test, it is concluded that glass fiber has a higher indentation resistance than carbon fiber [27]. It was also found that the typical fracture mode of additively manufactured carbon and glass fiber specimens is similar [27].

Alongside PLA, PETG is one of the materials most commonly used in 3D printing [28].

According to research by Mehtedi et al. [29], PETG is more flexible, tougher, less prone to fracture, and more resistant to external stresses compared to PLA [29].

The results of Martins et al. [30] also confirmed that PETG is a flexible material with high elongation and higher deformation capacity compared to PLA, which is more rigid and resistant to tensile forces. It can break more easily due to its rigidity [30].

Bembenek et al. [16] investigated FDM specimens, focusing on the infill pattern, and found that cubic infill for PLA and lines for PETG gave the best UTS (ultimate tensile strength) and STS (specific tensile strength) results [16].

Kadhum et al. [31] investigated the effect of 14 different filling patterns on the mechanical properties of 3D printed PLA and PETG objects with the same filling percentage [31]. Both materials' highest tensile strength values occurred for gyroid and concentric designs. PETG showed higher strength when comparing the two polymers [31].

Srinivasan et al. [32] investigated the effect of filling patterns on the tensile strength of parts fabricated by FDM printing with PETG filament. The results showed that the grid infill pattern resulted in the highest tensile strength (36.34 MPa) compared to the other patterns [32]. The lowest tensile strength values (13.54 MPa) were obtained for the concentric infill specimen [32].

For 3D printing, the appropriate choice of filling pattern can optimize the object's size and influence its strength, elasticity, and mechanical properties [33].

Guessasma et al. [33] developed a blood-element framework to investigate the effect of filling patterns and density and showed that honeycomb samples were the best choice for compression performance. It was shown that for the same fill ratio, among the gyroid, zigzag, and cross patterns, the gyroid pattern is the best option for improving mechanical strength, while the zigzag and cross are more suitable for promoting higher stretch–strain deformation, especially at low fill ratios [33].

Maqsood et al. [34] fabricated continuous carbon fiber-reinforced specimens with PLA matrix material using FDM 3D printing technology with grid and triangular infill patterns at 20%, 40%, and 60% infill densities. The grid specimen showed better mechanical properties than the triangular specimen. Based on the micrographs of the fractured composite, spallation was found to be the main cause of the failure [34].

The DIC system can be easily applied to various deformation tests, providing a versatile and reliable measurement technique for industrial and academic research [12].

In order to obtain a more accurate picture of material deformation during the Erichsen test (ECT, i.e., Erichsen Cupping Test), 3D Digital Image Correlation (DIC) can be used, which provides extensive information on stress and displacement distributions and fracture mechanisms and can calculate the mechanical properties of different materials [12].

The DIC technique is an efficient, non-contact, and highly accurate measurement method for Erichsen tests [12]. Research has shown that the DIC technique can be used to record stress distributions over the entire tested surface and identify localized thinning and material defects [12]. Compared to traditional single-point methods, this technique provides a more comprehensive picture of the material's ductility [12]. The integration of DIC into ECT significantly improves the evaluation of deformation and fracture characteristics of metal plates [12].

The challenge of DIC, for example, in larger structures, is to optimize camera placement and dissection [35]. In addition, optical noise, lighting conditions, and camera stabilization can significantly impact the accuracy of measurements [35]. Damage or movement of speckle patterns can affect measurement accuracy, especially in fracture experiments [35].

It provides full area coverage measurements, which is a significant advantage over conventional sensors [35]. DIC can be used in concrete, steel, and composite structures, for example, to analyze the mechanical properties of complex structural elements such as composite materials [35].

Guo et al. [36] applied Digital Image Correlation (DIC) technology in tension–tension fatigue tests of carbon/glass fiber hybrid rods, using the method to map damage propagation, stress concentrations, and stiffness degradation. Their results confirm that DIC is an effective tool for tracking the deformation behavior of hybrid composites [36].

The executed literature review can be summarized below.

- Additive manufacturing is a technology that builds materials layer by layer from 3D model data, allowing complex objects to be created [1,2].
- Additive manufacturing is particularly advantageous for performing complex geometric shapes, although disadvantages include limited surface quality and geometric accuracy [1,2].

- Increasing industrial competitiveness, raw material efficiency, and sustainability are driving the growing importance of additive manufacturing technologies [1,2].
- 3D printing is one of the best-known forms of additive manufacturing, and it includes many processes such as SLA, SLS, LOM, and FDM [1,2].
- In industrial applications, FDM technology is the most common due to the low cost of printers and the wide availability of raw materials [1–3].
- FDM printers are used in various industries, such as automotive, aerospace, and medical, mainly for rapid prototyping [1–3].
- During FDM printing, the individual layers do not always adhere perfectly to each other, which can affect the mechanical properties of the final product [15].
- Research by Bembenek et al. [16] has shown that the printing angle directly influences Young's modulus, so choosing the appropriate printing direction is key to optimizing mechanical performance [16].
- Pure polymers' mechanical and electrical properties are not always satisfactory, so various reinforcements such as carbon, glass, and aramid fibers can be used to increase their strength and functional properties [1,4].
- Research shows that continuous fiber-reinforced composites have better mechanical properties than their short fiber-reinforced counterparts [4].
- Carbon fiber reinforcement in PLA matrix increases the strength of the material and reduces its thermal expansion coefficient, making it suitable for aerospace and engineering applications [1,3,4,18].
- Glass fiber reinforcement dominates the composites industry worldwide, improving mechanical strength and durability and reducing brittleness [21].
- Research by Wang et al. [22] has shown that fiberglass reinforcement increases tensile strength and impact resistance and limits thermal deformation [22].
- Studies by Begum et al. [23] have shown that the printing orientation and the filling density significantly influence the strength of 3D printed GF/PLA composites.
- Based on the research of Liu et al. [24] among different fiber reinforcements, samples with a raster angle of $+45^\circ/-45^\circ$ showed the best mechanical performance, while composites printed in the vertical direction were the weakest [24].

While additive manufacturing has advanced considerably—particularly in the field of fiber-reinforced 3D printing—significant knowledge gaps remain regarding the mechanical behavior of these composite materials under different loading conditions. It is well established that adding carbon and glass fibers to polymer matrices enhances properties such as strength, stiffness, and impact resistance. However, most existing studies tend to examine isolated factors—such as fiber orientation, infill pattern, or interlayer adhesion—without fully considering how these parameters interact in practical, load-bearing applications. Furthermore, although Digital Image Correlation (DIC) is a widely adopted method for strain measurement, its potential to deliver high-resolution, full-field deformation data in the context of fiber-reinforced 3D printed components has yet to be fully exploited. The intrinsic anisotropy of FDM-manufactured parts, compounded by inconsistencies in layer bonding and filament deposition, adds complexity to predicting their mechanical performance. This study aims to address these challenges by leveraging DIC to perform a comprehensive, full-field analysis of the deformation and failure behavior of PLA- and PETG-based carbon- and glass fiber-reinforced 3D printing filaments.

The research systematically investigates the effects of printing parameters—including infill pattern, density, and build orientation—under both bending and compressive loads.

Through this approach, the study seeks to identify critical structure–property relationships that influence mechanical performance and to derive optimized printing strategies for enhanced material efficiency and functional reliability. The insights gained are intended to inform the design and manufacturing of high-performance, lightweight components in engineering applications where mechanical integrity and material efficiency are paramount.

The structure of the current paper is as follows. Section 2 contains the applied materials and methods, Section 3 introduces the results and gives explanations, and Section 4 provides the conclusions.

2. Materials and Methods

This section focuses on the materials and instruments used in the research and describes the measurement procedures and processes. The current study is based on three PLA-based filaments and one PETG filament from the Fillamentum, Filaticum, and eSun manufacturers.

2.1. Filaments

The four most common filaments available to the authors of the current study were selected for 3D printing the test specimens needed to perform the tests.

Fillamentum PLA is a conventional PLA that is easy to handle, has excellent print quality, and has excellent adhesion. Filaticum PETG is the ideal choice when a stable formulation and high heat resistance are required. Printed parts made from this material have the advantage of excellent resistance to chemical attack and retain their shape at temperatures up to 80 °C. The PLA-based filament eSUN ePLA with 15–20% glass fiber reinforcement is an easy-to-print filament. With glass fiber reinforcement, it exceeds conventional PLA's mechanical strength and impact resistance. Pieces made from ePLA-GF have high impact strength, resistance, and abrasion resistance. eSUN ePLA-CF is a high-quality, easy-to-print PLA-based filament. Incorporating impregnated short carbon fibers increases the strength and modulus of PLA. It offers exceptional printing performance and high-speed printing. The main properties of all the used filaments are shown in Table 1. Table 2 summarizes the mechanical properties of the materials.

Table 1. Properties of used filaments.

Specific	Basic PLA	PETG	Carbon Fiber PLA	Glass Fiber PLA
Nozzle [°C]	190–210	235–250	190–230	190–230
Bed [°C]	50–60	80–95	45–60	45–60
Layer height [mm]	0.4	0.4–1.0	0.4	0.4
Max speed [mm/s]	30–70	60	50–300	40–100
Color	Blue	White	Green	Natural

Table 2. Physical and mechanical properties.

Specific	Basic PLA	PETG	Carbon Fiber PLA	Glass Fiber PLA
Density [g/cm^3]	1.24	1.27	1.21	1.31
Melt Flow Index [g/10 min]	6 (210 °C/2.16kg)	-	5.37 (190 °C/2.16 kg)	6.36 (190 °C/2.16 kg)
Tensile Strength [MPa]	60	50	39	59.27
Elongation at Break [%]	6	120	4.27	7.99
Flexural Strength [MPa]	83	71	103	85.01
Flexural Modulus [MPa]	3800	2150	5003	4414.89
IZOD Impact Strength	16 [J/m]	85 [J/m]	5.08 [kJ/m^2]	10.16 [kJ/m^2]

2.2. 3D Printers and Printing Setups

To print the test specimens, two reliable, easy-to-use, affordable and readily available FDM 3D printers were applied, which were selected for their ability to print the selected fiber-reinforced composites; their properties are shown in Table 3.

Table 3. Properties of used 3D printers.

Specific	Bambu Lab X1 Carbon	Bambu Lab A1 Mini
Manufacturer (city, country)	Bambu Lab (Shenzhen, China)	
Nozzle size [mm]	0.4	0.4
Extruder system	Direct drive	Direct drive
Filament diam.	1.75 mm	1.75 mm
Hotend	All metal	All metal
Bed leveling	Dual auto	Automatic

For the 3D prints, the easy-to-use Bambu Studio v1.9.7.52 slicing software for the Bambu Lab printers was used, which provides a wide range of options for adjustment. The fill percentage, fill pattern, and construction orientation were adjusted for the research.

The density of the fill indicates the amount of material used for the fill. The fill values for this research were set between 5% and 40% with a 5% step size. The infill pattern represents the material's structure within the 3D printed object, whereby the printed pieces can have different strength, elasticity properties, and bulk properties. One of the aims of this research is to investigate different patterns; the authors have investigated the most popular grid, triangle, and gyroid patterns for this research (Figures 2 and 3).

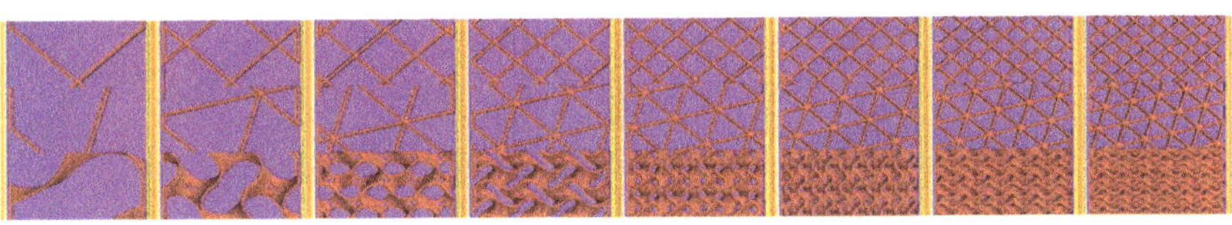

Figure 2. Filling settings in Bambu Studio for specimens in laid orientation; filling patterns from top to bottom: grid, triangle, gyroid; filling values from left to right: 5%–10%–15%–20%–25%–30%–35%–40%.

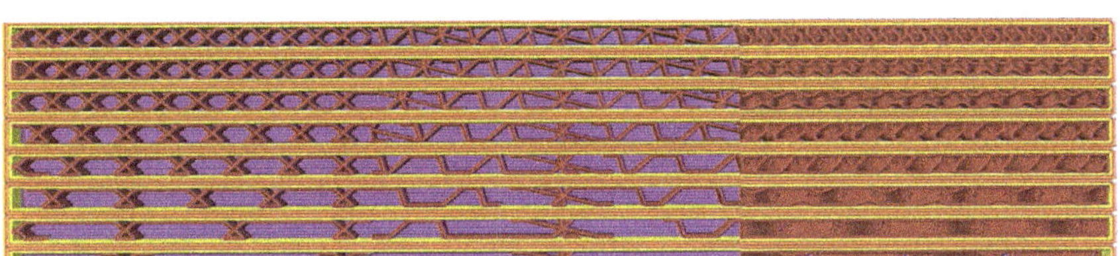

Figure 3. Filling settings in Bambu Studio for specimens with longitudinal edges; filling patterns from left to right: grid, triangle, gyroid; filling values from bottom to top: 5%–10%–15%–20%–25%–30%–35%–40%.

The test specimens were printed with two different construction orientations, laid flat and longitudinal edge-on (Figure 4). The orientation is important because the different ply alignments cause the forces through a ply to differ. The experimental matrix is presented in Table 4.

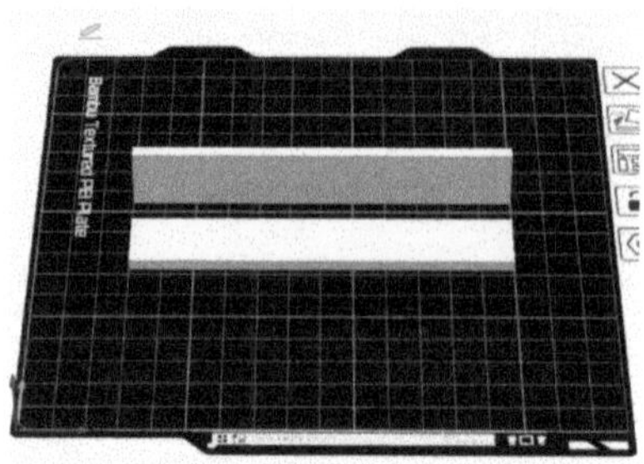

Figure 4. Demonstration of 3D printing orientations.

Table 4. Experimental matrix (1—PLA, 2—PETG, 3—Carbon Fiber PLA, 4—Glass Fiber PLA). The symbols "+" mean that the related tests were executed during the experiments of the current research.

Infill Percentage [%]	Grid (Gr)								Triangle (Tr)								Gyroid (Gy)							
	On-Edge (E)				Flat (F)				On-Edge (E)				Flat (F)				On-Edge (E)				Flat (F)			
	1	2	3	4	1	2	3	4	1	2	3	4	1	2	3	4	1	2	3	4	1	2	3	4
5	+	+	+	+	+	+	+	+	+	+	+	+	+	+	+	+	+	+	+	+	+	+	+	+
10	+	+	+	+	+	+	+	+	+	+	+	+	+	+	+	+	+	+	+	+	+	+	+	+
15	+	+	+	+	+	+	+	+	+	+	+	+	+	+	+	+	+	+	+	+	+	+	+	+
20	+	+	+	+	+	+	+	+	+	+	+	+	+	+	+	+	+	+	+	+	+	+	+	+
25	+	+	+	+	+	+	+	+	+	+	+	+	+	+	+	+	+	+	+	+	+	+	+	+
30	+	+	+	+	+	+	+	+	+	+	+	+	+	+	+	+	+	+	+	+	+	+	+	+
35	+	+	+	+	+	+	+	+	+	+	+	+	+	+	+	+	+	+	+	+	+	+	+	+
40	+	+	+	+	+	+	+	+	+	+	+	+	+	+	+	+	+	+	+	+	+	+	+	+

2.3. GOM Aramis DIC System

The measurement process was carried out using the GOM ARAMIS 5M non-contact, photogrammetric system. The GOM ARAMIS 2018 3D measurement system is part of the Digital Image Correlation (DIC) systems. The system allows the deformation of individual objects to be measured during the inspection. The principle of operation of the measurement system is based on photogrammetry. A sequence of photographs of the object, taken by two cameras in time-lapse mode, provides information on the 3D displacement of each point on the measured sample surface, independently of the material, by means of image correlation using random pattern recognition. The method can be used to inspect both small and large specimens [37–40].

The development of the approach for analyzing deformation behavior and structural integrity was supported by previous research, which demonstrated that the internal microstructural arrangement significantly impacts the mechanical properties and strength of materials [10].

The system uses two 12 MP cameras (GOM, Braunschweig, Germany), which enable high-resolution measurements, allowing small and large displacements to be detected [40].

2.4. Specimen Preparation and Speckle Pattern

The ARAMIS 3D measuring system (GOM, Braunschweig, Germany) determines the position of the points on the surface under test from a grey-scale image. Preparing the sample surface properly and using an adequate pattern is essential to obtain accurate measurement results. In order to achieve appropriate contrast and minimize reflections, the surface should first be coated with white matte paint. A gradient pattern of black spots

(Figure 5) is then applied and distributed so that the ratio of white to black areas remains 50/50. The size and distribution density of the patches will depend on the size of the area under study. When measuring the deformation of small objects, the gradient should consist of finer, more densely spaced patches, whereas for larger objects, it is recommended to use sparser, larger patches. Pattern contrast, spot size, and distribution are key factors for accurate measurement [38,40].

Figure 5. Speckle pattern on the specimen before the ECT.

The applied paint materials were Schuller Eh'klar PrismaColor Acryl RAL9016 (Schuller Eh'klar, Pécs, country: Hungary) for the white primer layer and United Sprays Matte Black Spray (United Paints and Chemicals S.A.E., Giza, Egypt) paint for the speckle pattern.

3. Results and Discussion

As can be seen in Table 4, 192 different combinations of tests could be performed, maintaining a repeatability of 3 per test, which means 576 tests per test (in ECT). Any conclusions or analyses presented in the results section are the average of the 3 measurements for ease of interpretation of the data.

The results are presented in two ways; the first is to conclude the conventional crack maximum and the force data measured at that time. The data presented here are derived from the hydraulic equipment's travel and force display by the operator at the moment of crack initiation. Of course, the resulting data are presumably more inaccurate since they depend to a large extent on the operator's speed, but trends can, of course, already be observed here. In the evaluation, the force-time and displacement-time data are mainly analyzed.

In the second case, the DIC tests were evaluated. The distance and forces are directly linked to the ARAMIS measurement card. In this case, no reading is performed by the operator, so this type of measurement result does not contain any reading uncertainty. The travel-time and displacement-time data are indirectly integrated into the measurement evaluation. In the evaluation, conclusions are drawn from these data. Thanks to the DIC evaluation, much data is also presented, and at the end of the Results chapter, the results of special or outlier measurements are analyzed.

3.1. Measurement Experience, Die Displacement Results

Table 5 shows the results of the tool displacement measurements. Several conclusions can be drawn from the analysis of the data. The PETG material shows the highest values overall, with the highest deformation measured. Additionally, in the case of PETG material, it can be observed that orientation is important, as specimens placed edge-on gave better results than those laid flat. This is interesting because, due to the layer order, the higher value would have been expected in the opposite case, presumably due to the effect of the filling. For the PLA material, the gyroid infill pattern gave the best results between 5 and

20%. Interestingly, this value decreases with increasing pattern density, while it increases with the other two patterns. This implies that good flexural strength results can be obtained for PLA material with a lower filling density, even with a gyroid filling pattern.

Table 5. Result of displacement.

PLA	Grid (Gr)		Triangle (Tr)		Gyroid (Gy)		PETG	Grid (Gr)		Triangle (Tr)		Gyroid (Gy)	
%	On-Edge (E)	Flat (F)	On-Edge (E)	Flat (F)	On-Edge (E)	Flat (F)	%	On-Edge (E)	Flat (F)	On-Edge (E)	Flat (F)	On-Edge (E)	Flat (F)
5	27.5	27.3	26.7	28.1	30.8	30.6	5.0	40.1	27.9	28.2	31.2	30.0	41.3
10	26.8	27.4	26.5	28.7	35.3	30.6	10.0	40.6	41.0	41.5	41.0	41.2	41.2
15	29.3	27.3	28.3	27.9	35.7	29.8	15.0	37.3	37.6	41.3	41.3	30.4	41.4
20	29.0	26.8	27.3	27.8	35.8	27.7	20.0	40.1	39.8	41.3	39.6	41.3	40.5
25	27.9	27.3	27.1	27.3	29.7	27.7	25.0	40.1	37.2	41.3	34.7	41.3	40.4
30	29.2	27.1	27.7	27.4	29.0	27.3	30.0	40.2	41.4	41.0	37.3	41.4	33.8
35	29.5	27.4	27.9	27.4	28.3	27.5	35.0	40.5	41.0	41.0	40.0	41.3	37.5
40	29.7	27.5	28.3	27.5	28.5	27.1	40.0	40.6	40.1	36.7	36.3	41.3	37.8
GLASS	Grid (Gr)		Triangle (Tr)		Gyroid (Gy)		CARBON	Grid (Gr)		Triangle (Tr)		Gyroid (Gy)	
%	On-Edge (E)	Flat (F)	On-Edge (E)	Flat (F)	On-Edge (E)	Flat (F)	%	On-Edge (E)	Flat (F)	On-Edge (E)	Flat (F)	On-Edge (E)	Flat (F)
5	25.6	28.7	25.6	26.9	25.7	26.0	5.0	25.5	27.1	26.1	26.8	25.8	27.5
10	25.4	27.3	25.8	27.0	25.8	25.9	10.0	25.5	26.8	25.9	26.8	25.6	27.8
15	25.4	26.8	25.8	26.7	25.2	25.8	15.0	26.0	26.7	26.4	26.6	26.1	27.4
20	25.8	26.9	26.2	26.8	25.2	25.7	20.0	26.5	28.5	26.1	26.8	25.7	27.5
25	25.9	26.2	26.1	26.6	25.2	25.7	25.0	27.1	26.9	26.8	26.7	25.8	27.2
30	25.7	26.5	25.9	26.6	25.6	25.8	30.0	27.8	27.3	27.1	26.8	26.4	27.4
35	25.7	26.4	26.1	26.9	25.7	25.8	35.0	29.1	27.5	27.5	26.3	26.0	27.0
40	25.6	26.7	26.0	26.7	25.7	25.8	40.0	28.7	28.8	26.8	27.1	26.8	27.1

For the glass and carbon composites, the measurement results are stable, with no significant differences in filling rates or patterns. It is observed that the grid and triangle patterns showed similar performance for all materials. The effect of layout can significantly affect the results, especially for PETG and PLA materials. For glass and carbon materials, the values are relatively uniform and less sensitive to the fill pattern and ratio, which also indicates that using these materials can achieve similar mechanical properties at lower fill values, which can result in significant weight savings for a structure.

3.2. Results of DIC Measurements

The results of the DIC measurements are shown in diagram X. In such measurements, although the sub-applicant manufacturer has long been a leading automotive and research supplier, it is worthwhile to check, i.e., validate the results. Validation ensures that the data, models, or systems are reliable, accurate, and a good representation of reality, thus minimizing the risks of errors. By evaluating the displacement-time data of the Aramis system, the maximum displacement values were compared with the values in Table 4 for each material. It was found that the results of the Aramis system reproduce the values measured in the conventional measurement.

From the results in Figure 6, it is clear that the displacement results are also quite close. The results of the DIC measurements add a lot to the results of the conventional tests. If the slopes of the curves are almost identical, the following conclusion can be drawn. The curves' steepness is related to the stiffness (k), which characterizes the resistance of the material or structure to deformation. If the slope is the same, then the stiffness of the specimens is the same, i.e., the initial resistance to bending of the materials and geometries tested is similar. The different displacement maxima show that, despite the same stiffness, the materials may behave differently at fracture due to internal structural differences. These differences are reflected in differences in infill density and pattern. The conclusions that can be drawn here are the same as in Section 3.1. The additional information provided by the DIC tests is that the whole forming process can be traced.

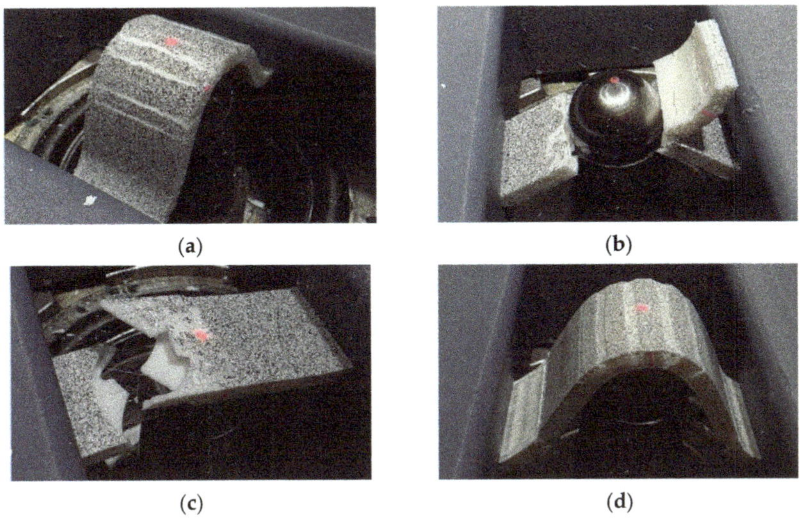

Figure 6. Different infill types in PETG specimens: (**a**) PETG_GrE40%, (**b**) PETG_GrE25%, (**c**) PETG_GrE5%, (**d**) PETG_TrE35%.

A further advantage of the DIC test is that the individual failure states and the deformation of the specimens can be clearly observed. More specific results are shown in Figures 6–9.

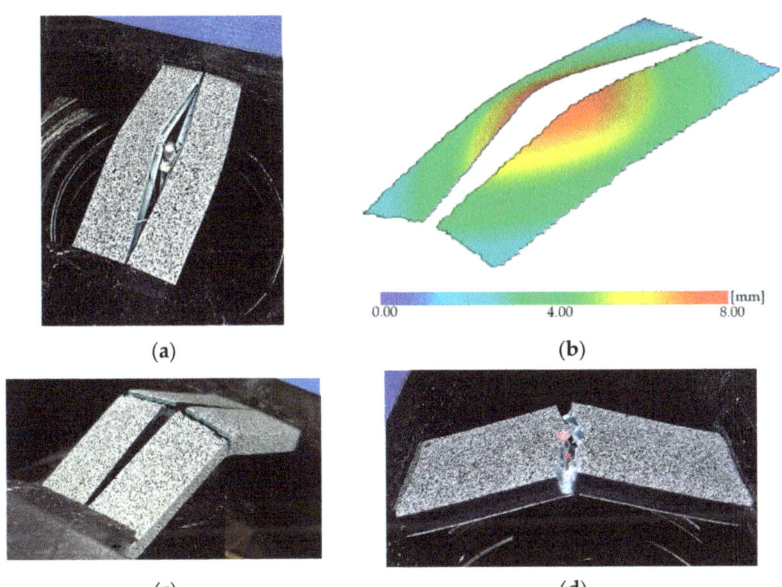

Figure 7. Different infill types in PLA specimens: (**a**) PLA_GrE5%, (**b**) PLA_GrE5% with DIC, (**c**) PLA_GrF15%, (**d**) PLA_GyF20%.

Figure 8. Brittle behavior in the carbon fiber specimen (i.e., CF_TrE25%).

Figure 9. Elongation in the PETG gyroid-filled specimen (i.e., PETG_GyF35%).

In Figure 6, it can be seen that the PETG material stretches in the inter-pattern area as a function of the fill, i.e., local stretching occurs at several points, which may also be responsible for better results. This type of phenomenon suggests that the material has a better load-bearing capacity in the patterns, while the material fails faster in the "empty shell". It can be seen that, although the PETG test specimens achieved similar elongations, the filling density and the pattern had a significant effect on the failure state, so there is an apparent effect on the way the material will fail.

The difference in destruction between the set and laid samples is clearly observed in the case of PLA. The broken specimens clearly show that the set specimen only opened along the layers, while the laid specimen is wholly broken. In Figure 7, it can be seen how the PLA specimens broke.

The carbon-reinforced material showed higher forces but more brittle behavior, as can be seen in Figure 8. The advantage of using this material is that similar mechanical properties can be achieved from relatively few materials compared with other materials.

On the PETG gyroid-filled specimens, it can be seen that there is no localized elongation in this pattern, but the whole specimen follows the shape change together. This behavior is more favorable as the layers move together, making this pattern more suitable for this type of loading. Figure 9 shows one of the PETG gyroid-filled specimens.

When examining the force data, it was observed that the values were quite noisy. This is because the equipment used does not have a direct force measuring cell, but the force can be measured from the hydraulic pressure. In the case of fiber-reinforced composite specimens, it could be expected that small forces would occur due to their expected brittleness compared to the capabilities of the equipment, which is in the lower limit range of the hydraulic force gauge. Thus, in the evaluation, the force maxima were examined rather than the full curves because noise does not distort the results as much, and the results for the maxima are shown in Figure 10.

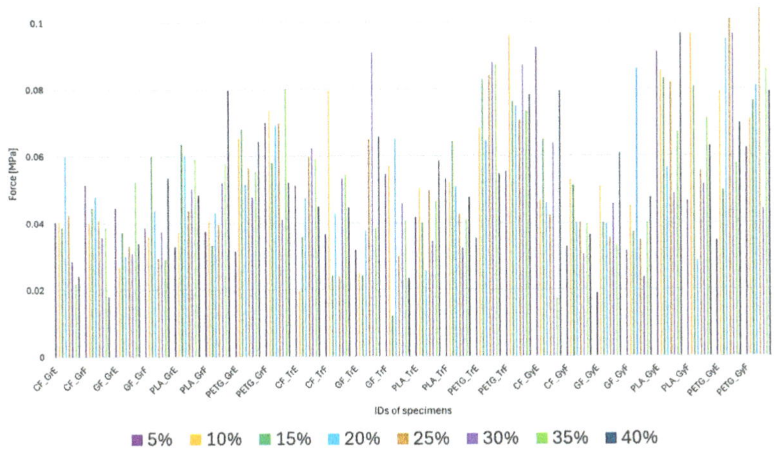

Figure 10. Maximum Force Values of the Specimens.

An overview of the results shows that specimens with 20-25-30% infill in most material groups achieved higher forces. When reviewing the materials, PETG, carbon fiber, and glass fiber with gyroid and triangle infill patterns gave the highest forces.

Among the composite materials, the laid arrangement gave better results. It is also interesting to observe that similar force maxima can be obtained for PLA material using a gyroid pattern. For PETG, the gyroid and triangle fill patterns gave better results in both the stationary and laid-down cases.

The boxplot (Figure 11) shows the statistical distribution of the maximum force values sustained by the tested materials (basic PLA, carbon fiber PLA, glass fiber PLA, and PETG).

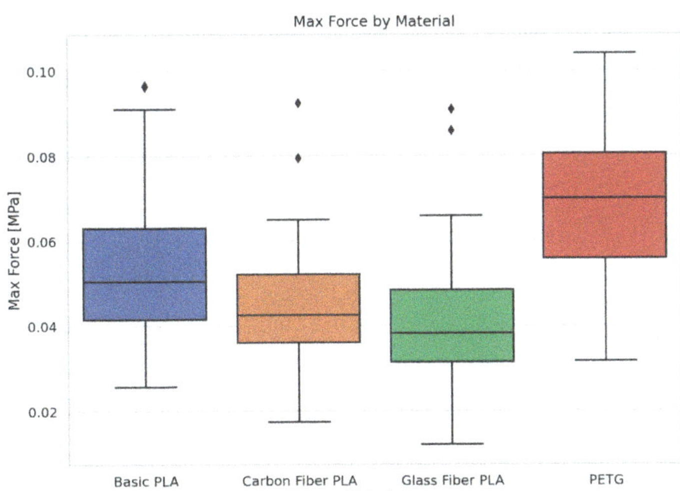

Figure 11. Boxplot of Maximum Force for Various Materials.

Samples made from PETG show the highest maximum force, meaning they withstand the highest load. Basic PLA ranks second in strength, with a broader spread suggesting more variability in performance. Carbon fiber PLA and glass fiber PLA display lower maximum force values. Interestingly, despite being fiber-reinforced, they do not outperform PETG or even basic PLA in strength. Based on the findings of this study, PETG demonstrated the highest load-bearing capacity among the tested materials.

The boxplot (Figure 12) illustrates the statistical distribution of displacement values for the tested materials (basic PLA, carbon fiber PLA, glass fiber PLA, and PETG). PETG samples exhibited more significant displacement, indicating a more elastic and flexible material. Carbon fiber PLA and glass fiber PLA showed low displacement, meaning they are stiffer. Basic PLA falls somewhere in between—more flexible than the fiber-reinforced variants but less flexible than PETG. In conclusion, PETG combines strength and flexibility, whereas fiber-reinforced PLA materials exhibit greater stiffness but tend to fail under lower loads.

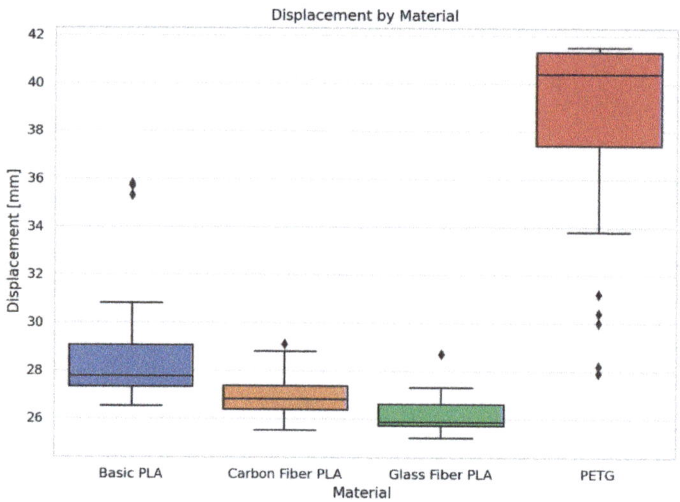

Figure 12. Boxplot of Displacement Across Different Materials.

Figure 13 illustrates a Pareto diagram with the maximum force values [MPa] for various printing settings using basic PLA material. The PLA_GrE20%, PLA_GrE25%, and PLA_GrE30% specimens deliver the best results. The top two account for about 75% of the total force. The trend again shows that 20–30% infill is optimal for basic PLA.

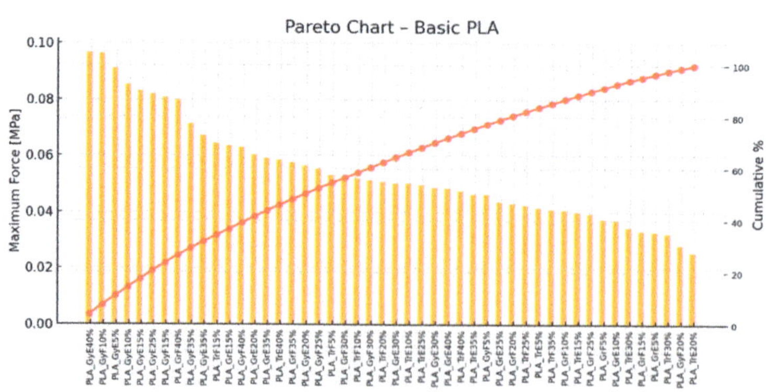

Figure 13. Pareto Chart—Basic PLA (orange color represents the maximum force values, hence red color line means the cumulative percentages).

Figure 14 presents a Pareto analysis of carbon fiber PLA printed samples, highlighting the ranked printing settings based on maximum force [MPa]. The chart shows that CF_GrE20% achieved the highest maximum force, followed by CF_GrE25% and CF_GrE30%. According to the red cumulative line, the top three specimens account for more than 80% of the total force. This suggests that the 20–30% infill range provides outstanding mechanical performance for this material.

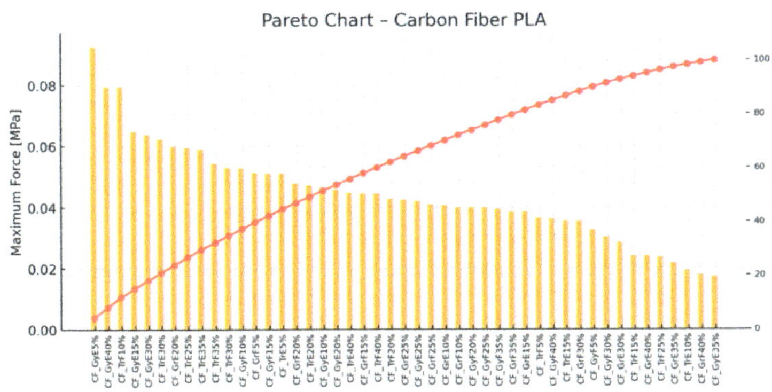

Figure 14. Pareto Chart—PLA Reinforced with Carbon Fiber (orange color represents the maximum force values, hence red color line means the cumulative percentages).

Figure 15 is a Pareto diagram illustrating the maximum force values [MPa] for various printing settings using PETG material. The most substantial specimen is PETG_GrE25%, followed by PETG_GrE20% and PETG_GrE30%. These three account for about 80% of the total force. For PETG, too, the 20–30% infill range delivers the best mechanical performance.

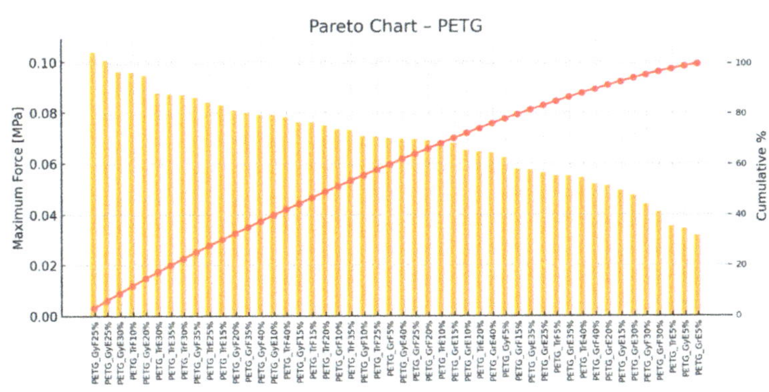

Figure 15. Pareto Chart—PETG (orange color represents the maximum force values, hence red color line means the cumulative percentages).

Figure 16 shows a Pareto analysis of glass fiber PLA printed samples, highlighting the ranked printing settings based on maximum force [MPa]. The most substantial specimen is GF_GrE20%, followed by GF_GrE25% and GF_GrE30%. Together, these three cover approximately 85% of the total performance. Higher infill percentages prove more beneficial, with values above 20% being particularly effective.

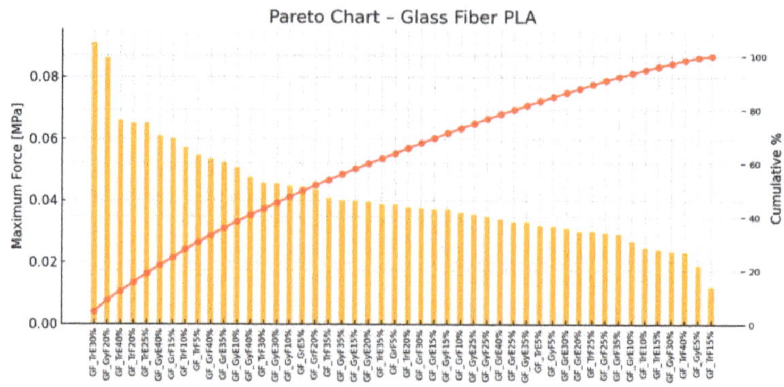

Figure 16. Pareto Chart—PLA Reinforced with Glass Fiber (orange color represents the maximum force values, hence red color line means the cumulative percentages).

Figure 17 is the time–distance diagram that visualizes the deformation behavior of the top 20% of 3D printed PLA specimens. The purpose of the analysis is to identify which combinations of material type, infill pattern, printing orientation, and infill percentage performed best under load.

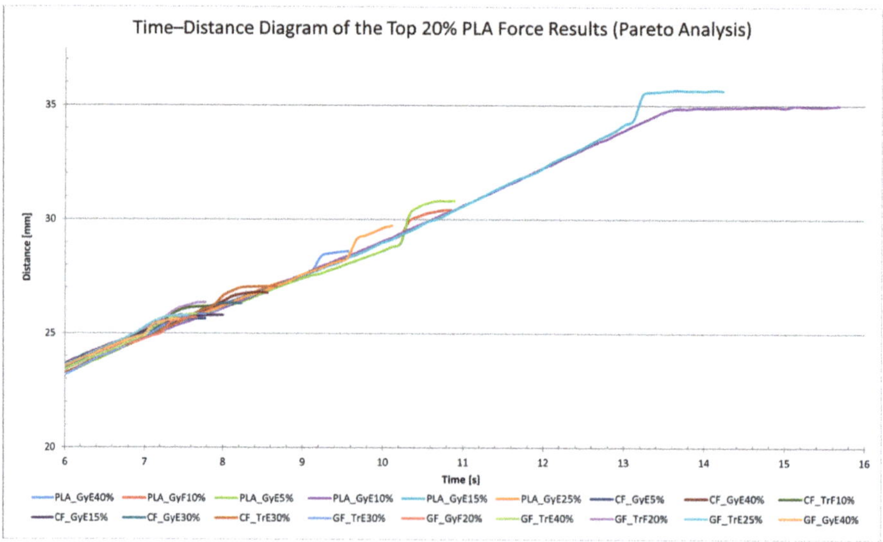

Figure 17. The top 20% PLA force results from Pareto analysis, shown in time-distance.

The specimen showing the highest elongation was PLA_GyE15% (35.66 mm), followed by PLA_GyE10% (34.98 mm) and PLA_GyE5% (30.81 mm). All three had gyroid infill and were printed in a vertical (E) orientation, indicating that this combination is particularly favorable for ductility. Interestingly, moderate infill percentages (10–15%) resulted in greater elongation than higher-density configurations such as PLA_GyE40%.

The diagram does not exclusively contain gyroid-patterned specimens; several triangular (Tr) infill samples also made it into the top 20%, particularly among CF (carbon fiber) and GF (glass fiber) materials. This highlights the relevance of the infill pattern, especially when working with high-strength composite materials.

In conclusion, the combination of gyroid infill, vertical orientation, and moderate infill density appears to be optimal for improving the deformation capability of PLA specimens.

4. Conclusions

The test results show PETG exhibits stronger and more stable mechanical properties than PLA, especially for grid and gyroid filling patterns, where the measured strength values are generally higher. In the case of PLA, the gyroid pattern initially shows more favorable results, especially at lower infill ratios, but with increasing infill ratios, the strength values start to decrease. It is important to note that increasing the infill ratio does not always result in an apparent increase in strength, so selecting the appropriate pattern and infill percentage is advisable to achieve optimal mechanical properties.

The PETG material was used as a reference in the experiments, which shows good behavior under bending stress. PLA and reinforced PLAs perform worse but are cheaper and can be easily printed with simpler printers.

Although carbon and glass composites provide a more stable structure, their mechanical properties are generally inferior to PLA and PETG. These materials are less sensitive to fill pattern and fill ratio variations, so their strength values are more uniform under different print settings. The tests showed that the carbon fiber and glass fiber PLA filaments behaved similarly; however, regarding both the tolerable force and the deformation achievable, the carbon fiber PLA filament performed slightly better. If bending is the expected stress for a structure, it is preferable to use basic PLA material, which shows better results than the fiber-reinforced filaments.

The experiments also demonstrated that the correct infill pattern and density are needed to achieve these results. The carbon fiber PLA and glass fiber PLA filaments show better results at grid fill patterns and lower fill densities. This may be of practical importance when low fill densities have to be chosen (for weight optimization) and when bending stresses are expected.

The practical significance of the results may be that the tests have shown an optimum infill value for these materials, which is between 20 and 30%, so for load-bearing elements, there is no reason to use higher infill. Further experiments will be conducted to optimize the exact fill rates for materials. The results obtained from this research will be applied to fabricating battery enclosures and other protective storage enclosures.

Author Contributions: Conceptualization, V.N., S.S., B.F.S., M.S., D.K. and S.F.; methodology, V.N., S.S., B.F.S., M.S., D.K. and S.F.; validation V.N., S.S., B.F.S., M.S., D.K. and S.F.; formal analysis, V.N., S.S., B.F.S., M.S., D.K. and S.F.; investigation, V.N., S.S., B.F.S., M.S., D.K. and S.F.; resources, V.N., S.S., B.F.S., M.S., D.K. and S.F.; data curation, V.N., S.S., B.F.S., M.S., D.K. and S.F.; writing—original draft preparation, V.N., S.S., B.F.S., M.S., D.K. and S.F.; writing—review and editing, V.N., S.S., B.F.S., M.S., D.K. and S.F.; visualization, V.N., S.S., B.F.S., M.S., D.K. and S.F.; supervision, V.N., S.S., B.F.S., M.S., D.K. and S.F.; project administration, V.N., S.S., B.F.S., M.S., D.K. and S.F.; funding acquisition, V.N., S.S., B.F.S., M.S., D.K. and S.F. All authors have read and agreed to the published version of the manuscript.

Funding: This research received no external funding.

Institutional Review Board Statement: Not applicable.

Data Availability Statement: All of the data are within the paper.

Acknowledgments: This paper was prepared by the research team "SZE-RAIL". This research was supported by the SIU Foundation's project 'Sustainable Railways—Investigation of the energy efficiency of electric rail vehicles and their infrastructure'. The publishing of the paper did not receive financial support or financing for the article process charge. The authors acknowledge the technical help of Hanna Csótár.

Conflicts of Interest: The authors declare no conflicts of interest.

List of Abbreviations

3D	Three-Dimension or Three-Dimensional
AM	Additive Manufacturing
CF	Carbon Fiber
DIC	Digital Image Correlation
ECT	Erichsen Cupping Test
FDM	Fused Deposition Modeling
GF	Glass Fiber
LOM	Laminated Object Manufacturing
PA	Polyamide
PETG	Polyethylene Terephthalate Glycol
PLA	Polylactic Acid
SLA	Stereolithography
SLS	Selective Laser Sintering

References

1. Valvez, S.; Santos, P.; Parente, J.M.; Silva, M.P.; Reis, P.N.B. 3D Printed Continuous Carbon Fiber-Reinforced PLA Composites: A Short Review. *Procedia Struct. Integr.* **2020**, *25*, 394–399. [CrossRef]
2. Heidari-Rarani, M.; Rafiee-Afarani, M.; Zahedi, A.M. Mechanical Characterization of FDM 3D Printing of Continuous Carbon Fiber-Reinforced PLA Composites. *Compos. B Eng.* **2019**, *175*, 107147. [CrossRef]
3. Tian, X.; Liu, T.; Yang, C.; Wang, Q.; Li, D. Interface and Performance of 3D Printed Continuous Carbon Fiber-Reinforced PLA Composites. *Compos. Part. A Appl. Sci. Manuf.* **2016**, *88*, 198–205. [CrossRef]
4. Maqsood, N.; Rimašauskas, M. Characterization of Carbon Fiber-Reinforced PLA Composites Manufactured by Fused Deposition Modeling. *Compos. Part. C Open Access* **2021**, *4*, 100112. [CrossRef]
5. Kohutiar, M.; Kakošová, L.; Krbata, M.; Janík, R.; Fekiač, J.J.; Breznická, A.; Eckert, M.; Mikuš, P.; Timárová, Ľ. Comprehensive Review: Technological Approaches, Properties, and Applications of Pure and Reinforced Polyamide 6 (PA6) and Polyamide 12 (PA12) Composite Materials. *Polymers* **2025**, *17*, 442. [CrossRef]
6. Huang, Y.; Sultan, M.T.H.; Shahar, F.S.; Grzejda, R.; Łukaszewicz, A. Hybrid Fiber-Reinforced Biocomposites for Marine Applications: A Review. *J. Compos. Sci.* **2024**, *8*, 430. [CrossRef]
7. Balakrishnan, T.S.; Sultan, M.T.H.; Shahar, F.S.; Basri, A.A.; Shah, A.U.M.; Sebaey, T.A.; Łukaszewicz, A.; Józwik, J.; Grzejda, R. Fatigue and Impact Properties of Kenaf/Glass-Reinforced Hybrid Pultruded Composites for Structural Applications. *Materials* **2024**, *17*, 302. [CrossRef]
8. Ézsiás, L.; Tompa, R.; Fischer, S. Investigation of the Possible Correlations Between Specific Characteristics of Crushed Stone Aggregates. *Spectr. Mech. Eng. Oper. Res.* **2024**, *1*, 10–26. [CrossRef]
9. Fischer, S. Investigation of the Settlement Behavior of Ballasted Railway Tracks Due to Dynamic Loading. *Spectr. Mech. Eng. Oper. Res.* **2025**, *2*, 24–46. [CrossRef]
10. Mousavi, S.M.R.; Rafe, A.; Yeganehzad, S. Textural, Mechanical, and Microstructural Properties of Restructured Pimiento Alginate-Guar Gels. *J. Texture Stud.* **2019**, *50*, 155–164. [CrossRef]
11. Fischer, S.; Harangozó, D.; Németh, D.; Kocsis, B.; Sysyn, M.; Kurhan, D.; Brautigam, A. Investigation of Heat-Affected Zones of Thermite Rail Weldings. *Facta Univ. Ser. Mech. Eng.* **2024**, *22*, 689–710. [CrossRef]
12. Aydin, M.; Wu, X.; Cetinkaya, K.; Yasar, M.; Kadi, I. Application of Digital Image Correlation Technique to Erichsen Cupping Test. *Eng. Sci. Technol. Int. J.* **2018**, *21*, 760–768. [CrossRef]
13. Szalai, S.; Szívós, B.F.; Kocsis, D.; Sysyn, M.; Liu, J.; Fischer, S. The Application of DIC in Criminology Analysis Procedures to Measure Skin Deformation. *J. Appl. Comput. Mech.* **2024**, *10*, 817–829. [CrossRef]
14. Bodnàr, Z. Unlocking Potential with AI-Designed PLA Compounds. In Proceedings of the 8th PLA World Congress 2024 Handbook, Munich, Germany, 28–29 May 2024; Volume 22.
15. Yang, L.; Li, S.; Zhou, X.; Liu, J.; Li, Y.; Yang, M.; Yuan, Q.; Zhang, W. Effects of Carbon Nanotube on the Thermal, Mechanical, and Electrical Properties of PLA/CNT Printed Parts in the FDM Process. *Synth. Met.* **2019**, *253*, 122–130. [CrossRef]
16. Bembenek, M.; Kowalski, Ł.; Kosoń-Schab, A. Research on the Influence of Processing Parameters on the Specific Tensile Strength of FDM Additive Manufactured PET-G and PLA Materials. *Polymers* **2022**, *14*, 2446. [CrossRef] [PubMed]
17. Ibrahim, N.I.; Sultan, M.T.H.; Łukaszewicz, A.; Shah, A.U.M.; Shahar, F.S.; Józwik, J.; Najeeb, M.I.; Grzejda, R. Characterization and Isolation Method of *Gigantochloa scortechinii* (*Buluh semantan*) Cellulose Nanocrystals. *Int. J. Biol. Macromol.* **2024**, *272*, 132847. [CrossRef]

18. Giani, N.; Mazzocchetti, L.; Benelli, T.; Picchioni, F.; Giorgini, L. Towards Sustainability in 3D Printing of Thermoplastic Composites: Evaluation of Recycled Carbon Fibers as Reinforcing Agent for FDM Filament Production and 3D Printing. *Compos. Part. A Appl. Sci. Manuf.* **2022**, *159*, 107002. [CrossRef]
19. Shahar, F.S.; Sultan, M.T.H.; Grzejda, R.; Łukaszewicz, A.; Oksiuta, Z.; Krishnamoorthy, R.R. Harnessing the Potential of Natural Composites in Biomedical 3D Printing. *Materials* **2024**, *17*, 6045. [CrossRef]
20. Rao, V.D.P.; Rajiv, P.; Geethika, V.N. Effect of Fused Deposition Modelling (FDM) Process Parameters on Tensile Strength of Carbon Fibre PLA. *Mater. Today Proc.* **2019**, *18*, 2012–2018.
21. Thomason, J.L. Glass Fibre Sizing: A Review. *Compos. Part. A Appl. Sci. Manuf.* **2019**, *127*, 105619.
22. Wang, G.; Zhang, D.; Wan, G.; Li, B.; Zhao, G. Glass Fiber-Reinforced PLA Composite with Enhanced Mechanical Properties, Thermal Behavior, and Foaming Ability. *Polymer* **2019**, *181*, 121803. [CrossRef]
23. Vasumathi, M.; Karupaiah, V.; Narayanan, V. Effect of Process Parameters on Mechanical and Tribological Characteristics of FDM Printed Glass Fiber-Reinforced PLA Composites. *Rapid Prototyp. J.* **2024**, *30*, 1859–1875. [CrossRef]
24. Liu, Z.; Lei, Q.; Xing, S. Mechanical Characteristics of Wood, Ceramic, Metal and Carbon Fiber-Based PLA Composites Fabricated by FDM. *J. Mater. Res. Technol.* **2019**, *8*, 3743–3753. [CrossRef]
25. Ismail, K.I.; Pang, R.; Ahmed, R.; Yap, T.C. Tensile Properties of In Situ 3D Printed Glass Fiber-Reinforced PLA. *Polymers* **2023**, *15*, 3436. [CrossRef] [PubMed]
26. Omar, N.W.Y.; Shuaib, N.A.; Hadi, M.H.J.A.; Azmi, A.I. Mechanical Properties of Carbon and Glass Fibre Reinforced Composites Produced by Additive Manufacturing: A Short Review. *IOP Conf. Ser. Mater. Sci. Eng.* **2019**, *670*, 012020.
27. Goh, G.D.; Dikshit, V.; Nagalingam, A.P.; Goh, G.L.; Agarwala, S.; Sing, S.L.; Wei, J.; Yeong, W.Y. Characterization of Mechanical Properties and Fracture Mode of Additively Manufactured Carbon Fiber and Glass Fiber-reinforced Thermoplastics. *Mater. Des.* **2018**, *137*, 79–89. [CrossRef]
28. Valvez, S.; Silva, A.P.; Reis, P.N.B. Compressive Behaviour of 3D-Printed PETG Composites. *Aerospace* **2022**, *9*, 124. [CrossRef]
29. El Mehtedi, M.; Buonadonna, P.; El Mohtadi, R.; Loi, G.; Aymerich, F.; Carta, M. Optimizing Milling Parameters for Enhanced Machinability of 3D-Printed Materials: An Analysis of PLA, PETG, and Carbon-Fiber-Reinforced PETG. *J. Manuf. Mater. Process.* **2024**, *8*, 131. [CrossRef]
30. Martins, R.F.; Branco, R.; Martins, M.; Macek, W.; Marciniak, Z.; Silva, R.; Trindade, D.; Moura, C.; Franco, M.; Malça, C. Mechanical Properties of Additively Manufactured Polymeric Materials—PLA and PETG—For Biomechanical Applications. *Polymers* **2024**, *16*, 1868. [CrossRef]
31. Kadhum, A.H.; Al-Zubaidi, S.; Abdulkareem, S.S. Effect of the Infill Patterns on the Mechanical and Surface Characteristics of 3D Printing of PLA, PLA+ and PETG Materials. *Chemengineering* **2023**, *7*, 46. [CrossRef]
32. Srinivasan, R.; Kumar, K.N.; Ibrahim, A.J.; Anandu, K.; Gurudhevan, R. Impact of Fused Deposition Process Parameter (Infill Pattern) on the Strength of PETG Part. *Mater. Today Proc.* **2020**, *27*, 1801–1805. [CrossRef]
33. Guessasma, S.; Belhabib, S. Infill Strategy in 3D Printed PLA Carbon Composites: Effect on Tensile Performance. *Polymers* **2022**, *14*, 4221. [CrossRef]
34. Maqsood, N.; Rimašauskas, M. Development and Fabrication of Continuous Carbon Fiber-Reinforced Thermoplastic Porous Composite Structures with Different Infill Patterns by Using Additive Manufacturing. *J. Thermoplast. Compos. Mater.* **2023**, *36*, 2050–2075. [CrossRef]
35. Mousa, M.A.; Yussof, M.M.; Hussein, T.S.; Assi, L.N.; Ghahari, S.A. A Digital Image Correlation Technique for Laboratory Structural Tests and Applications: A Systematic Literature Review. *Sensors* **2023**, *23*, 9362. [CrossRef] [PubMed]
36. Guo, R.; Xian, G.; Li, C.; Hong, B. Effect of Fiber Hybrid Mode on the Tension–Tension Fatigue Performance for the Pultruded Carbon/Glass Fiber-Reinforced Polymer Composite Rod. *Eng. Fract. Mech.* **2022**, *260*, 108208. [CrossRef]
37. Walotek, K.; Bzówka, J.; Ciołczyk, A. Examples of the Use of the ARAMIS 3D Measurement System for the Susceptibility to Deformation Tests for the Selected Mixtures of Coal Mining Wastes. *Sensors* **2021**, *21*, 4600. [CrossRef]
38. Guastavino, R.; Göransson, P. A 3D Displacement Measurement Methodology for Anisotropic Porous Cellular Foam Materials. *Polym. Test.* **2007**, *26*, 711–719. [CrossRef]
39. Samsami, S.; Pätzold, R.; Winkler, M.; Herrmann, S.; Augat, P. The Effect of Coronal Splits on the Structural Stability of Bi-Condylar Tibial Plateau Fractures: A Biomechanical Investigation. *Arch. Orthop. Trauma. Surg.* **2020**, *140*, 1719–1730. [CrossRef]
40. Szalai, S.; Szürke, S.K.; Harangozó, D.; Fischer, S. Investigation of Deformations of a Lithium Polymer Cell Using the Digital Image Correlation Method (DICM). *Rep. Mech. Eng.* **2022**, *3*, 116–134. [CrossRef]

polymers

MDPI

Article

Untapped Potential of Recycled Thermoplastic Blends in UD Composites via Finite Element Analysis

Pei Hao [1], Ninghan Tang [1], Juan Miguel Tiscar [2] and Francisco A. Gilabert [1,*]

1 Department of Materials, Textiles and Chemical Engineering (MaTCh), Mechanics of Materials and Structures (MMS), Tech Lane Ghent Science Park-Campus A, Ghent University (UGent), Technologiepark-Zwijnaarde 46, 9052 Ghent, Belgium; pei.hao@ugent.be (P.H.); ninghan.tang@ugent.be (N.T.)

2 Instituto de Tecnología Cerámica (ITC), Asociación de Investigación de las Industrias Cerámicas (AICE), Universitat Jaume I, Campus Universitario Riu Sec, 12006 Castellón, Spain; juanmiguel.tiscar@itc.uji.es

* Correspondence: fran.gilabert@ugent.be

Abstract: The increasing demand for fully recyclable composites has spurred extensive research on thermoplastics, valued for their recyclability and excellent mechanical properties. High-performance thermoplastics such as PEEK and PPS have been widely adopted in aerospace applications due to their outstanding load-bearing capabilities, which are well documented. Recently, thermoplastic polymer blends have gained attention for their enhanced recyclability and sustainability, as well as their ability to improve thermal stability, viscosity, and manufacturability. However, limited data are available on the mechanical characterization of composites that incorporate these blends, particularly when recycled thermoplastics are used. In this study, we first examine the stress–strain behavior of the following three polymer blends relevant for structural applications: PES/PEEK, PPS/PEEK, and HDPE/PP. We then perform a numerical analysis to predict the mechanical performance of unidirectional fiber-reinforced composites using each blend as the matrix. This involves a micromechanical Representative Volume Element (RVE) approach combined with an advanced polymer model previously validated against experimental data. The findings are discussed to critically assess the suitability of these blends for producing fully matrix-recycled composites.

Keywords: recycled thermoplastics; fiber-reinforced plastics; polymer blends; mechanical behavior; micro-mechanical modeling; stress–strain predictions

1. Introduction

The increasing demand for fully recyclable composite materials has driven significant interest in the use of thermoplastics, which offer both recyclability and high-performance characteristics [1,2]. In particular, high-performance thermoplastics have gained prominence in the automotive and aerospace industries due to their excellent mechanical properties, chemical resistance, and thermal stability [3–5]. However, the aerospace sector, like many others, faces growing sustainability concerns, necessitating the development of more environmentally friendly material solutions. One emerging approach to enhance the sustainability of these materials is the use of polymer blends, which can improve recyclability while maintaining or enhancing performance [6,7]. Additionally, in the broader context of recyclable polymers, polypropylene (PP) and polyethylene (PE) remain the most widely used recycled commodity thermoplastics [8]. There have also been efforts to develop natural fiber-reinforced thermoplastic composites as a sustainable alternative, combining renewable reinforcement materials with recyclable polymer matrices to further advance circular economy principles in composite manufacturing [9–13].

Despite the growing interest in recyclable thermoplastic composites, several research gaps must be addressed to fully realize their potential. One critical challenge is the lack of comprehensive mechanical characterization data for composites incorporating recycled thermoplastic blends. While these materials are typically tested following ASTM and ISO standards, the resulting datasheet properties provide only a basic understanding of their mechanical behavior [3,13–16]. This limited characterization makes it difficult to accurately assess key failure mechanisms, such as matrix cracking, fiber breakage, interface debonding, and delamination, which are crucial for ensuring structural integrity and reliability. Tiwari et al. [17] performed SEM analysis of the fracture surfaces, which showed that 50/50 polyphenylene sulfide (PPS)/polyether ether ketone (PEEK) blends were homogeneous, with dispersed sphere-shaped PEEK particles at various milled carbon fiber concentrations and good fiber–matrix adhesion. Although efforts have been made to integrate polymer blends into composite materials, there is still a significant lack of in-depth mechanical performance analysis from a composite engineering perspective. To bridge this gap, correlating experimentally obtained results with fundamental failure mechanisms is essential for the broader application of recycled composites in real-world scenarios. One of the most promising approaches to achieving this is the use of simulation-based analysis, particularly for unidirectional fiber-reinforced composites. Computational models such as the Representative Volume Element (RVE)-based approach can provide valuable insights into mechanical performance, failure behavior, and optimization strategies [18–20]. Ultimately, this approach facilitates the widespread adoption of sustainable composite materials.

However, a high-fidelity model for polymer blends is still lacking in research, limiting the credibility of simulation-based approaches. Pioneering work on modeling polymer blends was conducted by Seelig and van der Giessen [21]. They developed a numerical approach for polycarbonate (PC)/acrylonitrile–butadiene–styrene (ABS) blends, a common polymer blend primarily sourced from Waste Electrical and Electronic Equipment (WEEE) recycling. The material was modeled as a two-phase system, with ABS particles dispersed in a PC matrix, considering large viscoplastic strains, softening–rehardening behavior, and plastic dilation caused by rubber particle cavitation. Later, Hund et al. [22] used the finite element method (FEM) and constitutive models to analyze the large-strain deformation and fracture behavior of PC/ABS blends, demonstrating their accuracy under complex loading conditions, which also depended on the material composition. Eddhahak and Gaudy [23] conducted an experimental study on ABS/PC blends prepared directly from WEEE waste. The phenomenological G'Sell and Jonas constitutive law was used to predict the mechanical response, with parameter identification optimized through the nonlinear Generalized Reduced Gradient algorithm. Drozdov et al. [24,25] developed a constitutive model for the viscoplastic behavior of low-density polyethylene (LDPE) and metallocene-catalyzed polypropylene (mPP) polymer blends, treating the semicrystalline polymer as a two-phase composite medium. The model was used to examine the effect of annealing on the viscoplasticity and elastic moduli of the amorphous and crystalline phases. More recently, Reuvers et al. [26] developed a thermodynamically consistent model for polyamide 6 (PA6) blended with an amorphous copolymer, incorporating nonlinear viscoelastic and elastoplastic behavior at finite strains. The mechanical and thermal properties were also characterized.

The objectives of this paper are to extend an advanced polymer model from pure polymers to blends, with the necessary considerations, covering both high-performance and commodity thermoplastics. Section 2 provides an overview of thermoplastics, polymer blends, and blend-based composites. Section 3 analyzes the stress–strain response of three representative polymer blends (polyaryl ether sulfone) (PES)/PEEK, PPS/PEEK, and HDPE/PP) and presents the rate- and temperature-dependent advanced polymer model, along with the transformation for polymer blends, considering the effects of composition on elastic modulus, viscoplasticity, and rupture. Model validation is provided through predictions of mechanical behavior across the entire weight fraction spectrum. Section 4 discusses the RVE model for simulating recycled polymer blend-based composites under critical transverse tensile loading. Section 5 presents accurate mechanical predictions and insights into the nonlinear interactions between the matrix and fibers, based on high-fidelity finite element micromechanical simulations. Additionally, it explores the feasibility of using recycled resin matrices in UD composites by comparing their performance with that of pure polymer matrices. Discussion on the guidance for incorporating polymer blends into composites is presented with the aid of an RVE simulation, and their suitability is quantified using a tentative composite performance index in Section 6.

2. Overview on Thermoplastics in Pure, Blended, and Composite Forms: Current Applications and Challenges

Thermoplastics have emerged as indispensable materials across industries due to their recyclability and adaptability. These polymers are broadly categorized into high-performance polymers (HPPs) and commodity grades, each serving distinct roles in pure, blended, and composite forms, as shown in Figure 1.

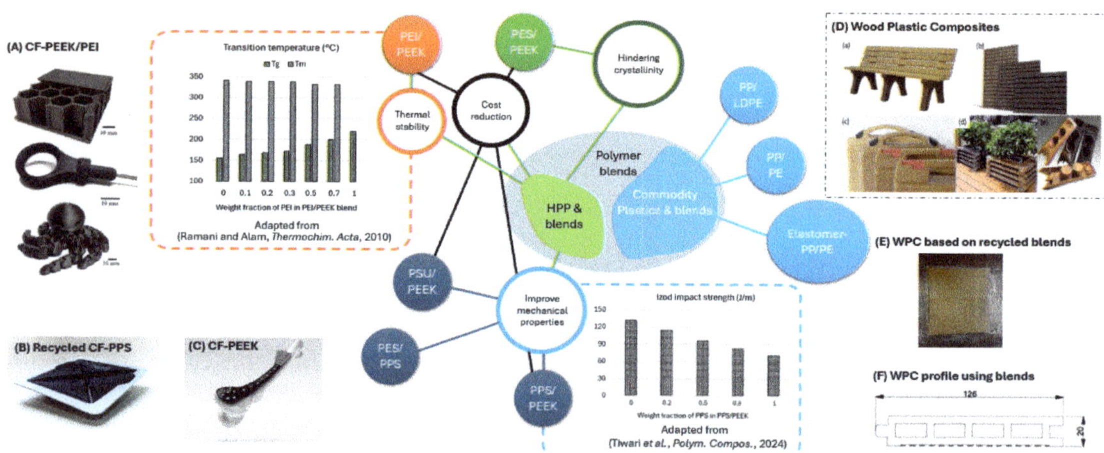

Figure 1. Thermoplastics and polymer blends as matrices for fiber-reinforced composites and their commercial applications in industry: (**A**) CF-PEEK/PEI used for additive manufacturing [27]; (**B**) Rotorcraft access panel from recycled carbon PPS (source: https://thermoplasticcomposites.nl/rotocraft-access-panel-from-recycled-carbon-pps/(accessed on 30 March 2025)); (**C**) Commercial CF reinforced PEEK-OPTIMA™ (source: https://invibio.com/en/application-areas/trauma (accessed on 30 March 2025)); (**D**) Commercial products made from wood plastic composites (WPC) ((a) Park furniture; (b) Screening; (c) Tool box; (d) Plant boxes; (e) Car interior) [28]; (**E**) WPC based on the recycled PE blends from municipal waste [29]; and (**F**) WPC profile manufactured from recycled plastic blends [30].

2.1. High-Performance Thermoplastics

In their pure state, high-performance thermoplastics such as PEEK, PPS, and polyaryletherketone (PAEK) dominate applications requiring exceptional mechanical strength, thermal stability (>200 °C), and chemical resistance [2]. These materials are widely used in aerospace, automotive, and biomedical sectors for load-bearing components, such as aircraft brackets or surgical implants. Their ability to undergo repeated melting and solidification without significant degradation aligns with circular economy goals, enabling reprocessing into high-value products. However, their high cost, energy-intensive processing, and limited compatibility with additives hinder broader adoption [3].

To address these limitations, high-performance polymer blends have gained increasing attention [6,7,31–33], owing to their enhanced recyclability, improved thermal stability, optimized viscosity, superior mechanical performance, and better processability (see Figure 1). Ramani and Alam [32] compared the glass transition temperature (T_g) and melting temperature (T_m) of various PEEK/poly(ether imide) (PEI) blend compositions and concluded that a 50/50 PEEK/PEI ratio offers optimal thermal stability. Blends such as PEEK/PEI [27] or PPS/recycled PEEK reinforced with carbon nanotubes (CNTs) balance stiffness and toughness while improving processability [14]. These systems are increasingly utilized in high-temperature aerospace components and biomedical load-bearing applications [4]. Yet, recycling these blends remains challenging due to phase separation and difficulties in maintaining homogeneity, limiting their use in closed-loop systems [5].

When reinforced with carbon or glass fibers, high-performance thermoplastics form composites with unparalleled strength-to-weight ratios. These composites excel in structural aerospace and automotive parts, offering damage tolerance and fatigue resistance. However, recycling these composites is complicated by the need to separate fibers from the matrix, often leading to downcycling.

2.2. Commodity Thermoplastics

Commodity polymers such as PP, PE, and polyethylene terephthalate (PET) are ubiquitous in packaging, consumer goods, and automotive interiors due to their low cost and ease of processing [8]. Their high recyclability supports sustainable, high-volume production, although their lower thermal and mechanical performance restricts use in structural applications.

Blending commodity polymers, such as elastomer-modified PP/PE matrices, enhances properties like impact strength and modulus while integrating recycled content [11]. These blends are increasingly paired with natural fibers (e.g., wood, banana fibers) to create sustainable wood plastic composites (WPCs) for automotive interiors and construction panels [9,12,13,29,30]. Innovations in 3D printing further enable the reuse of recycled commodity blends, although inconsistent properties in recycled materials often limit performance [34].

Commodity composites, typically reinforced with natural or recycled synthetic fibers, offer lightweight, low-cost solutions for non-structural applications [13]. However, recycled thermoplastics in these composites frequently exhibit degraded thermal stability and mechanical properties due to chain scission and contamination, reducing their viability for demanding uses. Poor fiber–matrix adhesion in recycled systems further complicates performance.

2.3. Challenge

The integration of thermoplastics into advanced composites necessitates robust modeling to predict interactions between the constituents with complicated nonlinear behaviors. While multiscale simulations based on the RVE approach provide insights, there is a gap in understanding how this simulation approach can be used to explore the effects of using blends as matrices. In part, this is due to the current recycling barriers that persist for both categories; high-performance systems struggle with phase separation, while commodity blends face variability in recycled feedstocks. Limited data exist to evaluate the viability of recycled polymer blends for composite applications, highlighting a gap that has motivated this research. This study represents an initial exploration into the feasibility of using recycled thermoplastics as composite matrices, combining insights from the existing literature with advanced constitutive modeling to assess their performance. By integrating established material data with computational frameworks validated through experimental studies, this work aims to bridge critical knowledge gaps and establish foundational guidelines for sustainable composite design.

3. Blend-Based Polymer Matrix and Constitutive Modeling

This section examines the mechanical response of polymers in their pure state. As previously noted, comprehensive experimental data on polymer blends, particularly stress–strain behavior across a broad compositional range, remains limited. This study focuses on PEEK- and PP-based polymers and their respective blends. The subsequent subsections provide the detailed specifications and properties of these materials.

An advanced polymer model is first briefly introduced to characterize the representative stress–strain response of pure polymers, emphasizing the critical mechanical attributes that must be accurately captured for reliable predictive capabilities. The model is then systematically extended to polymer blends, incorporating the variability in mechanical characteristics, including elastic modulus, peak yield stress, and failure behavior, to ensure robust and generalized applicability.

3.1. Mechanical Response of Pure Polymers

Figures 2–4 plot the stress–strain responses of six different polymers used in the blends, categorized into the following three groups: PES and PEEK, GF-PPS and GF-PEEK, and HDPE and PP. These polymers exhibit distinct mechanical responses in terms of both elastic and plastic properties [11,17,31].

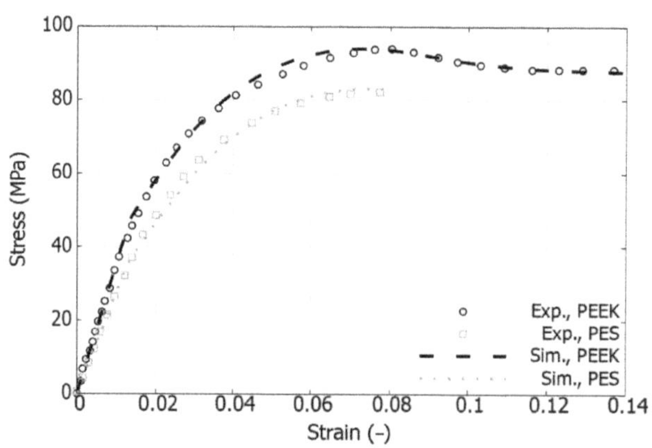

Figure 2. Comparison of experimental and simulation results of stress–strain curves for PES and PEEK in their pure state. Experimental data extracted from reference [31].

Figure 2 illustrates that both PES and PEEK exhibit similar stress–strain response trends; however, PES demonstrates a slightly lower elastic modulus and reduced peak stress compared to PEEK. Specifically, the peak stress of PES reaches 83 MPa, while that of PEEK attains 92 MPa, both occurring at a strain of 0.075. The failure strain of PES is 0.08, occurring just beyond its peak stress, whereas PEEK undergoes strain softening beyond its peak stress, ultimately failing at a strain of 0.15. Figure 3 shows that the incorporation of short glass fibers (GFs) significantly enhances the mechanical properties of commercial PPS and PEEK filled with 40% and 30% GF, respectively. Specifically, the elastic modulus exhibits a substantial increase compared to the pure polymer counterparts, exceeding 9644 MPa, as observed in the 30% GF-filled PEEK. Additionally, the ultimate strength notably improves; however, the materials exhibit increased brittleness, leading to early failure. The failure strain is 0.016 for GF-PPS and 0.027 for GF-PEEK, while in the case of GF-PPS, the material does not reach a fully saturated peak stress before failure.

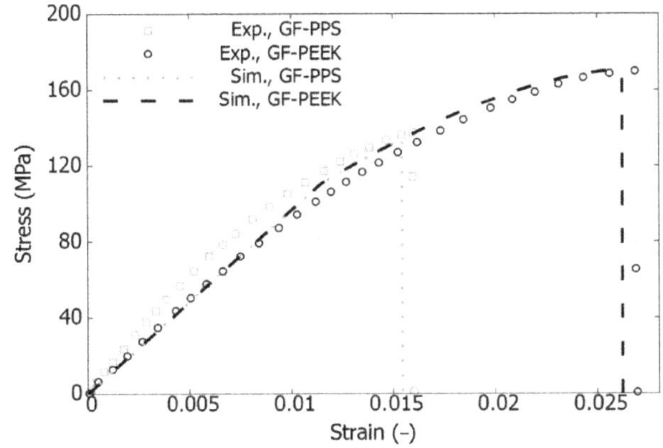

Figure 3. Comparison of experimental and simulation results of stress–strain curves for GF-PPS and GF-PEEK in their pure state. Experimental data extracted from reference [17].

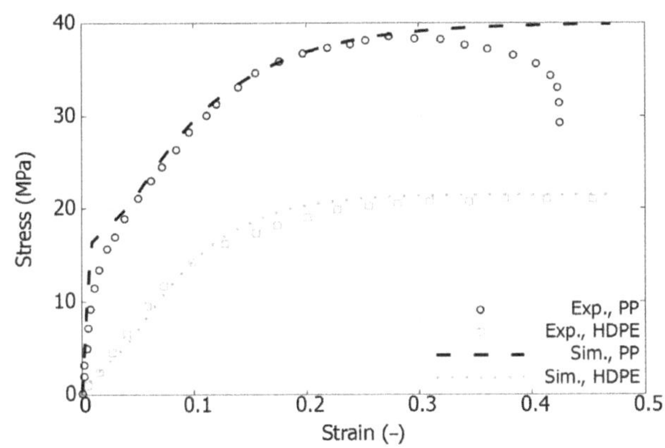

Figure 4. Comparison of experimental and simulation results of stress–strain curves for HDPE and PP in their pure state. Experimental data extracted from reference [11].

The stress–strain response of the commodity polymer group is presented in Figure 4. Compared to high-performance polymers, these materials exhibit significantly lower strength levels, with HDPE reaching 21 MPa and PP attaining 38 MPa. The post-yield behavior of HDPE and PP differs considerably; HDPE exhibits a saturated stress plateau, whereas PP undergoes a sudden drop beyond a strain of 0.3, attributed to the necking effect—a geometrically induced phenomenon that was not decoupled using conventional engineering stress–strain measures. The original data for HDPE extends up to a strain of 1200%, making it challenging to extract precise data points within the elastic region, as the curves are densely clustered in a narrow range. Consequently, the elastic phase is represented based on the elastic modulus values provided in the corresponding table within the reference [11].

3.2. Constitutive Model for Pure Polymers

Based on these characteristics, we present our Unified Semi-Crystalline Polymer (USCP) model, which is capable of providing accurate predictions. It is important to emphasize that this study primarily focuses on the composition of polymer blends. Moreover, research on the rate sensitivity of the plastic response is scarce in the existing literature. Consequently, rate sensitivity is excluded from this work, with particular attention given to the calibration strategy employed.

The original USCP model [35], formulated within the finite strain kinematic framework, is a generalization of the Boyce–Parks–Argon (BPA) model [36]. It incorporates a single viscoplastic law that unifies the amorphous and crystalline phases of the polymer, providing a comprehensive approach to modeling polymer behavior. Figure 5 illustrates the rheological analogue.

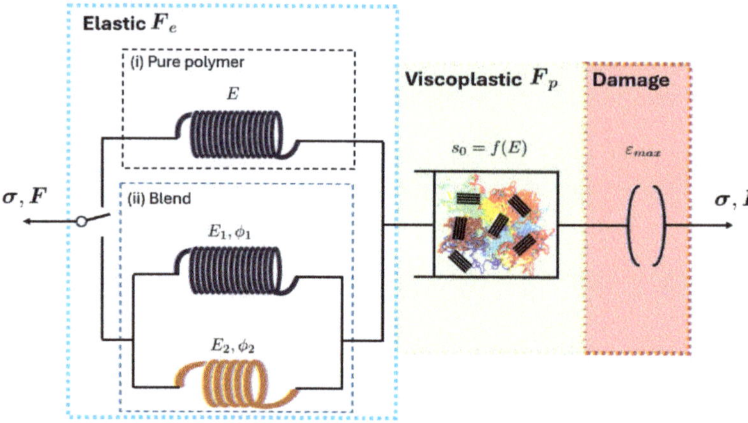

Figure 5. Schematic representation of elastic modulus, intermolecular resistance, and damage in corresponding rheological model.

The Cauchy stress tensor in the intermolecular branch σ is obtained by eliminating the plastic deformation gradient $F_e = FF_p^{-1}$, with $\dot{F}_p = F_e^{-1}\tilde{D}^{\mathrm{p}}F_eF_p$. The rate of inelastic deformation is written as

$$\tilde{D}^{\mathrm{p}} = \dot{\bar{\varepsilon}}\, N, \tag{1}$$

where N is the direction tensor, and $\dot{\bar{\varepsilon}}$ is the effective plastic strain rate.

$$\dot{\bar{\varepsilon}} = \dot{\varepsilon}_0 \exp\left[-\frac{A(s-\alpha\sigma_{\mathrm{h}})}{\theta}\left(1-\left(\frac{\sigma_{\mathrm{eq}}}{s-\alpha\sigma_{\mathrm{h}}}\right)^m\right)\right], \tag{2}$$

where σ_{eq} is the defined equivalent stress. The hydrostatic part of the Cauchy stress tensor is σ_{h}, and the description of model parameters are given in Table A1. The evolution law for the athermal effective stress is formulated using a smooth, Heaviside-like function to capture the pre-peak hardening, post-peak softening, and second yield behavior resulting from the crystalline phase contribution, as follows:

$$\begin{aligned} \dot{s} &= H_1(\bar{\varepsilon})\cdot\left(1-\frac{s}{s_1}\right)\cdot\dot{\bar{\varepsilon}} + H_2(\bar{\varepsilon})\cdot\left(1-\frac{s}{s_2}\right)\cdot\dot{\bar{\varepsilon}} + H_3(\bar{\varepsilon})\cdot\left(1-\frac{s}{s_3}\right)\cdot\dot{\bar{\varepsilon}}, \\ \text{with } s_0 &= \sqrt{3}\frac{8.5^{-1/m}}{(1-\nu)}\frac{E}{2(1+\nu)} \end{aligned} \tag{3}$$

where the athermal strength s_i (i = 1, 2, 3) corresponds to the preferred states at different stages of deformation, as described in reference [35]. The functions governing the hardening evolution are expressed as follows:

$$H_1(\bar{\varepsilon}) = -h_1\left\{\tanh\left(\frac{\bar{\varepsilon}-\bar{\varepsilon}_{\mathrm{p}}}{f\bar{\varepsilon}_{\mathrm{p}}}\right)-1\right\}, \tag{4}$$

$$H_2(\bar{\varepsilon}) = h_2\left\{-\tanh\left(\frac{\bar{\varepsilon}-\bar{\varepsilon}_p}{f\bar{\varepsilon}_p}\right)\tanh\left(\frac{\bar{\varepsilon}-\bar{\varepsilon}_c}{f\bar{\varepsilon}_c}\right)+1\right\}, \quad (5)$$

$$H_3(\bar{\varepsilon}) = h_3\left\{\tanh\left(\frac{\bar{\varepsilon}-\bar{\varepsilon}_c}{f\bar{\varepsilon}_c}\right)+1\right\}, \quad (6)$$

where $\bar{\varepsilon}_p$ is the plastic strain at the peak yielding point, and $\bar{\varepsilon}_c$ is the characteristic plastic strain when the crystalline nano-block initiates the yielding process. This formulation involves three hardening (or softening) parameters h_1, h_2, and h_3. The parameter f is the smoothing factor.

As shown in Figures 2–4, the model accurately captures the behavior and has been validated for pure polymers. It is now extended to polymer blends.

3.3. Extension Version for Blends

3.3.1. Elastic Modulus

It is intuitive to incorporate the mix ratio-dependent elastic modulus into the model (see Figure 5). PES was procured from Amoco Performance Products under the trade name Radel (grade A-300), while PEEK was synthesized in the laboratory as described by [37]. The HDPE used was HD 6605, a homopolymer with a melt flow index of 5 g/10 min (ExxonMobil Chemical), and the PP was Certene PHM-20AN, a homopolymer with a melt flow index of 20 g/10 min (Muehlstein and Co., Inc.) [11]. Both PES/PEEK [37] and HDPE/PP [11] systems followed the ASTM D 638 method [38] for conducting the tensile test.

The blend moduli were reported to exhibit a composition-dependent behavior, as described by the general equation proposed by Nielsen and Landel [39] for one-phase binary mixtures, and further specified by the form presented by Kleiner [40], as follows:

$$E = E_1\phi_1 + E_2\phi_2 + \beta_{12}\phi_1\phi_2 \quad (7)$$

where β_{12} is the term that represents the magnitude of the deviation from nonlinearity, with the best-fitted values being 1000 for the PES/PEEK system and 50 for the HDPE/PP system. Variable ϕ_i $(i = 1, 2)$ denotes the volume fraction of each component, where ϕ_1 can be calculated and converted from the given weight fraction w_1 of one component using the following equation:

$$\phi_1 = \frac{\rho_2}{\rho_1(1/w_1 - 1) + \rho_2} \quad (8)$$

where the density values are sourced from the literature and the corresponding datasheets of the commercial products, as provided in Table 1.

Table 1. Material properties for the investigated materials in their pure states [11,17,31].

Property	PES	PEEK	GF-PPS	GF-PEEK	HDPE	PP
ρ (g/cm^3)	1.370	1.263	1.68	1.51	0.948	0.905
E (MPa)	2755 *	3556 *	9644 *	9644 *	760	2060

* values obtained by extracting from the stress-strain curves.

Figures 6a and 7a show good agreement between the experimental results and the calculated ones. According to Equation (3), the composition dependency is naturally incorporated into the plastic formulation. For instance, Figure 6b shows a comparison between the experimental and simulation results using the extended version of the USCP model for the PES/PEEK blends. The stress–strain curves for various mix ratios are well captured, with the nonlinear portion following a smooth trend, reaching the peak yield point consistently. Shown in Figure 7b, good agreement is observed between the experimental results and the simulation for the HDPE/PP blends.

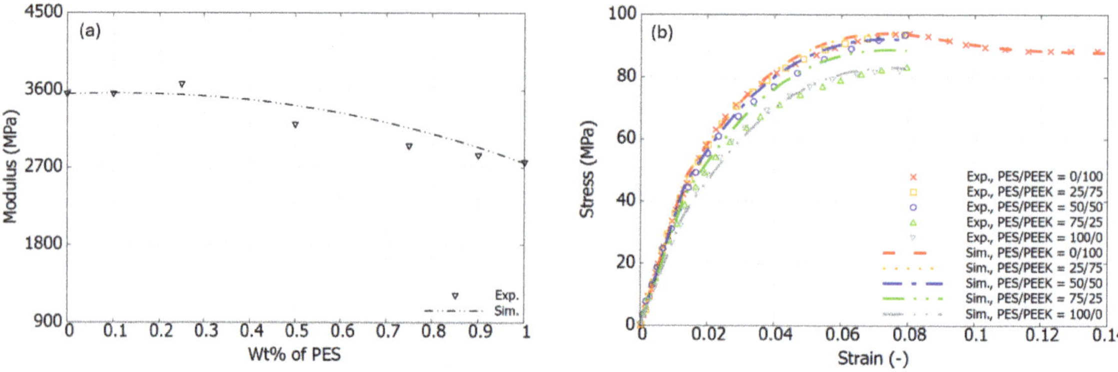

Figure 6. PES/PEEK blends: (**a**) Experimental data and best-fit values of elastic modulus; (**b**) Comparison between experimental and simulation results of stress–strain response.

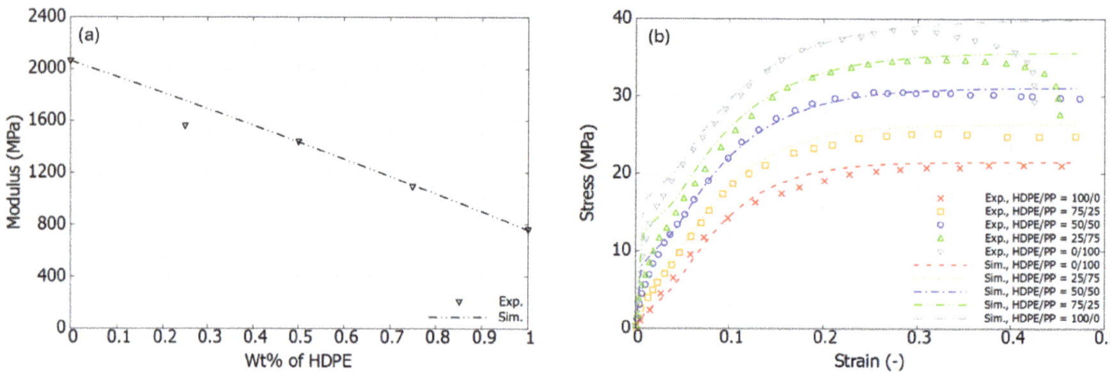

Figure 7. HDPE/PP blends: (**a**) Experimental data and best-fit values of elastic modulus; (**b**) Comparison between experimental and simulation results of stress–strain response.

3.3.2. Failure

Figure 5 shows that the proposed model is constructed based on multiple modules. To characterize failure, a damage module is introduced. The model's capability is demonstrated using a failure criterion based on the maximum failure strain, in accordance with the experimental observations of brittle fracture. Certainly, various energy-based damage theories can also be incorporated into the model.

Previous research by Tiwari et al. [17] investigated the mechanical behavior of PPS, PEEK, and their blends. PPS, a semi-crystalline polymer with a melting temperature of 279.68 °C, was studied in its reinforced form. Specifically, they examined 40% short glass fiber (GF)-filled PPS (Ryton®-R-4-230 NA) from Solvay and 30% short GF-filled PEEK (PEEK 450 GL 30, Victrex, Lancashire, UK). In this work, the USCP was adopted directly to the GF-filled thermoplastics, and the failure strain of blends were considered using the rule of mixture, as expressed in the following equation:

$$\varepsilon_f = \varepsilon_{f1}\phi_1 + \varepsilon_{f2}\phi_2 \tag{9}$$

Figure 8 presents the stress–strain response for various mix ratios of GF-PPS and GF-PEEK. The simulation results accurately capture the overall trend, with the predicted failure showing strong agreement with the experimental data.

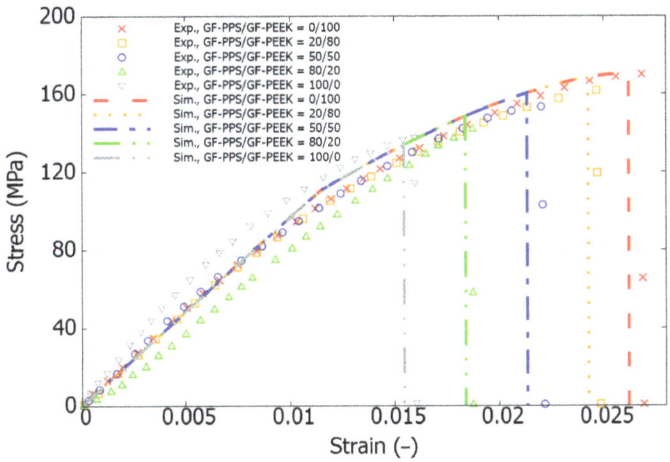

Figure 8. Experimental and simulation results of stress–strain curves for GF-PPS/GF-PEEK matrix.

We demonstrate that our model provides reliable predictions, and it is subsequently used as the matrix in the RVE simulations.

4. High-Fidelity FE Micromechanical RVE Model

To investigate the influence of matrix blend composition on the micromechanical behavior of a unidirectional (UD) composite ply, the development of an RVE model is crucial. Constructing this RVE involves defining the geometry and boundary conditions, as well as selecting the appropriate material constitutive models.

For this study, the chosen fiber material is UD carbon with a diameter of 7 µm, represented by randomly positioned cylinders embedded in a polymer matrix under periodic boundary conditions. Figure 9 illustrates the geometry of the model, which contains four fibers with a fiber volume fraction of 40%. The fiber positions are generated using a random algorithm previously applied in the molecular dynamics simulations of two-dimensional cohesive granular materials [41].

Both the fibers and matrix are independently meshed using three-dimensional (3D) six-node wedge elements from the element library, designated as C3D6 [42]. The element size is carefully analyzed to be sufficiently small, approximately 0.8 µm, ensuring the accurate resolution of microstructural features and consistency in the results. Mechanical Periodic Boundary Conditions (MPBCs) are implemented using the node-to-node coupling approach proposed by Garoz et al. [19]. These MPBCs enable the application of a mechanical load in the transverse direction while allowing Poisson's contraction.

It is assumed that the carbon fibers follow a linear elastic and transversely isotropic response. Properties of the mechanical and thermal constitutive behaviors are adopted from Arteiro et al. [18]. The properties are presented in Table 2. The fiber–matrix interface is modeled using surface-to-surface cohesive contact as an alternative to surface-based tie constraints. In this approach, there is no need to specify the stiffness or damage properties of the cohesive contact behavior.

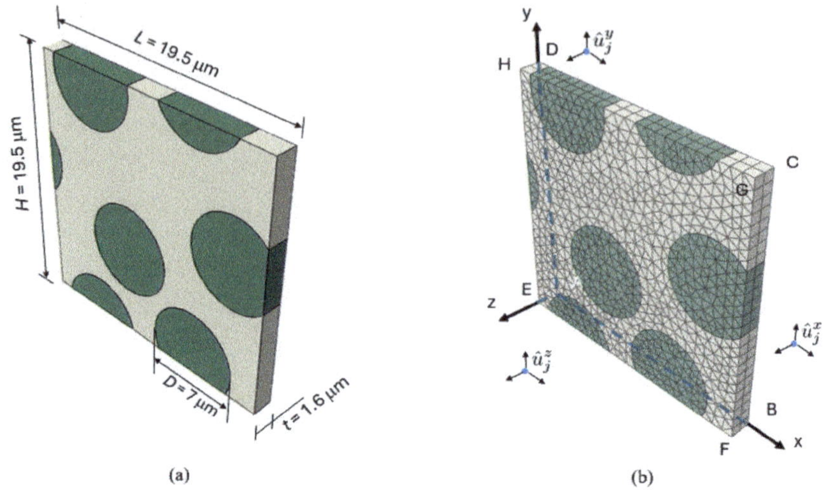

Figure 9. (**a**) Representative Volume Element (RVE) model and (**b**) node-coupling implementation of MPBCs.

Table 2. Transversely isotropic elastic properties [18] of carbon fiber.

Density ρ (kg/m^3)	Mechanical E_{11} (GPa)	E_{22} (GPa)	ν_{12}	G_{12} (GPa)	G_{23} (GPa)
1800	276	15	0.2	15	7

5. Numerical Analysis of UD Composite Using Blend-Based Matrices

This section details the performance response of UD composites using the previously developed RVE model, with an attempt to utilize different types of blends. The stress–strain response is straightforwardly obtained using engineering measures, where stress is calculated from the applied load and strain is the nominal strain. The predicted mechanical response of composites using each blend used as the matrix are presented in Figures 10–12.

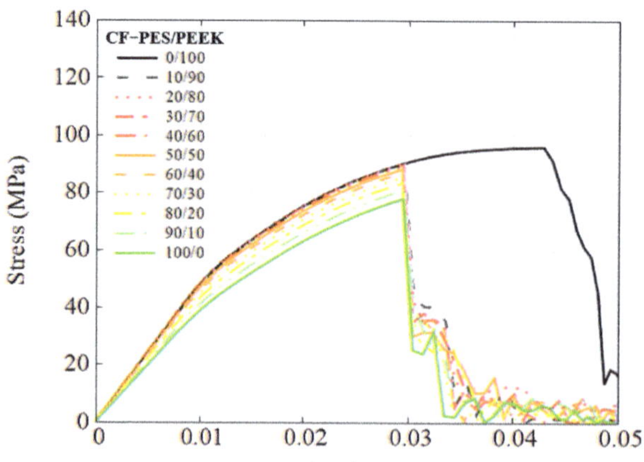

Figure 10. Engineering stress–strain response of CF-PES/PEEK composites using RVE simulation .

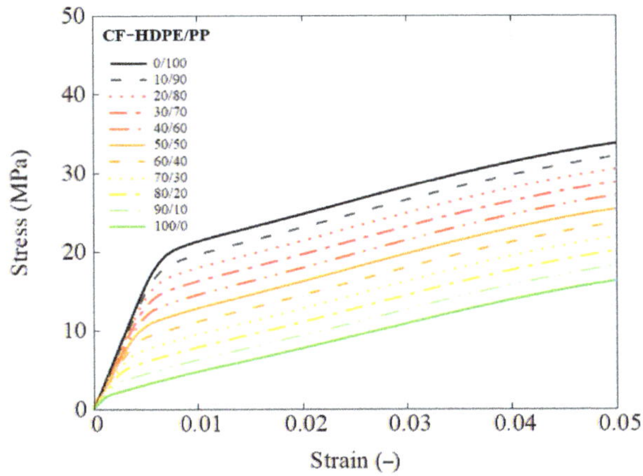

Figure 11. Engineering stress–strain response of CF-HDPE/PP composites using RVE simulation.

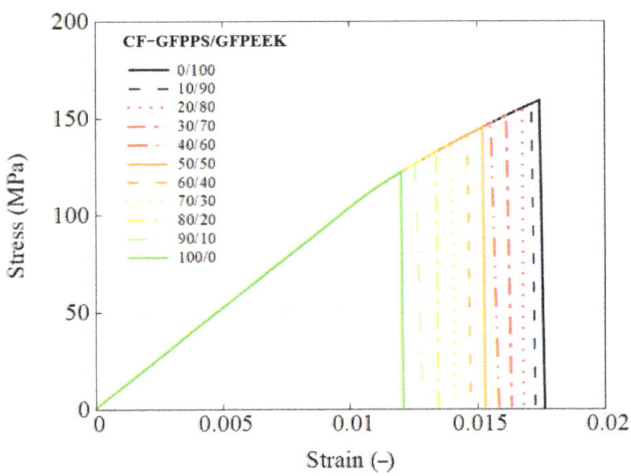

Figure 12. Engineering stress–strain response of CF-GF-PPS/GF-PEEK composites using RVE simulation.

The stress–strain behaviors of all the composites closely follow those of their corresponding blends. The incorporation of carbon fibers significantly enhances the overall mechanical performance, resulting in increased stiffness and strength. The key mechanical properties are summarized in Figure 13.

However, the addition of carbon fibers also introduces brittleness, leading to a substantial reduction in failure strain. For example, the PES/PEEK matrices in Figure 6 exhibit a failure strain of approximately 8%, whereas their composite counterparts (Figure 10) fail at around 3%. This reduction occurs because localized matrix pockets within the RVE of the composite reach their failure strain before the overall composite fails at the global scale.

Similarly, for the CF-GF-PPS/GF-PEEK composites, the failure strain ranges from 1.2% to 1.75%, while localized matrix regions experience failure at an even earlier stage. This phenomenon is further examined through fiber–matrix interaction investigation and load transfer analyses, supported by detailed results of RVE simulations. The findings are accompanied by corresponding stress–strain curves, visualizations of stress and strain distributions, and plasticity indicators to provide deeper insights into the composite failure mechanisms.

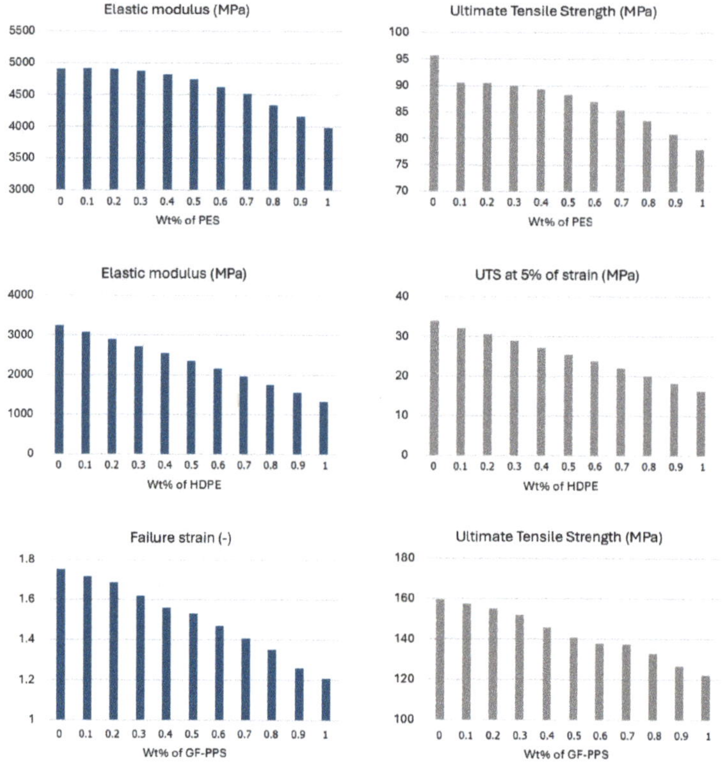

Figure 13. Summary of mechanical properties from RVE simulations for three different polymer blend composite systems.

A comparative analysis of the RVE performance using pure polymers and selected 50/50 blends is conducted across three blend groups, highlighting performance differences and the effectiveness of recycled matrices in maintaining UD composite integrity.

Figure 14 compares the following two different matrix systems: pure CF-PEEK as the baseline material and a 50/50 PES/PEEK blend. The global engineering stress–strain curves reveal minimal differences between the two, with a UTS of 95.7 MPa for the PEEK-based composite and 88.35 MPa for the PES-modified PEEK matrix. Strain and stress distributions are analyzed at critical loading points where matrix failure occurs. Both principal strain profiles exhibit similar failure locations but different strain values, influenced by fiber clustering effects. At global failure, the pure PEEK-based composite reaches a maximum strain of 13.9%, while the PES-modified composite fails at 7.4%. The peak matrix stress aligns with the UTS in Figure 6, indicating that incorporating 50 wt% PES effectively reduces the reliance on costly PEEK while maintaining mechanical performance, with only a minor reduction in failure strain of less than 1.5%.

Figures 15 and 16 demonstrate that the observations made for the PES/PEEK system are also applicable to the other two polymer blend systems. However, in the CF-HDPE/PP system, no failure was observed in the FEM simulations, so the chosen loading stage corresponds to the yielding point.

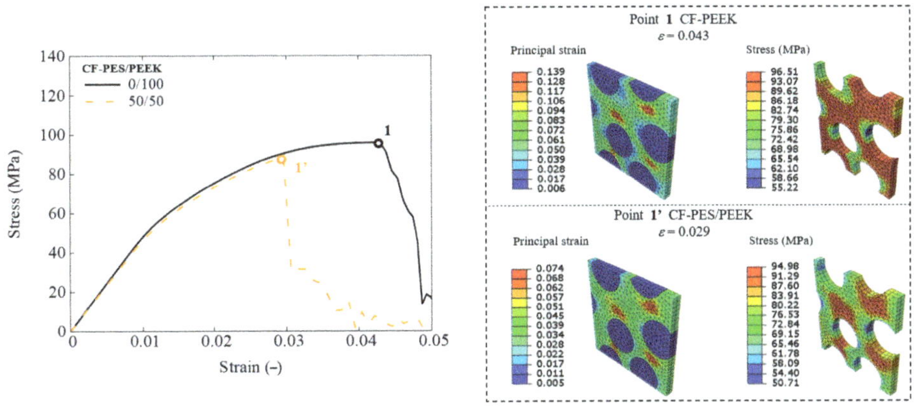

Figure 14. Global and local mechanical responses of CF-PES/PEEK composites, as demonstrated through RVE simulation.

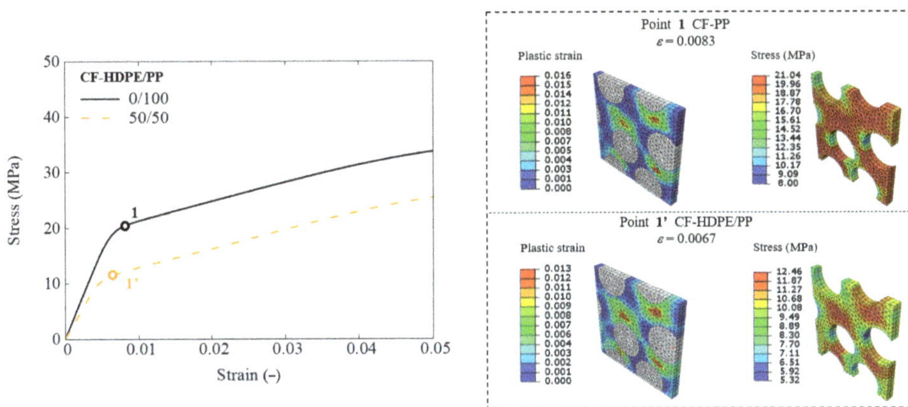

Figure 15. Global and local mechanical responses of CF-HDPE/PP composites, as demonstrated through RVE simulation.

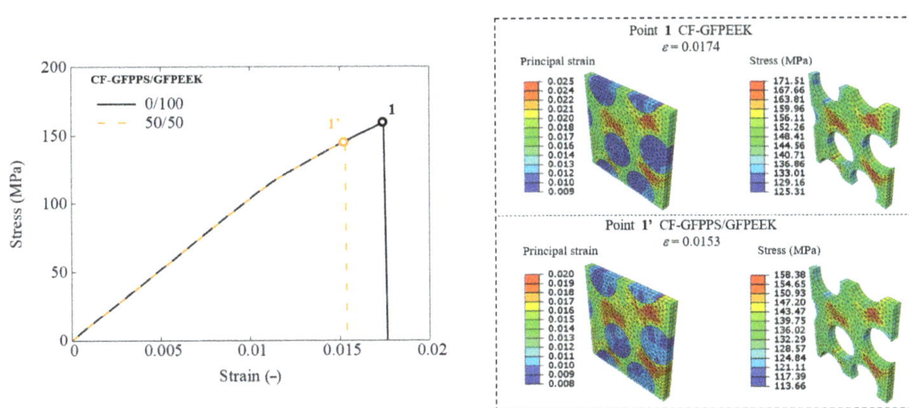

Figure 16. Global and local mechanical responses of CF-GFPPS/GFPEEK composites, as demonstrated through RVE simulation.

6. Discussion

The RVE simulations suggest that incorporating recycled polymer blend matrices into UD composites is a feasible approach, despite a few trade-offs. The obtained stress–strain curves indicate that while the addition of carbon fibers enhances overall stiffness and strength, it also leads to a reduction in failure strain compared to pure polymer systems. As mentioned, in the PES/PEEK system, the failure strain drops from approximately 8% in the neat blend to around 3% in the composite, primarily due to localized matrix failure. Despite this decrease in ductility, the prospect of reducing material costs and enhancing sustainability (by partially replacing expensive, high-performance polymers like PEEK with recycled materials) presents a compelling case for these blends in applications where extreme ductility is not critical.

The trade-off between performance and sustainability is a central consideration. Although thermoplastics in pure form may offer better ductility, recycled blends maintain comparable strength and stiffness, and their use could still bring lower costs and better environmental impact. When viewed from a life-cycle and sustainability perspective, the slight reduction in mechanical performance is often an acceptable compromise, especially in applications where maximum ductility is not the overriding requirement.

The simulations also suggest that the optimal usability of these blends depends on the operating conditions. They are best suited for environments demanding high load-bearing capacity under controlled loading conditions, such as those involving transverse tensile loads. It is critical, however, to account for the reduced failure strain and to optimize the microstructure by getting a more homogeneous fiber distribution. This may help mitigate early-stage failure due to localized stress concentrations.

To quantify the suitability of each blend, a tentative Composite Performance Index (CPI) could be formally introduced as follows:

$$CPI = \alpha\left(\frac{\mathrm{UTS}}{\mathrm{UTS}_{\mathrm{pure}}}\right) + \beta\left(\frac{\varepsilon_f}{\varepsilon_{f,\mathrm{pure}}}\right) + \gamma\, SF, \tag{10}$$

where the first terms between the parenthesis represents the normalized ultimate tensile strength, which reflects the load-bearing capacity of the composite. The second term is the normalized failure strain that indicates the relative level of ductility when using blends in the composite. The last one corresponds to a sustainability factor SF that might account for the cost reduction and environmental benefits of using recycled polymers, among other advantages. The coefficients α, β, γ are weights determined by the application requirements. With respect to SP, a tentative way to quantify this factor is to break it down into normalized, dimensionless components in which each one represents a key aspect or, more formally, the following:

$$SF = \sum_k c_k\, S_k \qquad \text{with} \qquad S_k = 1 - \frac{\chi_{k,\mathrm{blend}}}{\chi_{k,\mathrm{pure}}}, \tag{11}$$

where S_k is the relevant component to assess sustainability, and c_k are weighting coefficients that reflect the relative importance assigned to each component, fulfilling the condition $\sum c_k = 1$. Three plausible components to assess SF might be S_{env} that can quantify any relevant environmental impact metrics (e.g., CO_2 emissions) of the recycled and pure materials, S_{cst} for the cost of recycled blends and pure materials, and S_{rsc} could indicate the reduction in virgin material consumption when using recycled polymers. According to this proposal, as an example, for $k = \mathrm{cst}$, the descriptors $\chi_{\mathrm{cst,\,blend}}$ and $\chi_{\mathrm{cst,\,pure}}$ can be assigned to the cost of the recycled blend and pure polymers, respectively. It is worth mentioning that by adjusting the weights c_k, the industrial agent can tailor the index to reflect the specific priorities of a given application or product, whether they be environmental impact, cost savings, or resource conservation.

In terms of methodology, to ensure the effective implementation of the proposed CPI across industries, the following adaptation and customization guidelines are proposed: First, industries must define clear objectives by identifying primary application requirements, such as mechanical performance (e.g., strength, ductility), cost constraints, and environmental impact thresholds, to establish a baseline for index alignment. Next, the appropriate weights should be assigned to the coefficients of the Index to ensure that priorities are transparently reflected; for instance, sectors like packaging, governed by stringent EU sustainability mandates, might prioritize environmental impact over short-term cost savings. Third, the relevant sustainability indicators—such as CO_2 reduction metrics (S_{env}), waste reducibility targets (S_{rcs}), or cost-effectiveness ratios (S_{cst})—should be selected, and they must align with the industry-specific goals, ensuring measurable and actionable outcomes. Finally, contextual calibration through industry-specific benchmarking is critical to validate the accuracy of the index, leveraging empirical data (e.g., Packaging and Packaging Waste Regulation (PPWR) [43] compliance thresholds or Plastics Recyclers Europe (PRE) [44] emission estimates) to refine its components. This structured approach not only bridges theoretical frameworks with practical industrial needs but also empowers sectors to balance performance, sustainability, and economic viability in a quantifiable and adaptable manner.

On the other hand, simulations point out that the early onset of localized failure, particularly under high strain, represents a key limitation. This shortfall may restrict the use of recycled blends in applications that require both high strength and high ductility, though they remain attractive for applications where the highest level of ductility is not essential. In that sense, it is important to note that some of these limitations might be associated with the assumptions inherent in the simulations. The RVE model assumes ideal fiber–matrix bonding and a perfectly homogeneous matrix (no defects like voids), factors that might lead to an overestimation of composite performance relative to real-world production. A fundamental aspect of the recycling process, it significantly influences the final mechanical performance of the resulting material. In this study, our focus begins at the stage where recycled materials have already been produced and are ready for use, similar to the characterization of other pristine materials. However, a broader quantification of recycling effects, such as changes in molecular weight, impurity content, or crystallinity, is not included. Consequently, discrepancies between the simulated and experimental results are expected, and these assumptions underscore the need for further experimental validation to refine the predictive models.

Nevertheless, the prospects for additional blend formulations and innovative composite architectures appear promising. Experimenting with different blend ratios, incorporating secondary additives or nanofillers, and exploring hybrid composite designs that combine recycled and conventional matrices can potentially overcome current limitations. Moreover, adjusting the fiber orientations and the stacking sequences in UD-based laminates may help mitigate localized failures and further enhance mechanical performance.

7. Conclusions

In this work, an advanced constitutive model originally developed for pure thermoplastic polymers was successfully extended to capture the elasto-viscoplastic behavior of polymer blends, specifically demonstrating its applicability to both high-performance (PEEK-based) and commodity (PP-based) polymer systems. The model accurately accounts for composition-dependent variations in mechanical properties, such as elastic modulus and strength, and includes a damage module employing a strain-based failure criterion, effectively characterizing material failure.

Using Representative Volume Element (RVE)-based finite element simulations, the developed model provided local and global predictions of the mechanical performance under critical transverse tensile loading conditions. These simulations showed that the stress–strain behaviors of composite materials that incorporated recycled thermoplastic blends as matrices closely aligns with those of their corresponding neat polymer blends. However, the simulations also revealed that employing recycled blends as a matrix involved a reduction in ductility. It was observed that localized strain concentrations caused premature matrix cracking at the microscale, particularly in recycled blend matrix-rich regions between fibers.

The approach outlined here aims to minimize the reliance on extensive trial-and-error by providing a computational framework for evaluating recycled polymer blends in industrial composite applications. This method could lead to significant reductions in material costs, particularly when substituting costly polymers like PEEK, without compromising mechanical performance.

To evaluate the feasibility of using recycled polymer blends in composites, this study introduces a performance index that integrates both mechanical and sustainability metrics. Mechanical metrics, specifically the normalized ultimate tensile strength and normalized failure strain, are derived directly from the RVE simulation results. Although a detailed quantitative sustainability assessment remains an area for future investigation, the study proposes a preliminary structure for a sustainability factor. This factor can be effectively combined with mechanical metrics to enable a comprehensive evaluation of blend-based composites. Overall, the suggested performance index provides a clear, structured methodology to quantitatively determine the suitability of recycled polymer blends for targeted composite applications.

While the studied blends may exhibit reduced ductility relative to the pure thermoplastics, their comparable strength and the environmental benefits might represent attractive trade-offs for many industrial scenarios. However, careful consideration of the operating conditions and loading scenarios remains essential to fully exploit the advantages offered by recycled blend matrices. Future efforts should focus on refining the sustainability quantification indicators.

On the other hand, it is worth mentioning that this numerical investigation highlighted several aspects that require further exploration. Future work should address experimental validation under various loading conditions to refine model predictions, particularly addressing realistic fiber–matrix interface features. Additionally, accurately capturing and reconstructing realistic microstructures using advanced microscopy techniques such as SEM—including the presence of voids, heterogeneities, and variability in fiber distribution—is essential for enhancing the fidelity and predictive capability of models, going beyond the information provided by conventional stress–strain curves.

Nevertheless, this study provides valuable initial insights into the mechanical viability and sustainability advantages of using recycled thermoplastic blends as matrices for unidirectional fiber-reinforced composites. Continued research efforts in model refinement, experimental validation, and microstructural optimization are recommended to fully leverage the economic and environmental benefits of these innovative composite materials.

Author Contributions: Conceptualization, P.H. and F.A.G.; methodology, P.H., N.T. and F.A.G.; software, P.H., N.T. and F.A.G.; validation, P.H., N.T. and F.A.G.; formal analysis, P.H. and F.A.G.; investigation, P.H., N.T., J.M.T. and F.A.G.; resources, P.H. and F.A.G.; data curation, P.H. and N.T.; writing—original draft preparation, P.H. and F.A.G.; writing—review and editing, P.H., N.T., J.M.T. and F.A.G.; visualization, P.H. and N.T.; supervision, F.A.G.; project administration, F.A.G.; funding acquisition, P.H. and F.A.G. All authors have read and agreed to the published version of the manuscript.

Funding: P.H. acknowledges the funding support from the Special Research Fund ("Bijzonder Onderzoeksfonds", BOF) program by Ghent University (grant number BOF23/PDO/037). N.T. acknowledges the funding support from the China Scholarship Council (CSC) program (grant number 202106710030). The authors would like to thank the Commission for Scientific Research ("CommissieWetenschappelijk Onderzoek", CWO) of Ghent University for their financial support in establishing international research collaboration (grant number EA11-CWO-013).

Institutional Review Board Statement: Not applicable.

Data Availability Statement: Data are contained within the article.

Conflicts of Interest: The authors declare no conflicts of interest.

Appendix A. Thermomechanical Properties and Identified Model Parameters

The appendix summarizes the input parameters for the thermomechanical analyses of UD composites. Tables are provided separately for each constituent, including elasto-viscoplastic response of matrices, elastic response of carbon fiber and the traction-separation law of fiber–matrix interface. Readers are referred to the calibrated parameters for USCP model in Table A1 with the corresponding thermal properties for the investigated polymers used in this study.

Table A1. Material parameters for the investigated PES/PEEK [17], PPS/PEEK [31] and HDPE/PP [11].

Material Parameter	Unit	PES/PEEK	PPS/PEEK	HDPE/PP	Description
E_{ref}	MPa	3556	9644	2060	Modulus of base polymer
ν	-	0.40	0.40	0.43	Poisson's ratio
α	-	0	0	0	Pressure sensitivity
$\dot{\varepsilon}_0$	1/s	3266.36	3.4×10^{26}	1.049×10^{12}	Rate sensitivity
m	-	0.50	0.50	0.50	Rate sensitivity
A	K/MPa	1347	1272	1027	Rate sensitivity
s_0	MPa	50.74	137.6	2.29	Initial equivalent strength
s_1	MPa	98.61	210	5.1	Athermal peak strength
s_2	MPa	98.61	185.6	11.2	First saturation strength
s_3	MPa	93.02	184	27	Second saturation strength
h_1	MPa	1950	26,619	50.36	Pre-peak hardening
h_2	MPa	1837	32,546	339.6	Post-peak softening
h_3	MPa	160	160	340	Second yield hardening
$\bar{\varepsilon}_p$	-	0.052	0.0088	0.0012	Peak plastic strain
$\bar{\varepsilon}_c$	-	0.254	0.0188	0.0319	Activation plastic strain
f	-	0.3	0.3	0.3	Smooth factor

References

1. Barile, M.; Lecce, L.; Iannone, M.; Pappadà, S.; Roberti, P. Thermoplastic Composites for Aerospace Applications. In *Revolutionizing Aircraft Materials and Processes*; Pantelakis, S., Tserpes, K., Eds.; Springer International Publishing: Cham, Switzerland , 2020; pp. 87–114. [CrossRef]
2. Osti De Moraes, D.V.; De Melo Morgado, G.F.; Magnabosco, R.; Rezende, M.C. Study of the influence of PPS, PAEK, and PEEK thermoplastic matrices on the static mechanical behavior of carbon fiber-reinforced composites. *Polym. Compos.* **2024**, *45*, 1726–1736. [CrossRef]
3. Mitschang, P.; Blinzler, M.; Wöginger, A. Processing technologies for continuous fibre reinforced thermoplastics with novel polymer blends. *Compos. Sci. Technol.* **2003**, *63*, 2099–2110. [CrossRef]
4. Díez-Pascual, A.M.; Díez-Vicente, A.L. Nano-TiO_2 Reinforced PEEK/PEI Blends as Biomaterials for Load-Bearing Implant Applications. *Acs Appl. Mater. Interfaces* **2015**, *7*, 5561–5573. [CrossRef] [PubMed]
5. Zhao, Z.; Zhang, J.; Bi, R.; Chen, C.; Yao, J.; Liu, G. Study on the Overmolding Process of Carbon-Fiber-Reinforced Poly (Aryl Ether Ketone) (PAEK)/Poly (Ether Ether Ketone) (PEEK) Thermoplastic Composites. *Materials* **2023**, *16*, 4456. [CrossRef]
6. Simunec, D.P.; Jacob, J.; Kandjani, A.E.; Trinchi, A.; Sola, A. Facilitating the additive manufacture of high-performance polymers through polymer blending: A review. *Eur. Polym. J.* **2023**, *201*, 112553. [CrossRef]

7. Korycki, A.; Carassus, F.; Tramis, O.; Garnier, C.; Djilali, T.; Chabert, F. Polyaryletherketone Based Blends: A Review. *Polymers* **2023**, *15*, 3943. [CrossRef]
8. Plastics Europe. *Plastics—The Fast Facts*; Plastics Europe: Brussels, Belgium, 2024. Available online: https://plasticseurope.org/knowledge-hub/plastics-the-fast-facts-2024/ (accessed on 30 March 2025).
9. Maldas, D.; Kokta, B.V. Recycled Thermoplastic–Woodflour Composites. *Int. J. Polym. Mater.* **1994**, *26*, 69–78. [CrossRef]
10. Yuan, Q.; Wu, D.; Gotama, J.; Bateman, S. Wood Fiber Reinforced Polyethylene and Polypropylene Composites with High Modulus and Impact Strength. *J. Thermoplast. Compos. Mater.* **2008**, *21*, 195–208. [CrossRef]
11. Clemons, C. Elastomer modified polypropylene–polyethylene blends as matrices for wood flour–plastic composites. *Compos. Part Appl. Sci. Manuf.* **2010**, *41*, 1559–1569. [CrossRef]
12. Englund, K.; Villechevrolle, V. Flexure and water sorption properties of wood thermoplastic composites made with polymer blends. *J. Appl. Polym. Sci.* **2011**, *120*, 1034–1039. [CrossRef]
13. Arifuzzaman Khan, G.M.; Alam Shams, M.S.; Kabir, M.R.; Gafur, M.A.; Terano, M.; Alam, M.S. Influence of chemical treatment on the properties of banana stem fiber and banana stem fiber/coir hybrid fiber reinforced maleic anhydride grafted polypropylene/low-density polyethylene composites. *J. Appl. Polym. Sci.* **2013**, *128*, 1020–1029. [CrossRef]
14. Deng, S.; Lin, Z.; Cao, L.; Xian, J.; Liu, C. PPS/recycled PEEK/carbon nanotube composites: Structure, properties and compatibility. *J. Appl. Polym. Sci.* **2015**, *132*, app.42497. [CrossRef]
15. Kumar, S.; Mishra, R.K.; Nandi, T. Experimental and theoretical investigations of the high performance blends of PEEK/PEI. *J. Polym. Eng.* **2018**, *38*, 351–361. [CrossRef]
16. Jiang, Z.; Liu, P.; Chen, Q.; Sue, H.; Bremner, T.; DiSano, L.P. The influence of processing conditions on the mechanical properties of poly(aryl-ether-ketone)/polybenzimidazole blends. *J. Appl. Polym. Sci.* **2020**, *137*, 48966. [CrossRef]
17. Tiwari, S.; Bag, D.S.; Mishra, P.; Lal, G.; Dwivedi, M. High-performance PPS/PEEK blend and its composites with milled carbon fiber: Study on their mechanical, thermal and dielectric properties. *Polym. Compos.* **2024**, *45*, 11845–11859. [CrossRef]
18. Arteiro, A.; Catalanotti, G.; Melro, A.; Linde, P.; Camanho, P. Micro-mechanical analysis of the effect of ply thickness on the transverse compressive strength of polymer composites. *Compos. Part A Appl. Sci.* **2015**, *79*, 127–137. [CrossRef]
19. Garoz, D.; Gilabert, F.; Sevenois, R.; Spronk, S.; Van Paepegem, W. Consistent application of periodic boundary conditions in implicit and explicit finite element simulations of damage in composites. *Compos. B Eng.* **2019**, *168*, 254–266. [CrossRef]
20. Laheri, V.; Hao, P.; Gilabert, F. Direct embedment of RVE-based microscale into lab size coupons to research fracture process in unidirectional and bidirectional composites. *Compos. Sci. Technol.* **2023**, *235*, 109949. [CrossRef]
21. Seelig, T.; Van Der Giessen, E. Localized plastic deformation in ternary polymer blends. *Int. J. Solids Struct.* **2002**, *39*, 3505–3522. [CrossRef]
22. Hund, J.; Naumann, J.; Seelig, T. An experimental and constitutive modeling study on the large strain deformation and fracture behavior of PC/ABS blends. *Mech. Mater.* **2018**, *124*, 132–142. [CrossRef]
23. Eddhahak, A.; Gaudy, A. Investigation of a constitutive law for the prediction of the mechanical behavior of WEEE recycled polymer blends. *J. Polym. Res.* **2025**, *32*, 43. [CrossRef]
24. Drozdov, A.; Christiansen, J.; Klitkou, R.; Potarniche, C. Viscoelasticity and viscoplasticity of polypropylene/polyethylene blends. *Int. J. Solids Struct.* **2010**, *47*, 2498–2507. [CrossRef]
25. Drozdov, A.; Christiansen, J.; Klitkou, R.; Potarniche, C. Effect of annealing on viscoplasticity of polymer blends: Experiments and modeling. *Comput. Mater. Sci.* **2010**, *50*, 59–64. [CrossRef]
26. Reuvers, M.C.; Kulkarni, S.; Boes, B.; Felder, S.; Wutzler, A.; Johlitz, M.; Lion, A.; Brepols, T.; Reese, S. A thermo-mechanically coupled constitutive model for semi-crystalline polymers at finite strains: Mechanical and thermal characterization of polyamide 6 blends. *Contin. Mech. Thermodyn.* **2024**, *36*, 657–698. [CrossRef]
27. Diouf-Lewis, A.; Farahani, R.D.; Iervolino, F.; Pierre, J.; Abderrafai, Y.; Lévesque, M.; Piccirelli, N.; Therriault, D. Design and characterization of carbon fiber-reinforced PEEK/PEI blends for Fused Filament Fabrication additive manufacturing. *Mater. Today Commun.* **2022**, *31*, 103445. [CrossRef]
28. Chan, C.M.; Vandi, L.J.; Pratt, S.; Halley, P.; Richardson, D.; Werker, A.; Laycock, B. Composites of Wood and Biodegradable Thermoplastics: A Review. *Polym. Rev.* **2018**, *58*, 444–494. [CrossRef]
29. Ayana, K.D.; Ali, A.Y.; Ha, C.S. Wood Polymer Composites Based on the Recycled Polyethylene Blends from Municipal Waste and Ethiopian Indigenous Bamboo (Oxytenanthera abyssinica) Fibrous Particles Through Chemical Coupling Crosslinking. *Polymers* **2024**, *16*, 2982. [CrossRef] [PubMed]
30. Turku, I.; Keskisaari, A.; Kärki, T.; Puurtinen, A.; Marttila, P. Characterization of wood plastic composites manufactured from recycled plastic blends. *Compos. Struct.* **2017**, *161*, 469–476. [CrossRef]
31. Nandan, B.; Kandpal, L.D.; Mathur, G.N. Poly(ether ether ketone)/poly(aryl ether sulfone) blends: Relationships between morphology and mechanical properties. *J. Appl. Polym. Sci.* **2003**, *90*, 2887–2905. [CrossRef]
32. Ramani, R.; Alam, S. Composition optimization of PEEK/PEI blend using model-free kinetics analysis. *Thermochim. Acta* **2010**, *511*, 179–188. [CrossRef]

33. Nara, S.; Oyama, H.T. Effects of partial miscibility on the structure and properties of novel high performance blends composed of poly(p-phenylene sulfide) and poly(phenylsulfone). *Polym. J.* **2014**, *46*, 568–575. [CrossRef]
34. Zander, N.E.; Gillan, M.; Burckhard, Z.; Gardea, F. Recycled polypropylene blends as novel 3D printing materials. *Addit. Manuf.* **2019**, *25*, 122–130. [CrossRef]
35. Hao, P.; Dai, Z.; Laheri, V.; Gilabert, F.A. A unified amorphous-crystalline viscoplastic hardening law for non-isothermal modelling of thermoplastics and thermosets. *Int. J. Plast.* **2022**, *159*. [CrossRef]
36. Boyce, M.C.; Parks, D.M.; Argon, A.S. Large inelastic deformation of glassy polymers. part I: Rate dependent constitutive model. *Mech. Mater.* **1988**, *7*, 15–33. [CrossRef]
37. Nandan, B.; Kandpal, L.D.; Mathur, G.N. Poly(acrylonitrile-butadiene-styrene)/Poly(ether ether ketone) Blends: An Attempt Towards a Polymer Reinforced Polymer Composite. *Macromol. Mater. Eng.* **2004**, *289*, 749–756. [CrossRef]
38. *ASTM D638-14(en)*; Standard Test Method for Tensile Properties of Plastics. ASTM: West Conshohocken, PA, USA, 2014.
39. Nielsen, L.E.; Landel, R.F. *Mechanical Properties of Polymers and Composites*, 2nd ed., rev. and expanded ed.; Number 90 in Mechanical Engineering; M. Dekker: New York, NY, USA, 1994.
40. Kleiner, L.W.; Karasz, F.E.; MacKnight, W.J. Compatible glassy polyblends based upon poly(2,6-dimethyl-1,4-phenylene oxide): Tensile modulus studies. *Polym. Eng. Sci.* **1979**, *19*, 519–524. [CrossRef]
41. Gilabert, F.A.; Roux, J.N.; Castellanos, A. Computer simulation of model cohesive powders: Influence of assembling procedure and contact laws on low consolidation states. *Phys. Rev. E* **2007**, *75*, 011303. [CrossRef]
42. *SIMULIA User Assistance 2025, ABAQUS*; Dassault Systemes Simulia Corp.: Providence, RI, USA, 2025. Available online: https://help.3ds.com/2025/english/DSSIMULIA_Established/SIMULIA_Established_FrontmatterMap/DSDocHome.htm?contextscope=all (accessed on 30 March 2025) .
43. *Regulation (EU) 2025/40 of the European Parliament and of the Council of 19 December 2024 on Packaging and Packaging Waste, Amending Regulation (EU) 2019/1020 and Directive (EU) 2019/904, and Repealing Directive 94/62/EC (Text with EEA relevance), 2024. Legislative Body: CONSIL, EP.*; European Parliament: Brussels, Belgium , 2024 . Available online: https://eur-lex.europa.eu/eli/reg/2025/40/oj/eng (accessed on 30 March 2025).
44. Plastics Recyclers Europe (PRE). Available online: https://www.plasticsrecyclers.eu/publications/ (accessed on 30 March 2025).

polymers

MDPI

Article

Influence of Resin Composition on the Photopolymerization of Zirconia Ceramics Fabricated by Digital Light Processing Additive Manufacturing

Ning Kuang [1], Hao Qi [1], Wenjie Zhao [2,*] and Junfei Wu [1,*]

[1] College of Electromechanical Engineering, Qingdao University of Science and Technology, Qingdao 266061, China; kuang.ning@mails.qust.edu.cn (N.K.); 19558242420@163.com (H.Q.)

[2] College of Sino-German Science and Technology, Qingdao University of Science and Technology, Qingdao 266061, China

* Correspondence: zwj@qust.edu.cn (W.Z.); jfw_1968@126.com (J.W.)

Abstract: Digital light processing (DLP) is widely recognized as one of the most promising additive manufacturing technologies for ceramic fabrication. Nevertheless, during the additive manufacturing of zirconia ceramics, debinding and sintering often lead to structural defects, which severely deteriorate the material properties and hinder their broader application. In this study, we added an oligomer into the photosensitive resin and systematically investigated the effects of oligomer content on the viscosity and curing properties of ceramic suspensions. The results demonstrated that the introduction of oligomers is conducive to enhancing the crosslinking density and reducing defects. Finally, a 45 vol% solid content zirconia ceramic slurry was prepared by adding 20 wt% oligomers to the resin system. After printing, debinding, and sintering, the final zirconia ceramics exhibited a uniform microstructure without delamination or cracks, its bending strength reached 682.4 MPa. This study demonstrates that zirconia ceramics fabricated by photopolymerization with oligomer photosensitive resin exhibit excellent mechanical properties, significantly expanding the potential applications for high-performance zirconia ceramic components with additive manufacturing.

Keywords: additive manufacturing; DLP; zirconia; photopolymerization

1. Introduction

Zirconia ceramics possess excellent properties, including high strength, toughness, wear resistance, and biocompatibility, which have facilitated their extensive use in both industrial and medical fields, such as in bearings, cutting tools, dental restorations, and bone implants [1–3]. Nevertheless, the significant hardness and inherent brittleness of zirconia ceramics create substantial difficulties when machining components with complex geometries using conventional manufacturing approaches. While methods like isostatic pressing, injection molding, and gel casting can be utilized to fabricate ceramics with moderately complex shapes [4], these techniques are highly dependent on molds and encounter several limitations, such as expensive mold production, complicated procedures, extended manufacturing times [5], and challenges in producing parts with highly intricate designs. These constraints have hindered the broader adoption of complex-shaped ceramic components.

Innovative rapid prototyping and manufacturing technologies have successfully addressed the inherent limitations of conventional fabrication methods, offering unprecedented flexibility in producing geometrically complex ceramic components [6]. Notably, additive manufacturing (AM), commonly referred to as 3D printing, is a layer-by-layer fabrication technology that enables direct digital-to-physical transformation [7]. This technology demonstrates multiple advantages, including rapid production rates, high forming accuracy, and low material costs [8,9]. Ceramic additive manufacturing primarily encompasses five key techniques: laminated object manufacturing (LOM), fused deposition modeling (FDM), selective laser sintering (SLS), direct ink writing (DIW), and photopolymerization-based methods. Among these, photopolymerization-based methods, including stereolithography (SL), liquid crystal display (LCD), and digital light processing (DLP), have emerged as the dominant approaches for fabricating ceramic components, owing to their superior dimensional accuracy and enhanced surface quality [10–12].

The working principle of DLP technology is to project images with high-resolution digital projectors for layer-by-layer curing. These images are then stacked to form objects, as shown in Figure 1. Since each layer is formed in a single exposure, DLP features high precision, high speed, and high cost-effectiveness [13]; therefore, it is more widely applied in ceramic additive manufacturing. DLP technology operates through a photochemical mechanism in which UV irradiation triggers the crosslinking of photosensitive resin components, resulting in controlled polymerization and layer-by-layer solidification, while simultaneously curing the resin-coated ceramic powder [14,15].

Figure 1. DLP printer and schematic diagram.

Given the characteristics of the photopolymerization process, the ceramic slurry must possess excellent fluidity, low viscosity, and outstanding photopolymerization performance. Additionally, to minimize defects such as cracks and achieve a dense ceramic body, the ceramic slurry must also have a high solid loading [16–18]. Typically, the photosensitive resin in the slurry comprises various monomers to lower viscosity and enhance the curing character for effective photocuring reactions [19,20]. Therefore, selecting an appropriate photosensitive resin formulation is crucial for ensuring slurry stability while simultaneously meeting the requirements of both high solid loading and low viscosity.

In our previous studies, a photosensitive resin formulation with relatively low viscosity and a high photopolymerization performance was successfully prepared [21]. However, previous studies have predominantly focused on monomer screening in photosensitive resin systems, overlooking the fact that small-molecule monomers exhibit significant contraction rates, which can lead to cracks and deformations in ceramic components [22]. Recent research has shown several advancements. Xu et al. [23] successfully optimized resin–ceramic adhesion by precisely controlling double bond density and minimizing curing shrinkage. Johansson et al. [24] added non-reactive components 2-[[2-(Benzoyloxy)ethyl]amino]ethanol (BEA) and polyethylene glycol (PEG-200) to the photosensitive resin, which helped reduce polymerization shrinkage and modified the thermal decomposition of the polymer matrix. This optimization effectively suppressed layer delamination and interlayer cracking during thermal processing, yielding sintered components with a near-theoretical relative density of 99%. In a parallel study, Dang et al. [4] strategically replaced a portion of the ethoxylated pentaerythritol tetraacrylate (PPTTA) monomer with ethoxylated trimethylolpropane triacrylate (ETPTA) and incorporated functional oligomers. This formulation redesign reduced the critical exposure energy to 14.88 mJ/cm^2 while achieving a 98.5% monomer conversion rate. Han et al. [5] studied the influence of different PEG-200 contents on the degreasing process and mechanical properties of zirconia parts, and it was proved that PEG-200 could effectively inhibit the defects during the degreasing process.

From these studies, it is clear that reducing the effects of shrinkage helps prevent cracks and structural defects, ultimately improving the mechanical and functional performance of ceramics. Oligomer is a component of partial photosensitive resins that, compared to monomers, polymerizes with far fewer double bonds being consumed. As a result, the photopolymerization of oligomers induces substantially less volume shrinkage [22]. However, there are relatively few studies on the influence of oligomers on the photosensitive resin system of zirconia ceramics.

In this study, based on our previous research on resins [21], oligomers were introduced into the photosensitive resin system, and the effects of different contents of oligomers on the viscosity and photopolymerization performance of zirconia ceramic slurry were investigated. Through a comprehensive investigation of microstructure evolution and its impact on mechanical properties, the optimal balance point for the monomer-to-oligomer ratio in the photosensitive resin system was successfully identified. The zirconia ceramic slurry prepared with the resin containing a definite amount of oligomer was subjected to printing, debinding, sintering, and testing, resulting in significant improvements in the properties of the ceramic finish part.

2. Materials and Methods

2.1. Materials

In this study, zirconia powder (D50 = 1 μm, Shandong Sitaili Metal Materials Co., Ltd., Jinan, China) was used. The resins used in this study, including monomers Acryloyl morpholine (ACMO, Shanghai Guangyi Chemical Co., Ltd., Shanghai, China), 1,6-Hexanediol diacrylate (HDDA, Shanghai Guangyi Chemical Co., Ltd., Shanghai, China), Trimethylolpropane triacrylate (TMPTA, Shanghai Guangyi Chemical Co., Ltd., Shanghai, China), and oligomer Bisphenol A epoxy acrylate (BAEA, Shanghai Yinchang New Materials Co., Ltd., Shanghai, China). A 2,4,6-trimethyl benzoyl diphenyl phosphine oxide (TPO, Shanghai Guangyi Chemical Co., Ltd., Shanghai, China) was used as the photoinitiator, while BYK110 (Dongguan Haoyouduo New Materials Co., Ltd., Dongguan, China) was selected as a dispersant.

2.2. Preparation of Zirconia Ceramic Slurry

As shown in Figure 2, first, the photosensitive resin was prepared by mixing different resin monomers and oligomers via a magnetic stirrer (RCT-Basic, IKA-Werke GmbH & Co. KG, Staufen im Breisgau, Germany) at 500 r/min for 60 min. The monomer and oligomer compositions of the different resin groups are shown in Table 1. Next, a specified amount of zirconia ceramic powder was added to the prepared photosensitive resin, different content of dispersant was added to the suspension, ranging from 1 wt% to 3 wt% of zirconia powder, and TPO with 3 wt% of resins was added as the photoinitiator. The suspension was then ball-milled using a planetary ball mill (BQM-1L, Changsha Miqi Instruments &

Equipment Co., Ltd., Changsha, China) at 350 rpm, with alternating forward and reverse rotations for 10 h each, resulting in a homogenously dispersed zirconia ceramic slurry.

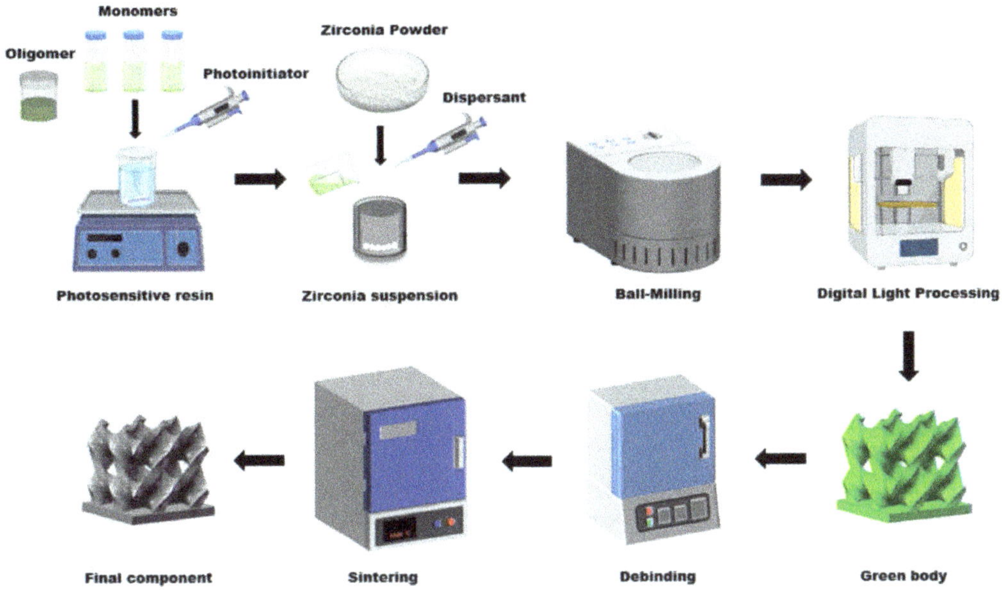

Figure 2. 3D printing zirconia ceramic fabrication process.

Table 1. Composition of resin.

Group	ACMO (wt%)	HDDA (wt%)	TMPTA (wt%)	BAEA (wt%)
1	50	10	40	0
2	50	10	30	10
3	50	10	20	20
4	50	10	10	30
5	50	10	0	40

2.3. Fabrication

The zirconia ceramic green body was printed using a DLP ceramic printer (BLD-25-C1, Qingdao Breuck 3D Additive Manufacturing Co., Ltd., Qingdao, China). The print thickness was selected as 30 μm, with the exposure power between 3 and 6 mW/cm^2, and an exposure time of 3–30 s. The green body was fabricated in a rectangular geometry measuring 35 mm × 4 mm × 3 mm. After printing, the green body was treated through debinding according to the thermogravimetric analysis. The zirconia green body was debound at 600 °C in air atmosphere in debinding furnace (FMJ-07/11, Hefei Facerom Intelligent Equipment Co., Ltd., Hefei, China) for 2 h and then sintered in the sintering furnace (FMJ-05/17, Hefei Facerom Intelligent Equipment Co., Ltd., Hefei, China) with a heating rate of 5 °C/min to 1600 °C for 2 h.

2.4. Characterization

The rheological property of zirconia slurries was determined using a digital viscometer (NDJ-8Spro, Shanghai Xiniulab Instruments Co., Ltd., Shanghai, China) at room temperature. The curing thickness was measured using a micrometer (Q2LF0025, Deqing Shengtaixin Electronic Technology Co., Ltd., Deqing, China). The thermal decomposition characteristics of the green body were investigated using thermogravimetric analysis (TGA; Labsys EVO, SETARAM Instrumentation, Caluire et Cuire, France) with a constant heating rate of 5 °C/min under a flowing air atmosphere. A scanning electron microscope (SEM, SU8010, HITACHI, Tokyo, Japan) characterization was performed to analyze the microstructural transformations from the green body through debinding to the sintering state. The sintered ceramic specimens were tested for three-point bending strength using a testing machine (Digital electronic universal testing machine, WH-70, Ningbo Weiheng Testing Instruments Co., Ltd., Ningbo, China), as shown in Figure 3, at a loading rate of 0.5 mm/min. To ensure repeatability, five samples were tested.

Figure 3. Three-point bending test.

3. Results and Discussion

3.1. Rheological Properties of Zirconia Ceramic Slurry

In order to prepare a slurry of zirconia ceramics with low viscosity and good dispersion properties, a dispersant was added to the slurry. Determining the optimal dispersant concentration represents a crucial parameter for attaining target rheological properties in high solid-loading slurries for ceramic additive manufacturing applications. To determine the optimal concentration of dispersant for the zirconia slurry, the rheological properties of suspensions containing varying dispersant concentrations (1, 1.5, 2.0, 2.5, and 3 wt% relative to zirconia powder) were investigated, as shown in Figure 4a; for the convenience of comparison, a slurry with resin Group 1 and 40 vol% solid content was used. It has been observed that the viscosity of all the slurries with different contents of dispersant decreases as the shear rate increases, and all the slurries exhibit non-Newtonian fluid characteristics. The viscosity decreases as the content of the dispersant increases from 1% to 2.5 wt%, but starts to increase when the content further increases to 3 wt%. This phenomenon is commonly ascribed to molecular flocculation induced by excessive dispersant dosage, resulting in impaired suspension fluidity [25,26]. Based on the rheological characterization results, a dispersant concentration of 2.5 wt% was determined to be optimal and was therefore selected for all subsequent processing steps in this investigation.

The solid loading of the ceramic suspension was further optimized by assessing its rheological behavior, as shown in Figure 4b. As the solid content increases, the slurry viscosity rises accordingly. At 50 vol% solid loading, a sharp increase in viscosity is observed. A good slurry fluidity is essential for smooth printing; normally, the viscosity should be less than 10 Pa·s at a shear rate of 10 s^{-1} [27]. When subjected to a shear rate of 10 s^{-1}, the slurry containing 50% solids demonstrates a viscosity of 20 Pa·s, and such a high viscosity is no longer suitable for printing. Therefore, the final solid loading of the zirconia ceramic slurry was set at 45 vol% to ensure the printing process.

Based on the dispersant and solid loading research, the influence of oligomer content on rheological behavior has been studied. As shown in Figure 4c, the viscosity of the slurry progressively increases with increasing oligomer concentration. This occurs because the oligomers exhibit higher viscosity compared to photosensitive resin monomers, thereby increasing the overall slurry viscosity. The viscosities at a typical shear rate of 10 s^{-1} were 5.18, 4.90, 7.06, and 9.25 Pa·s for the slurries of Groups 1, 2, 3, and 4, respectively, demonstrating that these slurries are suitable for the printing process. In contrast, Group 5 exhibited a viscosity of 23.36 Pa·s at a 10 s^{-1} shear rate, exceeding the printable range.

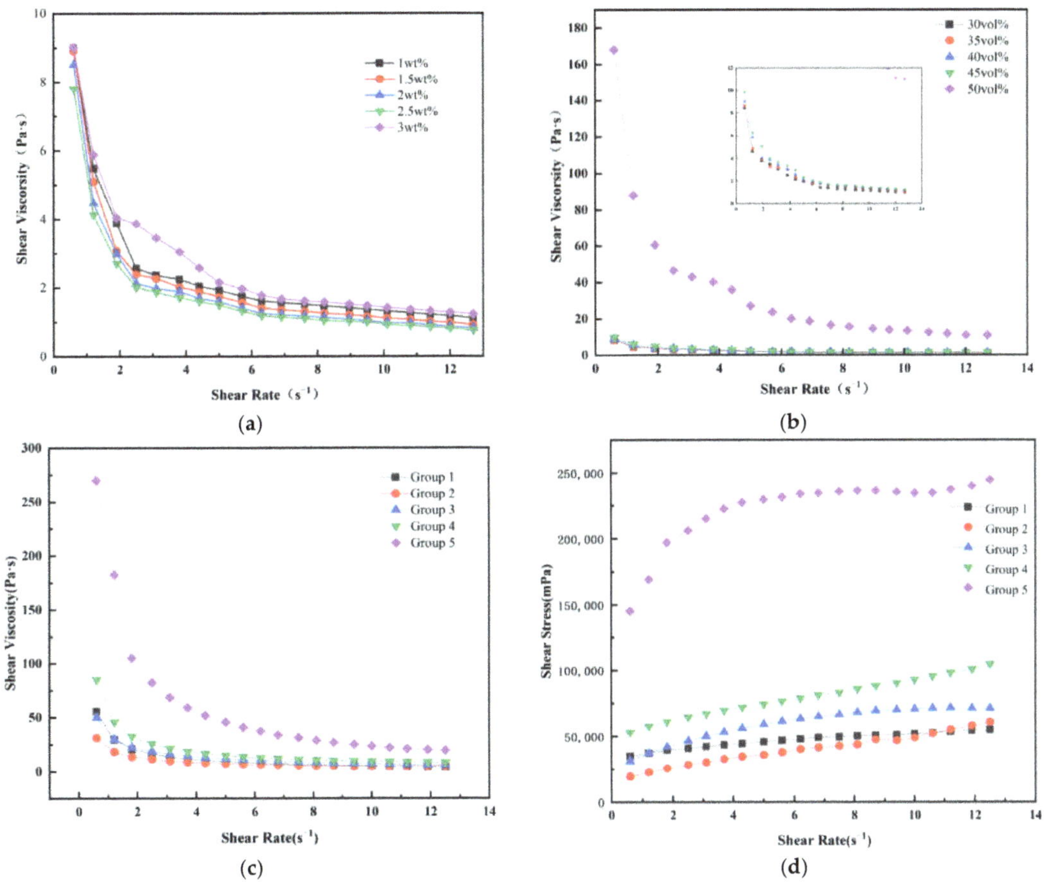

Figure 4. (**a**) Relationship between shear viscosity and shear rate of zirconia ceramic slurry with different dispersant content. (**b**) Rheological properties of zirconia ceramic slurry with different solid loading. (**c**) Relationship between shear viscosity and shear rate with different resin components. (**d**) Relationship between shear stress and shear rate with different resin components.

As shown in Figure 4d, the shear stress also increases with the increase in oligomer content. The shear stresses at a typical shear rate of 10 s^{-1} were 51,941, 48,994, 70,851, and 92,830 mPa for the slurries of Groups 1, 2, 3, and 4, respectively. Under the same conditions, the shear stress of the Group 5 was 234,655 mPa, which is significantly higher than that of the other groups. This indicates that when oligomers completely replace multifunctional monomers, the rheological properties of the slurry deteriorate significantly.

Due to the inadequate rheological properties of the slurry containing resin Group 5, it is no longer possible to complete the printing; therefore, this study subsequently utilized only slurries with resin Groups 1, 2, 3, and 4 for all further testing.

3.2. Photocuring Performance of Zirconia Ceramic Slurry

During light exposure, the photoinitiator in the photosensitive resin system generates free radicals that can initiate the polymerization of monomers or oligomers. This leads to the polymerization and crosslinking of active monomers or oligomers within the photosensitive materials, ultimately forming a three-dimensional network structure.

In the presence of a photoinitiator, the photosensitive resin premix undergoes photopolymerization under specific light intensity, leading to the formation of a crosslinked network structure. The cure depth C_d (μm) of the resin is influenced by the energy dose and can be described using the Beer–Lambert law [28], which relates light attenuation to material properties. According to this model, the energy imparted to the resin diminishes with increasing depth, resulting in a decrease in cure depth as the layer thickness increases. The relationship between cure depth and energy dose can be expressed as follows:

$$C_d = D_p \ln\left(\frac{E}{E_c}\right) \quad (1)$$

Here, projection depth D_p (μm) represents the penetration depth of the experimental setup, E_c ($mJ \cdot cm^{-2}$) is the critical exposure energy, and E is the actual exposure energy ($mJ \cdot cm^{-2}$) applied. This equation indicates that the cure depth increases logarithmically with the energy dose. In this study, we analyzed the relationship between C_d and lnE, as illustrated in Figure 5, and performed curve fitting to determine the variations in D_p and E_c, along with their corresponding R^2 values, as shown in Table 2, to assess the model's accuracy.

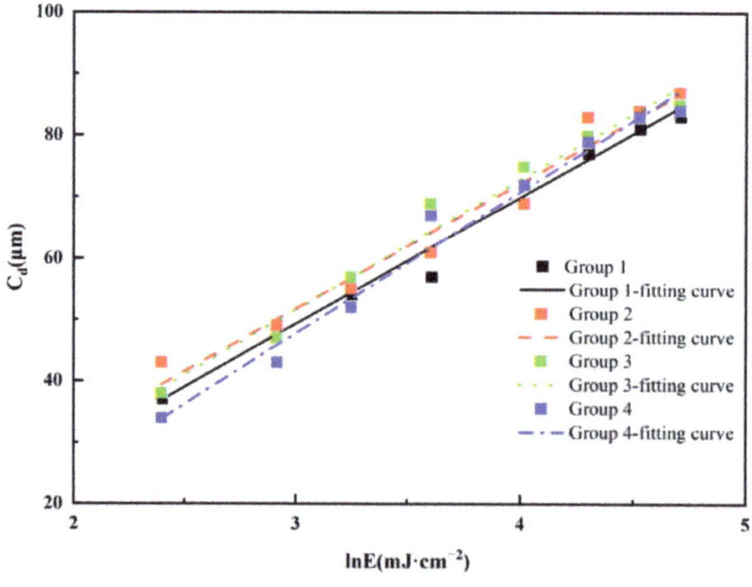

Figure 5. Linear fitting relationship between exposure energy and cure depth of the zirconia ceramic slurries with different resin compositions.

Table 2. Photocuring characteristic parameters of zirconia ceramic slurry with different resin compositions.

Group	D_p	E_c	R^2
1	20.64	1.840	0.97095
2	20.48	1.611	0.96357
3	21.26	1.783	0.97718
4	23.09	2.546	0.97713

All experimental curing depths exceeded the 30 μm layer thickness, fulfilling the fundamental printing specifications. As evidenced by the fitted data, both the projection depth (D_p) and critical exposure energy (E_c) exhibited a decreasing trend with increasing oligomer content. Notably, Group 4 displayed the highest values among all ceramic slurries, suggesting superior photosensitivity that enables curing at lower energy levels compared to Groups 1–3. Furthermore, the high correlation coefficients (R^2) observed across all groups indicate excellent fitting reliability.

3.3. TG-DTG Analysis

As evidenced by the TG-DTG results in Figure 6, the primary decomposition temperature ranges of organic components occurred between 400 °C and 450 °C. Near-complete thermal decomposition was achieved by 600 °C, after which the green body mass stabilized. Consequently, during the debinding process, a slow heating rate is maintained up to 100 °C to ensure thorough moisture removal without inducing cracks. Furthermore, an isothermal holding stage was implemented around 400 °C and 450 °C. The epoxy acrylate cured by photopolymerization has enhanced thermal stability due to its crosslinked network structure. The decomposition initiation temperature has risen to 300–350 °C, and the main decomposition stage is concentrated in the range of 350–450 °C [29]. The test results also verified this conclusion, with the increase in oligomer content, the resin decomposition process becomes more violent. As shown in Figure 6b, the debinding processes in Groups 1, 2, and 3 are much milder, among them, Group 3 exhibits the widest decomposition temperature range and shows a distinct multi-stage decomposition process, thereby effectively minimizing the formation of internal defects.

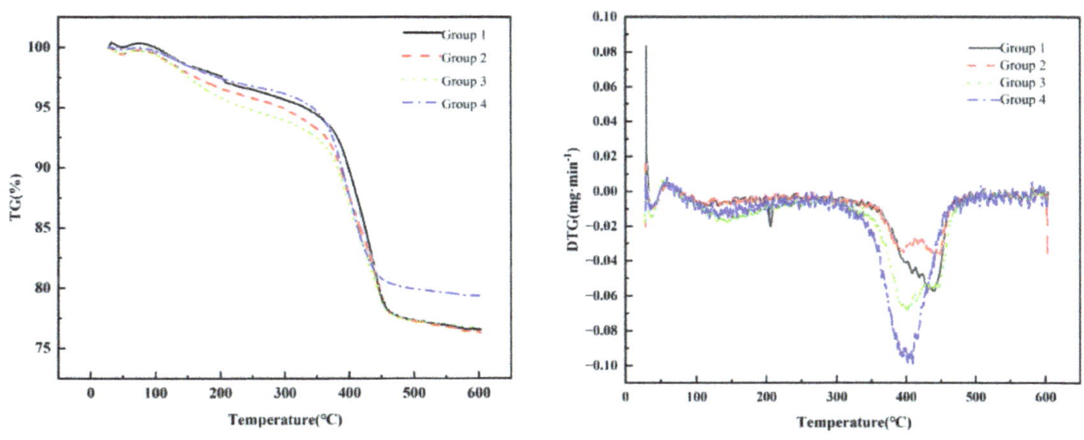

Figure 6. (**a**) TG curves of printed zirconia ceramic green body with different resin compositions. (**b**) DTG curves of printed zirconia ceramic green body with different resin compositions.

3.4. Properties of Zirconia Ceramics

Ceramic specimens were, respectively, printed using slurries with four different group resins. The SEM micrographs of the printed green body are shown in Figure 7. As can be seen from the figures of the green body with Group 1 and Group 2, the interlayer cracks between each layer can be clearly seen. This is likely caused by the low polymerization degree of the polymers in the slurry. In contrast, the green body of Groups 3 and 4 exhibited nearly indistinguishable interlayer cracks from the printing process; these results demonstrate that the addition of oligomers was found to promote more complete photopolymerization and increase crosslinking density.

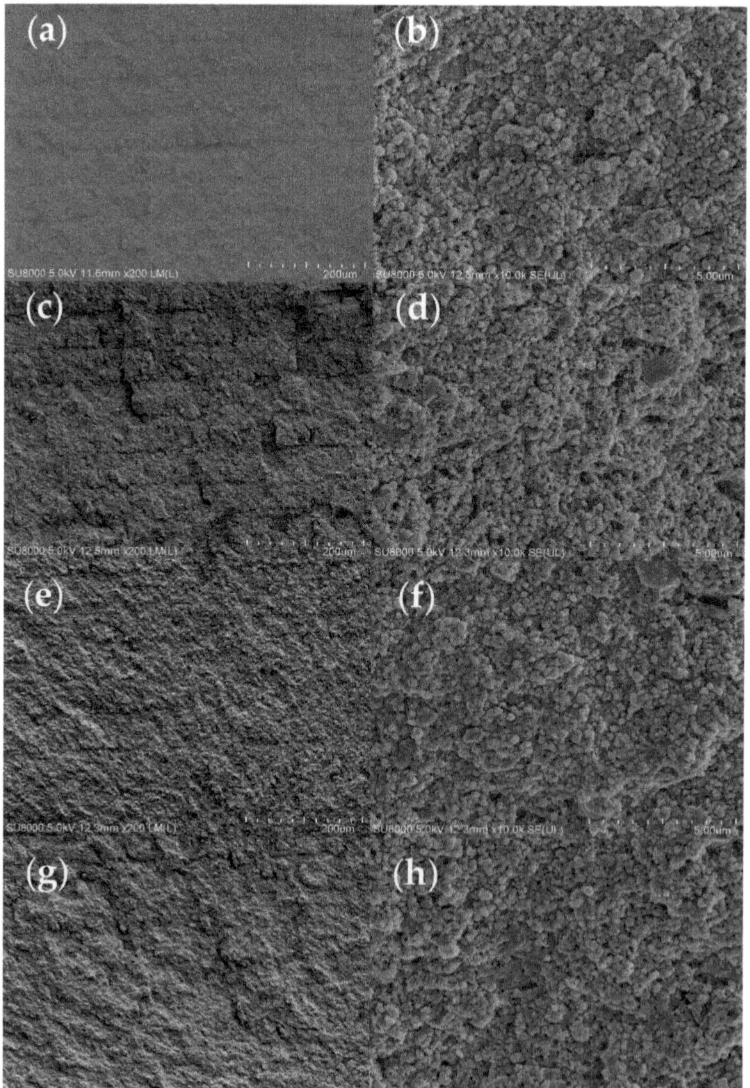

Figure 7. SEM images of green body fracture surfaces with different resin groups: (**a**,**b**) Group 1; (**c**,**d**) Group 2; (**e**,**f**) Group 3; (**g**,**h**) Group 4.

Figure 8 presents SEM images comparing the surface morphologies of debinded samples fabricated by different group resin composites. During the debinding process, polymer decomposition leads to significant volumetric shrinkage. The micrographs in Figure 8a,c,e clearly reveal interfacial delamination between layers, demonstrating insufficient interlayer bonding strength. However, no delamination is observed in Figure 8g, demonstrating that the oligomer addition effectively mitigates interlayer cracks and, consequently, enhances the mechanical properties of the components. Additionally, it can be observed from Figure 8f that the ceramic particles are distributed very uniformly, with no obvious defects detected. This is attributed to the relatively smooth debinding process of Group 3, which also aligns with the TG-DTG results.

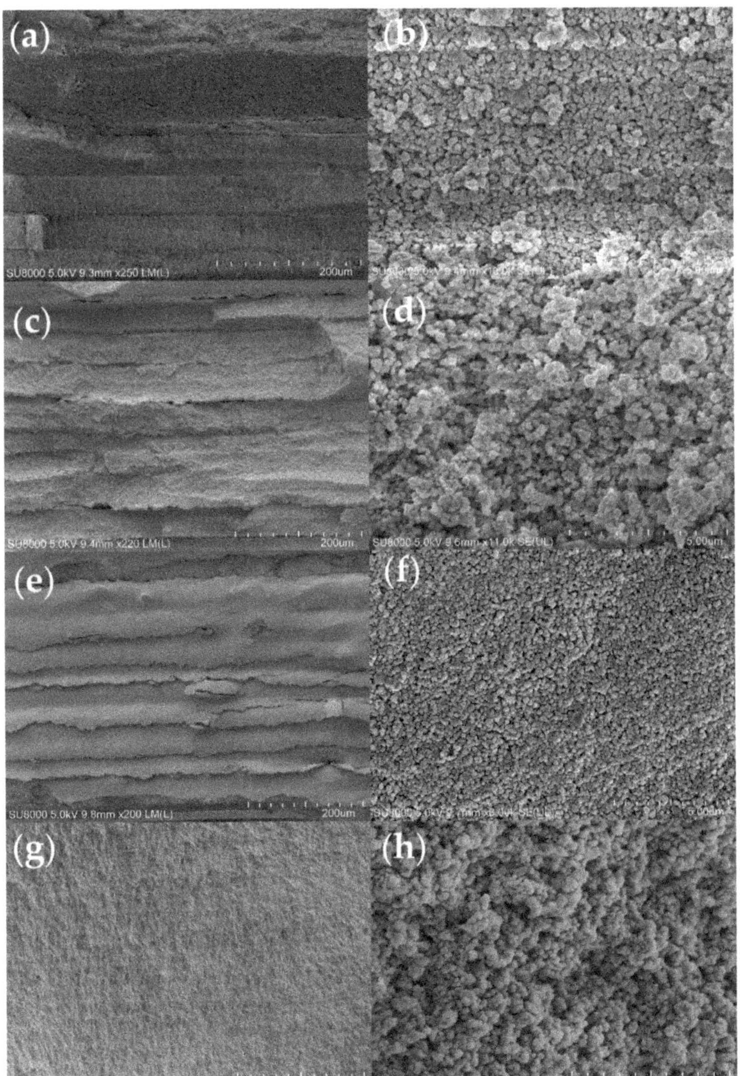

Figure 8. SEM images of debinded ceramic specimens with different resin groups: (**a**,**b**) Group 1; (**c**,**d**) Group 2; (**e**,**f**) Group 3; (**g**,**h**) Group 4.

Figure 9 presents the fracture surface morphologies of sintered ceramics specimens with different resin groups. It could be clearly observed that in samples of Groups 2, 3, and 4, interlayer cracks almost disappeared. Furthermore, higher-magnification microstructural examination demonstrates that the specimen with Group 3 resin has the least amount of micropores remaining inside the ceramics. The three-point bending test also confirmed this result. As shown in Figure 10, the bending test of specimens with oligomer (Groups 2, 3, and 4) exhibited significantly enhanced performance compared to Group 1, in which no oligomer was added. Among the tested groups, Group 3 demonstrated in all the tested resin groups the highest bending strength of 682.4 MPa.

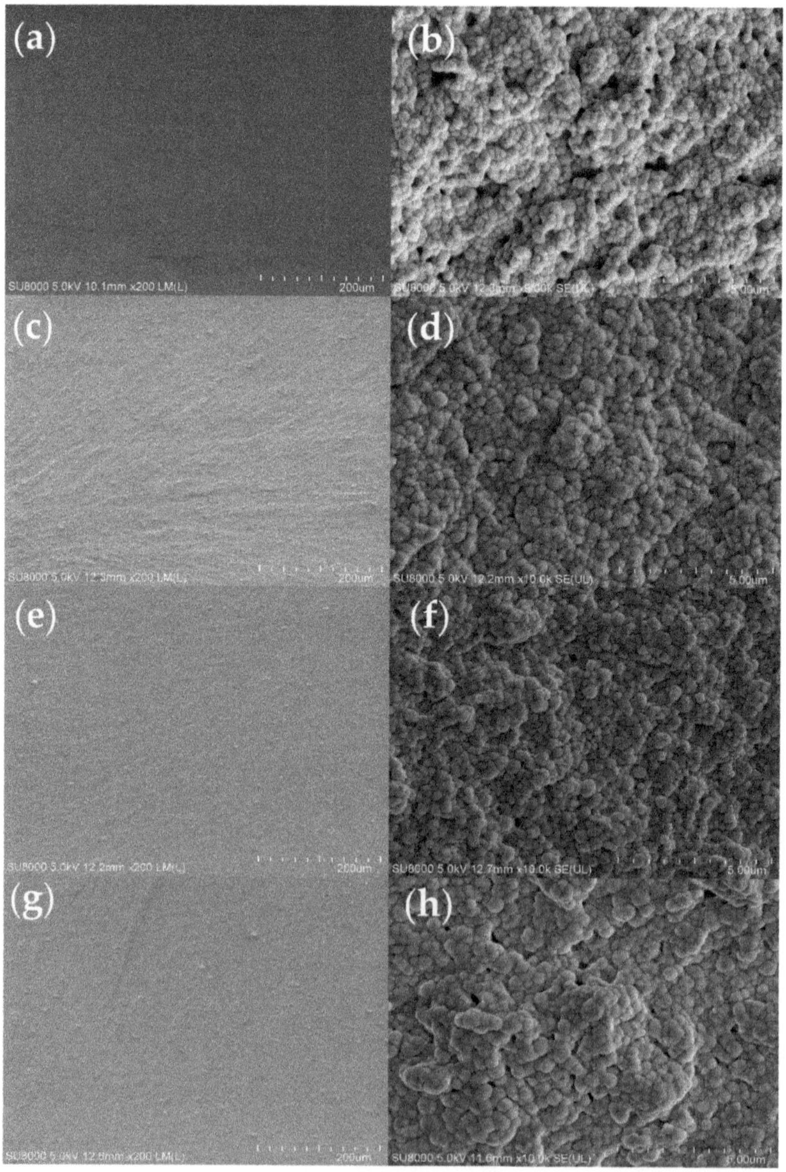

Figure 9. SEM images of the bending fracture surfaces of the sintered specimens with different resin groups: (**a**,**b**) Group 1; (**c**,**d**) Group 2; (**e**,**f**) Group 3; (**g**,**h**) Group 4.

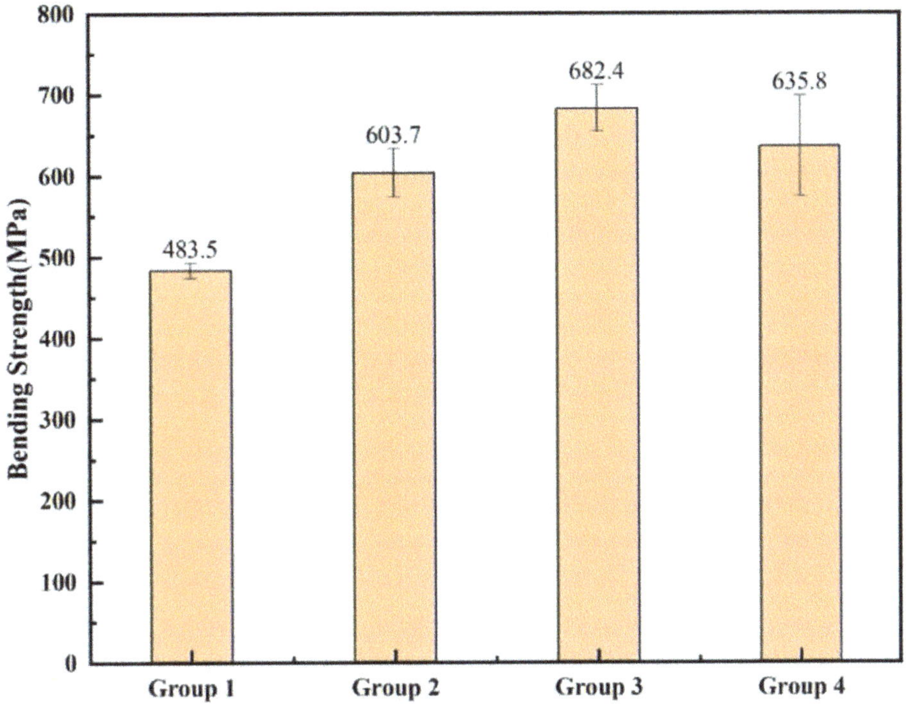

Figure 10. Bending strength of sintered zirconia ceramic specimens with different resin groups.

These results further indicate that the resin formulation explored in this study is suitable for DLP additive manufacturing of zirconia ceramics, and they are helpful in reducing defects that occur during the printing, debinding, and sintering processes, and enhancing the mechanical properties of zirconia ceramics.

4. Conclusions

In this study, the effects of different oligomer content on the rheological properties, the photopolymerization behavior of slurries, and the mechanical properties of printed ceramic specimens were systematically investigated, along with the influence of dispersant content and solid loading on the rheological properties of the zirconia ceramic slurry.

1. Dispersant Content Impact: This study elucidated the critical role of dispersants in controlling rheological behavior, with the 2.5 wt% dispersant formulation showing minimum viscosity among all tested compositions.
2. Solid loading Impact: The addition of a 2.5 wt% dispersant enabled the achievement of an optimal slurry solid loading of 45 vol%.
3. Oligomer Content Impact: This study revealed that although the addition of oligomers negatively affects the rheological properties of the slurry, it enhances the photopolymerization performance, increases the degree of polymerization and crosslinking density in DLP printing, and helps reduce interlayer cracks, thereby improving the strength of printed parts. The zirconia ceramic samples fabricated using a slurry containing 20 wt% oligomers of photosensitive resin exhibited a bending strength of 682.4 MPa after printing, debinding, and sintering processes.

Author Contributions: Conceptualization, N.K.; methodology, N.K.; software, N.K. and H.Q.; validation, N.K., W.Z. and J.W.; formal analysis, N.K.; investigation, N.K. and H.Q.; resources, W.Z. and J.W.; data curation, N.K. and H.Q.; writing—original draft preparation, N.K.; writing—review and editing, W.Z.; visualization, W.Z.; supervision, W.Z.; project administration, W.Z. and J.W.; funding acquisition, W.Z. and J.W. All authors have read and agreed to the published version of the manuscript.

Funding: This research was funded by the Shandong Provincial Natural Science Foundation (ZR2023MB086).

Institutional Review Board Statement: Not applicable.

Data Availability Statement: The original contributions presented in this study are included in the article. Further inquiries can be directed to the corresponding authors.

Conflicts of Interest: The authors declare no conflicts of interest.

References

1. Matsui, K.; Hosoi, K.; Feng, B.; Yoshida, H.; Ikuhara, Y. Ultrahigh Toughness Zirconia Ceramics. *Proc. Natl. Acad. Sci. USA* **2023**, *120*, e2304498120. [CrossRef]
2. Du, J.G.; Su, J.Z.; Geng, J.X.; Duan, L.Y.; He, W.B. Research Advances on Primary Machining Technologies of Zirconia Ceramics. *Int. J. Adv. Manuf. Technol.* **2024**, *130*, 33–55. [CrossRef]
3. Yao, Y.X.; Cui, H.B.; Wang, W.Q.; Xing, B.H.; Zhao, Z. High Performance Dental Zirconia Ceramics Fabricated by Vat Photopolymerization Based on Aqueous Suspension. *J. Eur. Ceram. Soc.* **2024**, *44*, 116795. [CrossRef]
4. Dang, M.T.; Yue, X.Z.; Zhou, G.H.; Hu, S. Influence of Resin Composition on Rheology and Polymerization Kinetics of Alumina Ceramic Suspensions for Digital Light Processing (DLP) Additive Manufacturing. *Ceram. Int.* **2023**, *49*, 20456–20464. [CrossRef]
5. Han, Z.Q.; Liu, S.H.; Qiu, K.; Liu, J.; Zou, R.F. The Enhanced ZrO_2 Produced by DLP via a Reliable Plasticizer and Its Dental Application. *J. Mech. Behav. Biomed. Mater.* **2023**, *141*, 105751. [CrossRef]
6. Zhang, K.Q.; Xie, C.; Wang, G.; He, R.J.; Ding, G.J.; Wang, M. High Solid Loading, Low Viscosity Photosensitive Al_2O_3 Slurry for Stereolithography-Based Additive Manufacturing. *Ceram. Int.* **2019**, *45*, 203–208. [CrossRef]
7. Chen, Z.; Li, Z.; Li, J.; Liu, C.; Lao, C.; Fu, Y.; Liu, C.; Li, Y.; Wang, P.; He, Y. 3D Printing of Ceramics: A Review. *J. Eur. Ceram. Soc.* **2019**, *39*, 661–687. [CrossRef]
8. Gonzalez, G.; Ignazio, R.; Candido, F.P.; Annalisa, C. Current and Emerging Trends in Polymeric 3D Printed Microfluidic Devices. *Addit. Manuf.* **2022**, *55*, 102773. [CrossRef]
9. Stampfl, J.; Martin, S.; Johannes, H.; Fritz, B. Lithography-Based Additive Manufacturing of Ceramics: Materials, Applications and Perspectives. *MRS Commun.* **2023**, *13*, 786–794. [CrossRef]
10. Zhang, G.X.; Zou, B.; Wang, X.F.; Yu, Y.; Chen, Q.H. Design, Manufacturing and Properties of Controllable Porosity of Ceramic Filters Based on SLA-3D Printing Technology. *Ceram. Int.* **2023**, *49*, 1009–1019. [CrossRef]
11. Jong, K.J.; Kong, J.H.; Fisher, J.G.; Park, S.W. Effect of the Volume Fraction of Zirconia Suspensions on the Microstructure and Physical Properties of Products Produced by Additive Manufacturing. *Dent. Mater.* **2019**, *35*, 97–106. [CrossRef] [PubMed]
12. Halloran, J.W. Ceramic Stereolithography: Additive Manufacturing for Ceramics by Photopolymerization. *Annu. Rev. Mater. Res.* **2016**, *46*, 19–40. [CrossRef]
13. Guo, J.; Zeng, Y.; Li, P.R.; Chen, J.M. Fine Lattice Structural Titanium Dioxide Ceramic Produced by DLP 3D Printing. *Ceram. Int.* **2019**, *45*, 23007–23012. [CrossRef]
14. Hammel, E.C.; Ighodaro, O.L.R.; Okoli, O.I. Processing and Properties of Advanced Porous Ceramics: An Application Based Review. *Ceram. Int.* **2014**, *40*, 15351–15370. [CrossRef]
15. He, R.; Liu, W.; Wu, Z.; An, D.; Huang, M.; Wu, H.; Jiang, Q.; Ji, X.; Wu, S.; Xie, Z. Fabrication of Complex-Shaped Zirconia Ceramic Parts via a DLP-Stereolithography-Based 3D Printing Method. *Ceram. Int.* **2018**, *44*, 3412–3416. [CrossRef]
16. Yu, Y.; Zou, B.; Wang, X.; Huang, C. Rheological Behavior and Curing Deformation of Paste Containing 85 wt% Al_2O_3 Ceramic During SLA-3D Printing. *Ceram. Int.* **2022**, *48*, 24560–24570. [CrossRef]
17. Zhang, L.; Huang, J.; Xiao, Z.; He, Y.; Liu, K.; He, B.; Xiang, B.; Zhai, J.; Kong, L. Effects of Debinding Condition on Microstructure and Densification of Alumina Ceramics Shaped with Photopolymerization-Based Additive Manufacturing Technology. *Ceram. Int.* **2022**, *48*, 14026–14038. [CrossRef]
18. Qin, L.; Sui, W.Q.; Wu, X.Q.; Yang, F.; Yang, S.P. Additive Manufacturing of ZrO_2 Ceramic Dental Bridges by Stereolithography. *Rapid Prototyp. J.* **2018**, *24*, 114–119.

19. Wang, X.; Zhou, Y.; Zhou, L.; Xu, X.; Niu, S.; Li, X.; Chen, X. Microstructure and Properties Evolution of Silicon-Based Ceramic Cores Fabricated by 3D Printing with Stair-Stepping Effect Control. *J. Eur. Ceram. Soc.* **2021**, *41*, 4650–4657. [CrossRef]
20. Li, T.; Chen, H.; Zhang, Y.; Gu, Y.; Hu, C.; Li, S.; Liu, B.; Duan, W.; Wang, G. A Comprehensive Evaluation of Vat-Photopolymerization Resins and Alumina Slurries for Ceramic Stereolithography. *Ceram. Int.* **2023**, *49*, 6440–6450. [CrossRef]
21. Kuang, N.; Xiao, M.; Qi, H.; Zhao, W.; Wu, J. Optimization of Resin Composition for Zirconia Ceramic Digital Light Processing Additive Manufacturing. *Polymers* **2025**, *17*, 797. [CrossRef] [PubMed]
22. Khudyakov, I.V. Fast Photopolymerization of Acrylate Coatings: Achievements and Problems. *Prog. Org. Coat.* **2018**, *121*, 151–159. [CrossRef]
23. Xu, X.H.; Zhou, S.X.; Wu, J.F.; Liu, Y.; Wang, Y.Y.; Chen, Z.C. Relationship between the Adhesion Properties of UV-Curable Alumina Suspensions and the Functionalities and Structures of UV-Curable Acrylate Monomers for DLP-Based Ceramic Stereolithography. *Ceram. Int.* **2021**, *47*, 32699–32709. [CrossRef]
24. Johansson, E.; Lidström, O.; Johansson, J.; Lyckfeldt, O.; Adolfsson, O. Influence of Resin Composition on the Defect Formation in Alumina Manufactured by Stereolithography. *Materials* **2017**, *10*, 138. [CrossRef] [PubMed]
25. Sun, J.; Binner, J.; Bai, J. 3D Printing of Zirconia via Digital Light Processing: Optimization of Slurry and Debinding Process. *J. Eur. Ceram. Soc.* **2020**, *40*, 5837–5844. [CrossRef]
26. Schwarzer-Fischer, E.; Zschippang, E.; Kunz, W.; Koplin, C.; Löw, Y.M.; Scheithauer, U.; Michaelis, A. CerAMfacturing of Silicon Nitride by Using Lithography-Based Ceramic Vat Photopolymerization (CerAM VPP). *J. Eur. Ceram. Soc.* **2023**, *43*, 321–331. [CrossRef]
27. Horri, B.A.; Ranganathan, P.; Selomulya, C.; Wang, H. 3D Printing with Ceramics. *Chem. Eng. Sci.* **2011**, *66*, 2798. [CrossRef]
28. Chartier, T.; Chaput, C.; Doreau, F.; Loiseau, M. Stereolithography of Structural Complex Ceramic Parts. *J. Mater. Sci.* **2002**, *37*, 3141–3147. [CrossRef]
29. Barkane, A.; Platnieks, O.; Jurinovs, M.; Gaidukovs, S. Thermal Stability of UV-Cured Vegetable Oil Epoxidized Acrylate-Based Polymer System for 3D Printing Application. *Polym. Degrad. Stab.* **2020**, *181*, 109347. [CrossRef]

polymers

MDPI

Article

Flame Retardance and Antistatic Polybutylene Succinate/Polybutylene Adipate-Co-Terephthalate/Magnesium Composite

Pornchai Rachtanapun [1,†], Jonghwan Suhr [2,†], Eunyoung Oh [2], Nanthicha Thajai [3], Thidarat Kanthiya [4], Krittameth Kiattipornpithak [4], Kannikar Kaewapai [5], Siriphan Photphroet [1], Patnarin Worajittiphon [6], Nuttapol Tanadchangsaeng [7], Pitiwat Wattanachai [5,8], Kittisak Jantanasakulwong [1,*] and Choncharoen Sawangrat [9,*]

[1] Faculty of Agro-Industry, Chiang Mai University, Chiang Mai 50100, Thailand; pornchai.r@cmu.ac.th (P.R.); siriphan_photp@cmu.ac.th (S.P.)
[2] School of Mechanical Engineering, Sungkyunkwan University, 2066 Seobu-ro, Jangan-gu, Suwon-si 16419, Gyeonggi-do, Republic of Korea; suhr@skku.edu (J.S.); ey0208@skku.edu (E.O.)
[3] Nanoscience and Nanotechnology (International Program/Interdisciplinary), Faculty of Science, Chiang Mai University, Chiang Mai 50200, Thailand; nanthicha581@gmail.com
[4] Office of Research Administration, Chiang Mai University, Chiang Mai 50200, Thailand; thidaratkanthiya05@gmail.com (T.K.); first200294@gmail.com (K.K.)
[5] Department of Civil Engineering, Faculty of Engineering, Chiang Mai University, Chiang Mai 50200, Thailand; kannikar@step.cmu.ac.th (K.K.); pitiwat@step.cmu.ac.th (P.W.)
[6] Department of Chemistry, Faculty of Science, Chiang Mai University, Chiang Mai 50200, Thailand; patnarin.w@cmu.ac.th
[7] College of Biomedical Engineering, Rangsit University, Pathumthani 12000, Thailand; nuttapol.t@rsu.ac.th
[8] Science and Technology Park (STeP), Chiang Mai University, Chiang Mai 50100, Thailand
[9] Department of Industrial Engineering, Faculty of Engineering, Chiang Mai University, Chiang Mai 50200, Thailand
* Correspondence: kittisak.jan@cmu.ac.th (K.J.); choncharoen@step.cmu.ac.th (C.S.)
† These authors contributed equally to this work.

Abstract: Antistatic and anti-flame biodegradable polymer composites were developed by melt-blending polybutylene succinate (PBS) with epoxy resin, polybutylene adipate-co-terephthalate (PBAT), and MgO particles. The composite films were prepared using a two-roll mill and an extrusion-blown film machine. Plasma and sparking techniques were used to improve the antistatic properties of the composites. The PBS/E1/PBAT/MgO 15% composite exhibited an improvement in V-1 rating of flame retardancy, indicating an enhancement in the flame retardancy of biodegradable composite films. The tensile strength of the PBS/PBAT blend increased from 19 MPa to 25 MPa with the addition of 1% epoxy due to the epoxy reaction increasing compatibility between PBS and PBAT. The PBS/E1/PBAT and PBS/E1/PBAT blends with MgO 0, 0.5, and 1% showed increases in the contact angle to 80.9°, 83.0°, and 85.7°, respectively, because the epoxy improved the reaction between PBS and PBAT via the MgO catalyst effect. Fourier-transform infrared spectroscopy confirmed the reaction between the epoxy groups of the epoxy resin and the carboxyl end groups of PBS and PBAT by new peaks at 1246 and 1249 cm^{-1}. Plasma technology (sputtering) presents better antistatic properties than the sparking process because of the high consistency of the metal nanoparticles on the surface. This composite can be applied for electronic devices as sustainable packaging.

Keywords: fiber; plasma; epoxy; reaction

1. Introduction

Plastics have become an integral part of modern life owing to their desirable properties, including durability, light weight, corrosion resistance, chemical resistance, low cost, and ease of manufacture [1–3]. Plastics are utilized in a wide range of applications, including films, sheets, fibers, yarns, adhesives, particles, and coatings, across various industries such as automotive, biomedical, pharmaceutical, construction, agriculture, electronics, and packaging [4,5]. However, the irreversible ecological damage caused by plastic waste is increasing. This increase in plastic waste has caused environmental harm to both land and marine life. One approach to address this issue is to develop biodegradable polymer blends (BPBs). There are three ways to classify environment-friendly materials: by the origin of the polymer matrix (e.g., natural and synthetic polymers), degradability (e.g., fully, partially, or non-biodegradable), and the content of renewable components (e.g., petroleum-based, partially, or fully bio-based components). Petroleum-derived plastics have led to enormous volumes of environmental pollution owing to their improper disposal. Only 173 million tons of plastic waste are collected for recycling and landfills [6], highlighting the urgent need for new, cost-effective, and degradable plastics that are currently a subject of interest for many academics and industries worldwide.

Poly(butylene succinate) (PBS) is a biodegradable polymer that can be generated from renewable resource-based succinic acid, obtained through the bacterial fermentation of sugars such as glucose, starch, and xylose [7]. Furthermore, the melting point of PBS is considerably lower than those of other commercially available biodegradable polymers, which can reduce the industrial processing time for blending with other materials [8]. The advantages of PBS include its semi-crystalline structure, thermal stability, and ease of forming processes (such as conventional film-casting and blowing techniques), making it an appropriate representative for producing biodegradable films [9]. However, the disadvantages of PBS are its poor ductility and high cost compared with other conventional plastics [10]. Therefore, there is still room for further studies on PBS blends with other polymers. Poly(butylene adipate-co-terephthalate) (PBAT) is a synthetic polymer based on fossil resources that is ductile at room temperature and has good processability [11,12]. PBAT can be blended with PBS to achieve desirable physical, mechanical, and barrier properties. In addition, a high potential for increasing material biodegradability can be achieved by blending PBAT and PBS [13]. Non-thermal plasma is one of the technologies to apply for blending technique [14]. The expectation of blending the high ductility of PBAT with the tensile strength of PBS is to balance the properties of biodegradable film production. However, the blending of polymers has issues regarding their compatibility. The incorporation of a compatibilizer into a polymer blend is pivotal for enhancing the compatibility of the biopolymer blend. Epoxy (E) is a crosslinker that improves the compatibility of polymer blends [15,16]. An epoxy is composed of epoxide groups, which are reactive groups used to crosslink with other functional groups such as hydroxyl and carboxylic groups [17]. This process leads to improved morphology, melting, and crystallization of the polymer blends [18]. Epoxies have the potential to enhance the compatibility of polymer blends.

Electrical and automotive industries require some materials to protect the products before using, such as packaging and coating. Plastic is a candidate material for these applications, while biopolymer has environmental impacts. The limitation of using biopolymers in packaging, electronics, and automotive applications is their potential fire risk and flammability [19]. The addition of various types of flame retardants is a promising and effective way to improve the anti-flaming performance of a blending polymer matrix [20–22]. An alternative flame retardant, magnesium oxide (MgO), has been widely used in building materials, fireproof coatings, and refractory materials [23]. MgO is used as an additive for antistatic property [24]. Although MgO can be used as a binder in ceramics and ceramic-matrix composites [25], its use in polymers has not yet been thoroughly investigated.

Atmospheric-pressure plasma is an alternative technique for surface modification that affects surface properties such as chemistry and morphology. Atmospheric-pressure plasma has recently been demonstrated to be an effective technique for modifying polymer surfaces [26]. Only surface properties are changed, but the bulk properties remain unchanged when applying plasma treatment [27]. Dhanumalayan et al. (2017) reported that the average surface roughness increased after plasma treatment [28]. As static charges can destroy sensitive electronic equipment, researchers must develop antistatic materials [29]. Polymeric matrices are the best candidates for antistatic packaging because of their inherent lightness and good processability [30]. Metal particles, such as copper, silver, iron, and carbon allotropes (carbon black, graphite, or carbon nanomaterials), are commonly used as antistatic coatings [31]. The sparking process is an alternative method for generating metal nanoparticles and altering the surface of materials. The advantages of the sparking method include its simplicity, low cost, rapidity, non-vacuum system, and use of non-toxic starting materials (i.e., metal wires) [32–34]. The sparking process has the potential to be used as an antistatic additive to improve surface materials. Strengthening the electrostatic resistance of bioplastics using various surface treatment processes, including plasma technology and sparking processes, are two key strategies employed to improve antistatic biodegradable polymer composites. However, there are no previous reports on their use as flame-retardant additives with bioplastics and a comparison of their antistatic abilities with plasma technology and sparking processes. The integrated strategy simultaneously enhances flame retardancy and antistatic performance in biodegradable polymer composites, which expands their potential for being used in advanced sustainable packaging, electronic devices, and automotive components.

Therefore, this study aimed to develop antistatic and anti-flaming biodegradable polymer composites from PBS blends with epoxy resin, PBAT, and MgO to improve their mechanical properties, water resistance, morphology, thermal stability, chemical structure, and flame retardancy. The antistatic properties were improved by plasma technology and a sparking process for coating the film surface. Antistatic and anti-flaming biodegradable polymer composites can be used in packaging, electronics, and automotive applications. MgO was found to improve the reaction and flame retardant of PBS/PBAT blend, while plasma sputtering was an effective technique to improve antistatic property.

2. Materials and Methods

2.1. Materials

Polybutylene succinate (PBS) pellets (BioPBS™ FD92PM, density 1.24 g/cm^3, MFI 4 g/10 min at 190 °C, melting point 84 °C, molecular weight of 516.54 g/mol) were purchased from PTT Global Chemical Public Co., Ltd. (Bangkok, Thailand). Polybutylene adipate terephthalate (PBAT) pellets (ECOFLEX F Blend C1200 grade, MFI 2.7–4.9 g/10 min at 190 °C, density 1.26 g/cm^3, molecular weight of 52.1 kg/mol) were purchased from Unic Technology Co., Ltd. (Chonburi, Thailand). Epoxy resin (E) (diglycidyl ether of bisphenol part A, EPO-TEK 302 grade, molecular weight of 340 g/mol, viscosity 5000–10,000 cPs at 23 °C, glass transition temperature $\geq$ 40 °C, shore D hardness 73) was purchased from Easy Resin Co., Ltd. (Nonthaburi, Thailand). MgO powder with particle sizes of 0.1–0.5 µm was purchased from Cernic International Co., Ltd. (Nakhon Pathom, Thailand). Titanium, copper, and aluminum wires (0.5 mm diameter) were purchased from Advent Research Materials Co., Ltd. (Oxford, UK). Copper, zinc, aluminum, and titanium oxide (TiO_2) nanoparticles were purchased from Guangguang Jing Yan Material Co., Ltd. (Xinfeng, China).

2.2. Preparation of Biodegradable Polymer Composite

Biodegradable polymer composites were prepared using a melt-blending process due to high shear mixing to disperse metal particles in polymer matrix. Samples were prepared by melt-blending PBS (80%) mixed with 1% epoxy (without hardener), 19% PBAT, and MgO particles at concentrations of 0, 0.5, 1, 2, 5, 10, and 15g, using a two-roll mill (Squeeze 6520 Precision resource Co., Ltd., Bangkok, Thailand) at 130 °C for 10 min. Table 1 lists the composition and nomenclature of the PBS/E1/PBAT/MgO blends. The samples were placed in a mold and compressed into sheets for property assessment using a hot compress at 130 °C for 10 min. All samples were prepared in sheets to evaluate their mechanical properties, thermogravimetric analysis (TGA), differential scanning calorimetry (DSC), water resistance properties, morphological properties, and chemical structure. The best formulation was selected to prepare uniform compound by twin screw extruder, and then prepared to be film by blown film extruder.

Table 1. Composition of PBS/E1/PBAT/MgO blends.

Samples	Composition (g)	
	PBS/E1/PBAT	**MgO**
1. PBS/E1/PBAT	100	-
2. PBS/E1/PBAT/MgO 0.5	99.5	0.5
3. PBS/E1/PBAT/MgO 1	99	1
4. PBS/E1/PBAT/MgO 2	98	2
5. PBS/E1/PBAT/MgO 5	95	5
6. PBS/E1/PBAT/MgO 10	90	10
7. PBS/E1/PBAT/MgO 15	85	15

2.3. Preparation of Biodegradable Polymer Composite Blown Film

The anti-flaming PBS/E1/PBAT/MgO15 composites were mixed using a twin-screw extruder (CTE-D22L32 model, Chareon Tut Co., Ltd., Samutprakarn, Thailand). Before mixing, all materials were dried at 60 °C for 24 h to evaporate moisture content. The extruder temperatures for samples from the hopper to the die were set in sequence as follows, 100 °C, 110 °C, 140 °C, 140 °C, 150 °C, 130 °C, 130 °C, and 120 °C, corresponding to torques of 50% and 70%, with a screw speed of 80 rpm. After mixing, films were produced using an extrusion-blown film machine (Laboratory Film Blowing Machine, Precision resource Co., Ltd., Nonthaburi, Thailand) for antistatic study. The temperature profile was set at 115 °C, 150 °C, 145 °C, 115 °C, and 110 °C from hopper to die, and the screw speed was set at 300 rpm. The film thickness ranged from 0.1 to 0.2 mm. Nanocoated films were produced from these film samples using plasma technology and a sparking process. The antistatic properties were then compared.

2.4. Preparation of Plasma Technology (Sputtering) on Biodegradable Polymer Composite Films for Antistatic Properties

The PBS/E1/PBAT/MgO15 composite films were individually functionalized using copper, zinc, aluminum, and titanium oxide (TiO_2) nanoparticles through DC magnetron sputtering using an MSLV-20020330 system. A high-purity (99.99%) target was secured to the cathode end, placed 5 mm from the specimen's surface. The vessel was evacuated prior to sputtering and filled with gas. The argon gas pressure was maintained at 1.0×10^{-1} Torr/m^3 in the chamber, and the argon gas flow was stabilized at 5 sccm. The oxygen (O_2) gas pressure was set at 3.0×10^{-1} Torr/m^3. The substrate underwent smooth deposition of metal nanoparticles using DC magnetron sputtering powered by a 100 W DC power supply (RPG-50 reactive plasma generator) for the sputtering process. The coating duration was 3 min, and the coating process was applied to only one side of the PBS/E1/PBAT/MgO15 composite film. After sputtering, the final PBS/E1/PBAT/MgO15 composite film was extracted from the chamber to measure its physical, chemical, and electrical characteristics.

2.5. Preparation of Nano-Metal-Particles (NMP) Sparking Process on Biodegradable Polymer Composite Films for Antistatic Properties

Ti, Cu, and Al wires (0.5 mm diameter) were purchased from Advent Research Materials Co., Ltd. (Oxford, UK). The two sharp tips (20 mm long) of the metal wire were connected to the anode and cathode of the sparking machine. The tips were aligned 2 mm above the PBS/E1/PBAT/MgO composite film (5 cm × 10 cm × 0.1 cm) with a 1 mm gap between the anode and cathode. The PBS/E1/PBAT/MgO composite film was individually sparked with Ti, Cu, and Al at 3 kV and 3 mA, and the tip holders of the wires were moved along the XY axis at a speed of 100 mm/min. The number of repetitions for each metal wire is listed in Table 2.

Table 2. Repeated time of each metal wire.

No.	Metal Wire: Repeated Times
1	Control (Untreated)
2	Titanium (Ti:10)
3	Titanium (Ti:20)
4	Titanium (Ti:30)
5	Titanium (Ti:40)
6	Copper (Cu:40)
7	Aluminum (Al:40)

2.6. Flame-Retardant Characterization

The flame retardancy of the samples was tested using the UL94 standard [35] (size: W12.7 mm × L127 mm × T1 mm) for vertical burns (V-0 to V-2 Rating). A burner flame was applied to the free end of the specimen for two 10 s intervals, separated by the time it took for the flaming combustion to cease after the first application. Five specimens were tested. The classifications of the UL94 standards [35] are presented in Table 3.

Table 3. Classification of UL94V standards.

UL 94 Test (Vertical Burning Test)			
Test Criteria	**V-0**	**V-1**	**V-2**
Burning time for each individual test specimen (s) (after first and second flame applications T_1 or T_2)	≤10 s	≤30 s	≤30 s
Total burning time (s) ($T_1 + T_2$)	≤50 s	≤250 s	≤250 s
Dripping of burning specimen (ignition of cotton batting)	No	No	Yes
Combustion up to holding clamp (specimens completely burned)	No	No	No

2.7. Mechanical Properties

The mechanical and tensile properties of the anti-flaming biodegradable polymer composites were evaluated. The tensile properties were measured using a tensile tester (Model MCT–1150, Tokyo, Japan) at a distance of 10 mm and a crosshead speed of 20 mm/min. The specimens, prepared as bone-shaped samples according to the JIS K 6251-7 standard [36], were tested with a minimum of five specimens for each composition. The specimens were conditioned at 25 °C with 50 ± 2% RH for 24 h. The maximum force and elongation at breaking were recorded.

2.8. Thermogravimetric Analysis (TGA)

Thermogravimetric analysis of all samples was performed using a thermogravimetric analyzer (Mettler Toledo STARe system TGA/DSC3+, Greifensee, Switzerland). Each sample, weighing approximately 5–10 mg, was heated from 25 °C to 600 °C at a heating rate of 15 °C/min under N_2.

2.9. Differential Scanning Calorimetry (DSC)

Samples weighing 5–10 mg were placed in an alumina pan. The temperature program involved heating from −50 °C to 200 °C, cooling from 200 °C to −50 °C, and further heating from −50 °C to 200 °C, all at a constant heating rate of 10 °C/min under a nitrogen atmosphere. The crystallinities (%X_c) of the samples were calculated using Equation (1).

$$\text{Crystallinity \% (\%Xc)} = \left(\frac{\Delta H_m - \Delta H_c}{\omega H_0^m}\right) \times 100 \quad (1)$$

where ΔH_m and ΔH_c are the enthalpies of melting and cold crystallization, respectively, and ω and $\Delta H^0{}_m$ are the weight fraction of PBS and PBAT and the melting enthalpy of 100% pure polymer (110 and 114 J/g), respectively.

2.10. Morphological Properties

The cross-sectional morphologies of the films were examined using an SEM (FE-SEM, JSM-IT800 series, JEOL Co., Ltd., Tokyo, Japan). The samples, measuring 5 mm × 30 mm × 1 mm (width × length × thickness), were placed in liquid nitrogen for 2 min and then fractured to obtain the cross-sectional surfaces. All the fractured sample surfaces were vacuum-coated with a thin gold layer and observed at 15 kV. The morphology was characterized based on the fracture surfaces of the samples.

2.11. Water Contact Angle

Water contact angle measurements were performed using a drop shape analyzer (DSA30E, Krüss Co., Ltd., Hamburg, Germany). The samples were prepared as sheets measuring 20 × 20 × 1 mm (width × length × thickness) and fixed on a glass slide. Water was dropped onto the surface of the sheets, and images were automatically recorded every minute for 10 min. The water contact angle was then measured using image analysis after droplet formation.

2.12. Chemical Structure Characterization by Fourier-Transform Infrared (FTIR) Spectroscopy

The chemical structure was characterized using FTIR spectroscopy (FT/IR–4700, Jasco Corp., Tokyo, Japan). The samples were prepared as films. FTIR spectra were recorded between 4000 and 400 cm^{-1} with a resolution of 4 cm^{-1} and 64 scans. The chemical structure was identified by IR spectroscopy.

2.13. Determine the Antistatic Properties of the Plasma Technology and Nano-Metal-Particles (NMP) Sparking Method on Biodegradable Polymer Composite Films

The surface resistance of the biodegradable polymer composite films, improved by two types of surface treatments (plasma technology and NMP sparking method), was determined using an electrostatic field meter (ARS-H002ZA series, Korea DONG IL TECHNOLOGY Co., Ltd., Hwaseong-si, Gyeonggi-do, Korea). The surface static electricity of the films was tested under conditions of 24 ± 2 °C and 40 ± 2% RH, at a distance of 25 mm.

2.14. Statistical Analysis

The results were analyzed using one-way analysis of variance (ANOVA) with SPSS software (version 17, Armonk, NY, USA). Differences in tensile strength and water contact angle were estimated using Duncan's test, with significance set at $p < 0.05$.

3. Results and Discussion

3.1. Chemical Structure

Figure 1 presents the FTIR spectra of PBS, PBAT, PBS/PBAT, PBS/E1/PBAT, and PBS/E1/PBAT/MgO at concentrations ranging from 0.5 to 15%. The chemical structures of the blends were analyzed using FTIR spectroscopy. The epoxy exhibited characteristic peaks at 914 and 1670 cm^{-1}, corresponding to the absorption of epoxy groups and the C=C stretching bands of aromatic rings, respectively [37]. The peaks at 1455, 1508, and 1606 cm^{-1} are attributed to the CH_3 asymmetrical bending and the C–C stretching vibrations of the aromatic rings [38]. The PBS peak at 918 cm^{-1} corresponds to –C–OH bending in the carboxylic acid groups. The band at 1045 cm^{-1} was attributed to the –O–C–C– stretching vibrations in PBS. Peaks in the range of 1153 cm^{-1} were linked to the stretching of the –C–O–C bonds in the ester linkages of PBS. The band at 1713 cm^{-1} was associated with the C=O stretching vibrations of the ester groups, while the peaks at 1332 and 2944 cm^{-1} were assigned to the symmetric and asymmetric deformational vibrations of the –CH_2– groups in the main PBS chains [39–41]. PBAT exhibits a peak at 726 cm^{-1}, associated with the vibrations of the –CH_2– groups adjacent to the methylene groups of the polymer backbone. The peaks at 2955 cm^{-1} are attributed to the symmetric stretching vibrations of the –CH– groups, while the peaks at 1249 and 1710 cm^{-1} were linked to the stretching of the C–O bonds from aliphatic groups and carbonyl groups (C=O) in the ester linkage, respectively [42,43]. MgO displayed peaks at 409 and 666 cm^{-1}, indicating the vibration of Mg–O in the MgO particles. The band at 1418 cm^{-1} was due to C=C stretching vibrations, while the peak at 3367 cm^{-1} was assigned to the OH groups in the MgO particles [44,45]. The PBS/PBAT with 1% epoxy exhibited combined spectra of PBS, PBAT, and epoxy resin. A peak intensity at 1249 cm^{-1} (C–O stretching groups) decreased, indicating a change in the C–O vibration due to the reaction between the carboxyl end groups of PBS and PBAT with epoxy. A decrease in the peak intensity of the C–O vibration due to the epoxy resin has been previously reported [46]. The PBS/E1/PBAT with 0.5–15% MgO exhibited combined characteristics of PBS, PBAT, epoxy, and MgO spectra. In PBS/E1/PBAT/MgO0.5, two new peaks at 1246 and 1249 cm^{-1} were observed, with a decrease in the peak at 1210 cm^{-1}. The intensity of the peak at 1249 cm^{-1} decreased, followed by a reduction in the 1246 cm^{-1} peak intensity with increasing MgO content. These two new peaks (1246 and 1249 cm^{-1}) and the reduction in peak intensity (1210, 1246, and 1249 cm^{-1}) indicated the reaction between the COOH groups of PBS and PBAT with the epoxy groups of epoxy, accelerated by the MgO catalyst. Crosslinking within the PBS and PBAT phases and interfacial crosslinking of PBS/PBAT was observed, which improved its properties.

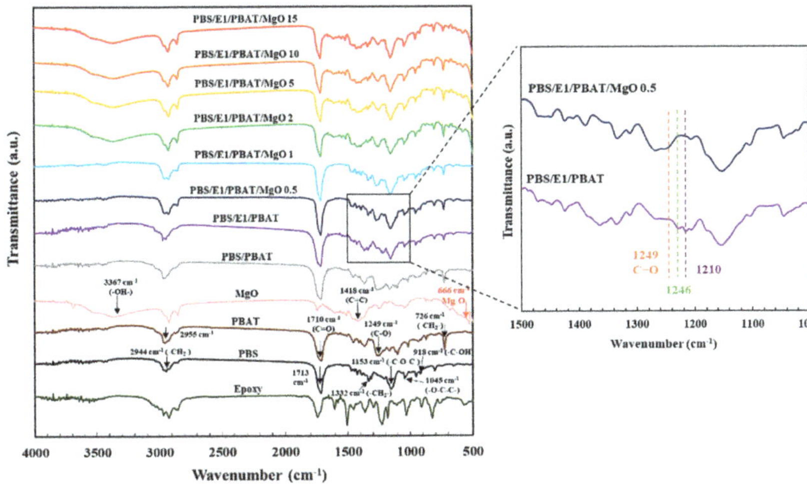

Figure 1. FTIR spectra of PBS, PBAT, PBS/PBAT, PBS/E1/PBAT, and PBS/E1/PBAT/MgO 0.5–15%.

3.2. Morphology

Figure 2 displays the SEM images of the fracture surfaces of PBS, PBAT, PBS/PBAT, PBS/E1/PBAT, and PBS/E1/PBAT with 0.25–5% MgO blends. PBAT exhibited a smooth fracture surface, while PBS showed a slightly rough fracture surface. The PBS/PBAT blend exhibited increased roughness due to phase separation incompatibility between PBS and PBAT. This phase separation results in non-uniform cross-sections in polymer blends [13]. Meanwhile, PBS/E1/PBAT demonstrated an increase in the smoothness of the fracture surface and improved dispersion of epoxy around the materials because the epoxy enhanced the compatibility of the PBS/PBAT blend. This compatibility was enhanced owing to increased interfacial adhesion between the two phases [47]. Adding MgO at concentrations of 0.5–5% in PBS/E1/PBAT blends resulted in increased roughness of the fracture surfaces with the MgO content, due to the aggregation of MgO particles in the samples. Epoxy of samples was indicated to form interactions with MgO, which induced agglomeration of MgO. These aggregates can affect to mechanical properties as a big defect for polymer matrix. Aggregation of metal nanoparticles has been previously reported [48]. However, adding MgO at concentrations of 10 and 15% resulted in smoother surfaces due to the well-dispersed MgO and smaller-sized particles, which improved the elongation at break of the PBS/E1/PBAT/MgO 10 and 15% blends. As previously reported, the mechanical properties of a polymer composite are enhanced by the dispersion and size of the particles in the matrix [49]. Figure 2 also includes SEM-EDS images showing the dispersion of MgO on the fracture surfaces of PBS/E1/PBAT with 0.5–15% MgO blends. In these SEM-EDS images (Figure 3), the PBS/E1/PBAT with 0.5–5% MgO blends showed agglomeration of large MgO particles within the matrix, which adversely affected the distribution of MgO particles and decreased the mechanical properties of the blends. The addition of 10–15 wt.% MgO demonstrated increased dispersion within the matrix, indicating an improvement in the elongation at break of the blends. Big aggregates were indicated during the first state of mixing, while aggregates crushing by high shear force induced fine MgO particle distribution in PBS/E1/PBAT/MgO 10 and 15% blends. Furthermore, the interaction between the carboxylic groups and Mg^{2+} ions increased the elongation at break. As previously reported, enhanced dispersion of MgO and interaction of metal ions contribute to an increase in elongation at break [50,51].

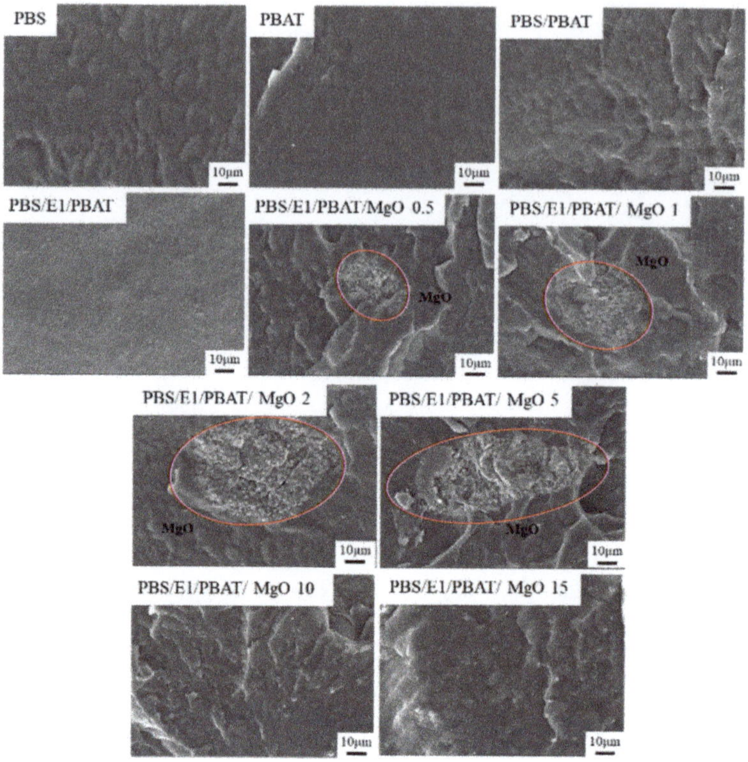

Figure 2. Scanning electron micrographs of PBS, PBAT, PBS/PBAT, PBS/E1/PBAT, and PBS/E1/PBAT/MgO 0.5–15%.

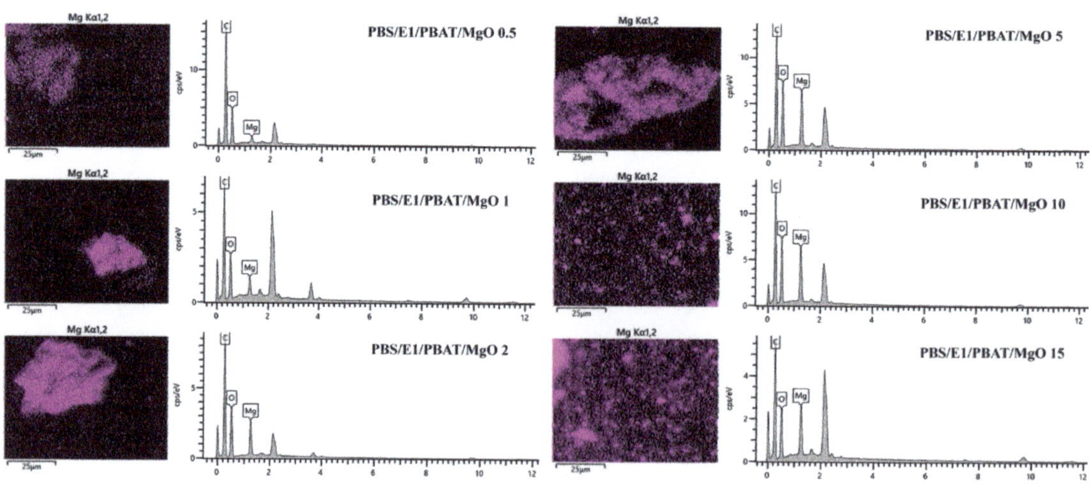

Figure 3. EDS mode of PBS/E1/PBAT/MgO 0.5–15%.

3.3. Mechanical Properties

The tensile properties of PBS, PBAT, PBS/PBAT, PBS/E1/PBAT, and PBS/E1/PBAT/MgO at concentrations ranging from 0.5 to 15% are presented in Figure 4. The tensile strength of PBS (25 MPa) was higher than that of PBAT (20 MPa). However, the elongation at break of PBS (593%) was lower than that of PBAT (701%). The PBS/PBAT blend showed tensile strength and elongation at break of 19 MPa and 367%, respectively. The mechanical properties of the PBS/PBAT blend decreased due to the phase separation between PBS and PBAT, indicating the incompatibility of the polymer blend. Phase separation in polymer blends leading to reduced tensile strength has been previously reported [52,53]. Upon adding 1% epoxy to PBS/PBAT, the tensile strength increased to 25 MPa, and the elongation at break decreased to 358%, due to increased compatibility between PBS and PBAT facilitated by the epoxy reaction. The functional groups of the polymer blends reacted between the carboxyl end groups of PBS and PBAT and the epoxy groups. Improvement in the tensile strength of polymer blends due to the reaction with epoxy groups has been previously documented [54,55]. The addition of 0.5 to 15% MgO to PBS/E1/PBAT resulted in reduced tensile strength compared to the PBS/E1/PBAT blend. PBS/E1/PBAT/MgO 0.5 exhibited the highest tensile strength (22 MPa), which was similar to low density polyethylene (10–18 MPa) [56–58]. The addition of 0.5–5% MgO to PBS/E1/PBAT decreased the tensile strength and elongation at break due to the presence of agglomerates and large particles, which reduced compatibility and acted as defects in the PBS/E1/PBAT/MgO blends. The agglomerated particles affected the interfacial adhesion and reduced the tensile strength [59]. The addition of 5–15% MgO showed the decreasing tensile strength due to the excessive amount of nanoparticles with low interfacial interaction between MgO and polymers surface, which induced MgO particle agglomeration as a big defect to limit the polymer stress distribution [60]. However, the elongation at break of PBS/E1/PBAT/MgO 10–15% increased to 415.9% and 505.2%, respectively, because of better dispersion of MgO. Furthermore, the high dispersion of MgO can be attributed to the high shear forces during blending, which promotes effective particle dispersion. The interaction of the carboxylic and hydroxyl groups with Mg^{2+} ions and/or epoxy groups causes plasticization properties. An increase in elongation at break due to the uniform dispersion of MgO and the interaction of metal ions in the polymer matrix has been previously reported [50,51].

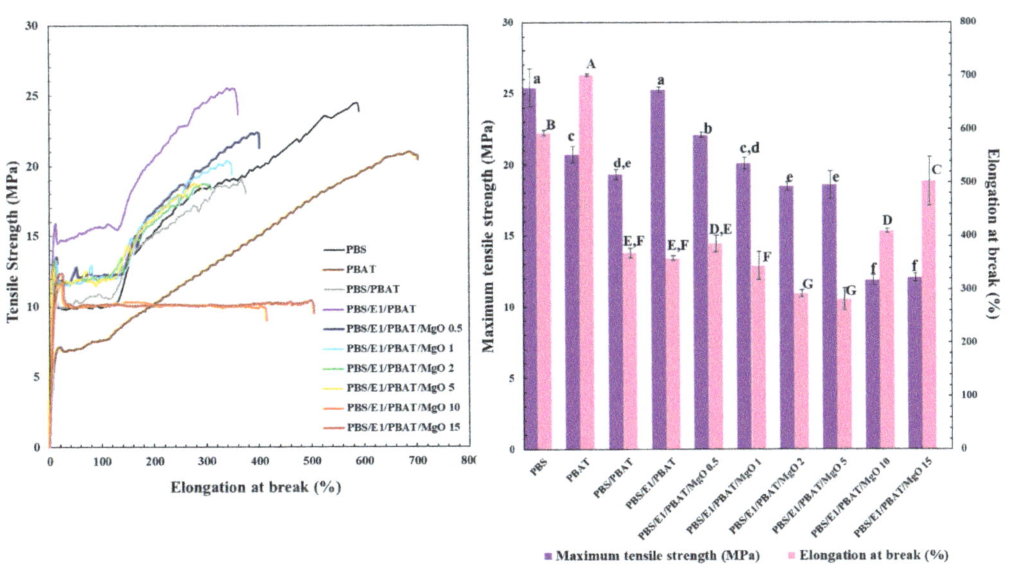

Figure 4. Tensile properties of PBS, PBAT, PBS/PBAT, PBS/E1/PBAT, and PBS/E1/PBAT blends with 0.5–15% MgO. Mean values of elongation at break (lowercase letters) and maximum tensile strength (uppercase letters) differ significantly ($p < 0.05$).

3.4. Differential Scanning Calorimetry (DSC)

DSC curves were utilized to determine the thermal properties of the PBS/E1/PBAT/MgO blends. Figure 5 displays the second scans of the DSC curves for PBS, PBAT, PBS/PBAT, PBS/E1/PBAT, and PBS/E1/PBAT/MgO15. PBAT exhibited a glass transition temperature (T_g) of −28 °C and a melting temperature (T_m) of 121 °C. The T_m of PBS was recorded at 87 °C, while its T_g was not observed during this measurement (Table 4). The glass transition and melting temperatures of PBS and PBAT have been previously reported [9,61]. The T_m of PBS/PBAT, PBS/E1/PBAT, and PBS/E1/PBAT/MgO15 were recorded at 87, 87, and 86 °C, respectively. The blends exhibited a decrease in T_m compared to pure PBS. The crystallinities of PBS, PBAT, PBS/PBAT, PBS/E1/PBAT, and PBS/E1/PBAT/MgO15 were 43%, 11%, 26%, 29%, and 19%, respectively. Crystallinities of PBS/PBAT, PBS/E1/PBAT, and PBS/E1/PBAT/MgO15 were PBS crystal which were calculated from the melting peak of PBS. However, recrystallization was not observed in any of the blends due to the high content of existing crystals without nucleation sites to induce new crystallization.

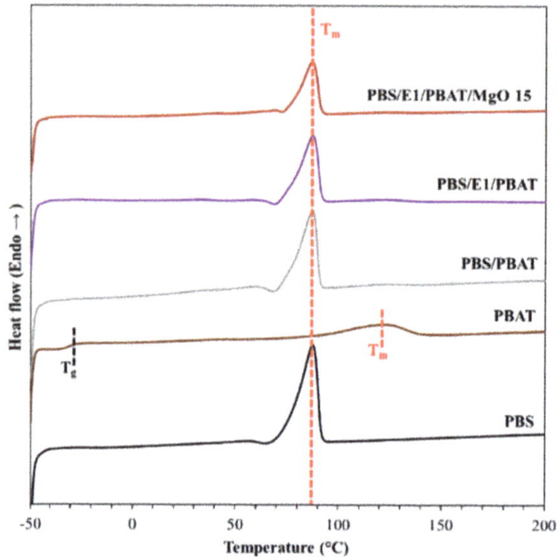

Figure 5. DSC curves of the second scan for PBS, PBAT, PBS/PBAT, PBS/E1/PBAT, and PBS/E1/PBAT/MgO15.

Table 4. Second scan DSC results for PBS, PBAT, PBS/PBAT, PBS/E1/PBAT, and PBS/E1/PBAT/MgO 15.

Sample	T_g (°C)	T_c (°C)	T_m (°C)	ΔH_m (J/g)	ΔH_c (J/g)	ΔX_c (%)
PBS	-	-	87.5	47.1	-	42.8
PBAT	−28.2	-	121.5	12.9	-	11.3
PBS/PBAT	-	-	87.1	33.9	-	26.3
PBS/E1/PBAT	-	-	87.1	31.4	-	28.6
PBS/E1/PBAT/MgO15	-	-	86.8	21.0	-	19.1

3.5. Thermogravimetric Analysis (TGA)

Figure 6 presents the TGA thermograms of PBS, PBAT, PBS/PBAT, PBS/E1/PBAT, and PBS/E1/PBAT/MgO ranging from 0.5 to 15%. PBS, PBAT, PBS/PBAT, and PBS/E1/PBAT exhibited one main weight-loss stage and a corresponding heat-flow peak. The primary thermal degradation temperatures for PBS and PBAT were approximately 400 and 413 °C, respectively. This thermal degradation has been documented in previous studies [62,63]. The PBS/PBAT blend showed a weight-loss stage at around 400 °C, similar to pure PBS. However, the weight-loss stage for PBS/E1/PBAT slightly increased to 410 °C, indicating enhanced thermal stability due to the crosslinking of PBS, PBAT, and epoxy resin. The degradation temperature of epoxy resin ranged between 400 and 600 °C, which contributes to the increased degradation temperature of the blends by cross-linking [54,64]. In the TGA thermogram of PBS/E1/PBAT/MgO 0.5–15%, the main step in weight loss occurred in the range of 398–400 °C, which is associated with the depolymerization of the polymer catalyzed by MgO. However, the addition of MgO at 2–15% introduced multiple-stage weight loss, observed at two stages for PBS/E1/PBAT/MgO 2 and 5%. The first stage, occurring between 310 and 320 °C, indicates the dehydroxylation of the surface of MgO particles within the polymer composites. The dehydroxylation of MgO surfaces has been previously been reported [44]. For PBS/E1/PBAT/MgO 10–15%, there were three stages: the first at 310–320 °C, the second at 390–400 °C, and the third at 450–460 °C, the latter being a heat transfer stage attributed to the thermal degradation of the MgO particles. The degradation temperature of the MgO particles has been reported in earlier studies [65]. Additionally, the residue percentages at 600 °C for PBS/E1/PBAT/MgO 10–15% were recorded at 19% and 23%, respectively. The increase in residue percentage could be attributed to the presence of inorganic MgO particles and their high thermal stability.

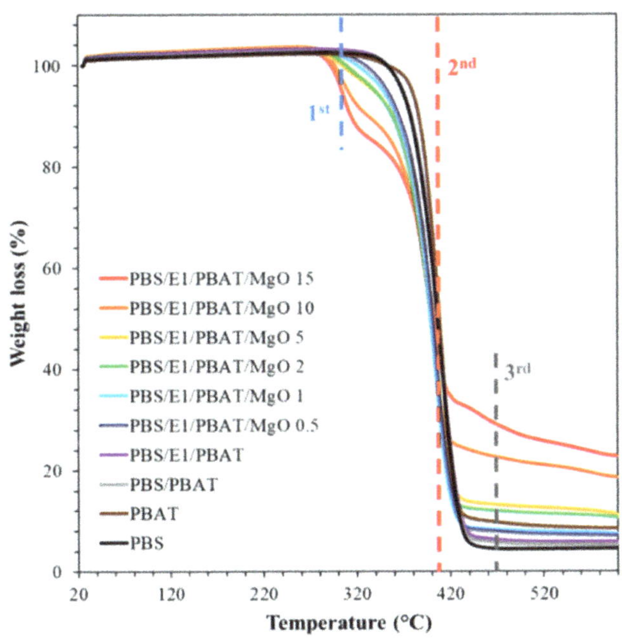

Figure 6. TGA thermograms of PBS, PBAT, PBS/PBAT, PBS/E1/PBAT, and PBS/E1/PBAT/MgO ranging from 0.5 to 15%.

3.6. Water Contact Angles

The water contact angle is commonly used to determine the hydrophilic and hydrophobic properties of surfaces. Typically, hydrophobic materials exhibit large water contact angles. Figure 7 shows the water contact angles for PBS, PBAT, PBS/PBAT, PBS/E1/PBAT, and PBS/E1/PBAT/MgO at 0.5–15%. The water droplet was absorbed onto the surface, and the contact angle was automatically recorded every minute for 10 min. PBS exhibited a contact angle of 77.8° after 10 min, whereas PBAT exhibited a higher angle of 82.7°, reflecting its more hydrophobic nature despite PBS generally being more hydrophilic [66]. The contact angle for the PBS/PBAT blend was 80.7°, higher than that of pure PBS, indicating that the hydrophobicity of PBAT enhanced the water resistance of the blends. Adding epoxy 1% to PBS/PBAT increased the contact angle to 80.9°, suggesting improved compatibility of the blend, which has been linked to increasing water contact angles in previous studies [67]. For the PBS/E1/PBAT blends with 0.5% and 1% MgO, the water contact angles at 10 min increased to 83.0° and 85.7°, respectively. This increase indicates that the addition of 1% MgO to the PBS/E1/PBAT blend increased the surface tension and hydrophobicity due to the metal particles and the increased density of the composites [68,69]. However, adding MgO at 2%, 5%, 10%, and 15% decreased the water contact angle at 10 min to 75.0°, 72.9°, 65.8°, and 63.6°, respectively, owing to increase in roughness of the composite films' surfaces and the hydrophilic nature of alkaline MgO [70]. The increasing MgO content exhibited hygroscopic properties and reacts with water to form $Mg(OH)_2$, which reduced hydrophobic nature of composites [71]. The decreasing water contact angle in polymer blend with MgO content has also been reported [72–75].

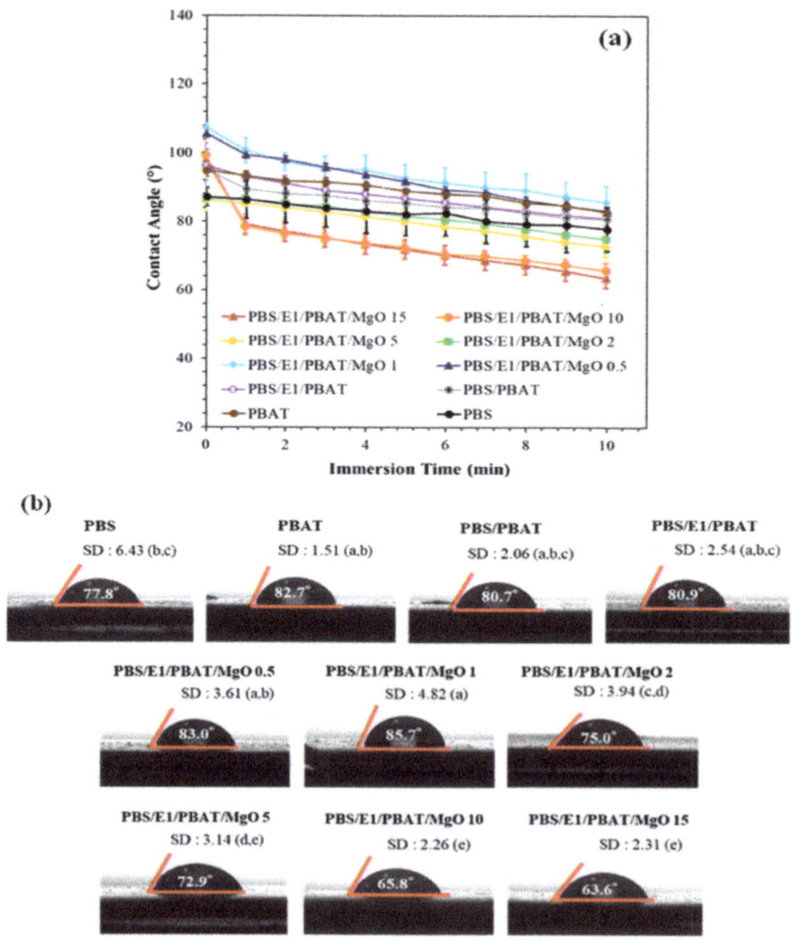

Figure 7. Water contact angles of PBS, PBAT, PBS/PBAT, PBS/E1/PBAT, and PBS/E1/PBAT/MgO 0.5–15% (**a**) plots of samples and (**b**) values at 10 min.

3.7. Flame-Retardant Characterization

The flammability of the samples was evaluated using the UL94 standard Vertical Burning test. The UL-94 ratings for PBS, PBAT, PBS/PBAT, and PBS/E 0.5–20/PBAT samples are presented in Table 5. PBS, PBAT, PBS/PBAT, and PBS/E1/PBAT all exhibited a V-2 flame retardancy rating, where flaming drips were observed after burning but did not reach the holding clamp. The addition of 0.5% to 2% MgO to PBS/E1/PBAT extended the burning times to over 30 s for T1, and 250 s for T1 + T2. Adding 5% and 15% MgO continuously improved flammability, with T1 times of 12 s and T1 + T2 times of 128, 125, and 122 s. Specifically, PBS/E1/PBAT/MgO 15 achieved a V-1 degree of flame retardancy, indicating that the drips from the samples did not ignite the cotton batting below, which reflects a significant improvement in flame retardancy of the biodegradable composite films due to the addition of MgO, enhancing self-extinguishing properties after the removal of the flame. V-1 rating classifications that indicate enhanced flame retardancy have been documented in previous studies [72,76]. MgO acted as both a catalyst and a flame-retardant agent, helping to form a crosslinked char that was firmly fixed on top of the composite, subsequently restricting and extinguishing the flame [77].

Table 5. Anti-flame ability of PBS, PBAT, PBS/PBAT, PBS/E1/PBAT, and PBS/E1/PBAT blend with MgO at 0.5–15%.

Samples	Class UL 94 (V0–V2)	Ignition (Flaming Drip) (Yes/No)	Specimen Burns up to Holding Clamp. (Yes/No)
PBS	V-2	Yes	No
PBAT	V-2	Yes	No
PBS/PBAT	V-2	Yes	No
PBS/E1/PBAT	V-2	Yes	No
PBS/E1/PBAT/MgO 0.5	-	Yes	No
PBS/E1/PBAT/MgO 1	-	Yes	No
PBS/E1/PBAT/MgO 2	-	Yes	No
PBS/E1/PBAT/MgO 5	V-2	Yes	No
PBS/E1/PBAT/MgO 10	V-2	Yes	No
PBS/E1/PBAT/MgO 15	V-1	No	No

3.8. Antistatic Properties of the Plasma Technology (Sputtering) and Nano-Metal-Particles (NMP) Sparking Process Method for Coating on Biodegradable Polymer Composite Films

Figure 8 shows static electricity voltage of the coated biodegradable polymer composite films using sparking process and plasma technology (sputtering). Plasma technology involved DC magnetron sputtering at 100, 110, and 120 W for sputter-coating with Ti, Cu, Zn, and Al nanoparticles. The uncoated (control) PBS/E1/PBAT/MgO15 film exhibited a high voltage of 1.56 kV, indicating poor antistatic properties. The voltages of all coated films decreased with each method applied. For the PBS/E1/PBAT/MgO15 films coated using the sparking process, the surface voltages were measured at 0.364 kV (Ti-10 min), 0.864 kV (Ti-20 min), 1.32 kV (Ti-30 min), 0.79 kV (Ti-40 min), 0.948 kV (Cu-40 min), and 0.83 kV (Al-40 min). Moreover, the surface voltages of PBS/E1/PBAT/MgO15 films coated with plasma technology were significantly reduced to 0.05 kV (Cu-100 W), 0.062 kV (Al-100 W), 0.036 kV (Zn-100 W), 0.05 kV (Zn-110 W), 0.05 kV (Ti-100 W), and 0.192 kV (Ti-120W). The film's antistatic properties are enhanced due to the conducting properties of the metal particles and oxides coated on it. Conductive additives, typically used with polymers, contribute to antistatic properties [78–81] by facilitating the movement of ions on the substrate surface [82]. Films coated using plasma technology exhibited lower voltage levels compared to those coated using the sparking process. Plasma sputtering involves accelerating electrons under an electric field until they collide with the atoms of the target material, leading to the breakdown of these atoms into positive ions. These ions are then accelerated towards the cathode, where they collide with the coating material particles, causing deposition onto the substrate positioned near the anode. The plasma process, resulting from the ionization of neutral gases [83–85], operates at the ion level, which enhances the nanoparticle dispersion ability on the surface of the film. In contrast, the sparking process produces metal nanoparticles by bombardment with electrons and ions, leading to a less uniform nanoparticle distribution on the surface compared to the plasma process [34]. Therefore, plasma technology is more effective in improving the antistatic properties due to the finer particle distribution and higher uniformity of the metal nanoparticles on the film surface.

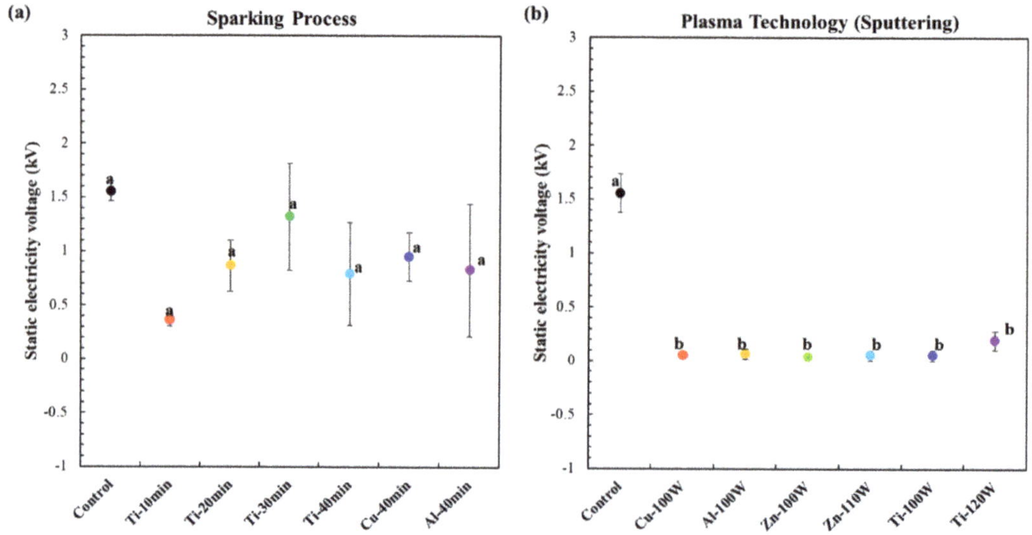

Figure 8. Effects of coating method on antistatic properties of the biodegradable polymer composite films: (**a**) sparking process; (**b**) plasma technology (sputtering).

4. Conclusions

Antistatic and anti-flaming biodegradable polymer composite films were successfully developed through the melt blending of PBS/PBAT and epoxy with MgO. FTIR analysis confirmed the reaction between the –COOH groups of PBS and PBAT and the epoxy groups. The PBS/E1/PBAT/MgO 15 blends exhibited a fine MgO particle distribution within the PBS/E1/PBAT matrix. The tensile strength of PBS/E1/PBAT was increased to 25.3 MPa, while the elongation at break of PBS/E1/PBAT/MgO 15 was raised to 505.2% compared to PBS/PBAT. This improvement in properties was due to the reaction of the epoxy in the PBS/E1/PBAT blends. The addition of MgO to the PBS/E1/PBAT blends enhanced their thermal decomposition behavior and water resistance, attributable to the high thermal stability and hydrophobicity of the metal particles. MgO acted as a catalyst for the –COOH and epoxy to accelerate reactions. The flame-retardant characteristics of the PBS/E1/PBAT/MgO 15 blends improved to a V-1 degree of flame retardancy. MgO served as a flame retardant, enhancing the strength of the residual char from the crosslinked matrix. The antistatic properties of the films were improved using both plasma technology and sparking methods. The antistatic effect of the plasma sputtering process proved superior to that of the sparking process because the sputtering method enhanced the uniform dispersion of metal nanoparticles on the film's surface. Anti-flaming and antistatic biodegradable polymer composites of PBS/E1/PBAT/MgO with enhanced properties hold potential for use in packaging, electronics, and automotive applications. It can be used as an environmentally friendly insulation or electrical circuit board packaging application. These polymer composites can replace petroleum polymer insulators and electronic devices packaging film.

Author Contributions: Designed the research study, K.J., P.R., C.S., J.S., and P.W. (Pitiwat Wattanachai) Prepared and investigated the samples, N.T. (Nanthicha Thajai), K.K. (Kannikar Kaewapai), K.K. (Krittameth Kiattipornpithak), S.P., E.O., and T.K. Contributed to the data analysis, K.J., K.K. (Kannikar Kaewapai), P.W. (Pitiwat Wattanachai), J.S., P.W. (Patnarin Worajittiphon), N.T. (Nuttapol Tanadchangsaeng), and P.R. Suggested, discussed, and concluded the results, N.T. (Nuttapol Tanadchangsaeng), J.S., P.R., K.K. (Kannikar Kaewapai), P.W. (Pitiwat Wattanachai), and K.J. Drafted the manuscript, K.J., P.R., P.W. (Patnarin Worajittiphon), J.S., and N.T. (Nanthicha Thajai). All authors have read and agreed to the published version of the manuscript.

Funding: The present study was partially supported by the Thailand Research Fund (TRF) Research Team Promotion Grant, RTA, Senior Research Scholar (N42A671052). This research project was supported by Fundamental Fund 2025, Chiang Mai University. This research has received funding support from the NSRF via the Program Management Unit for Human Resources & Institutional Development, Research and Innovation (grant number B13F670056).

Institutional Review Board Statement: Not applicable.

Data Availability Statement: The datasets used and/or analyzed during the current study are available from the corresponding author on reasonable request.

Acknowledgments: This material is based upon work supported by the Air Force Office of Scientific Research under award number FA2386-22-1-4064. This research work was partially supported by Chiang Mai University. The authors wish to express their sincere thanks to the Bio Plasma Technology Center (ABPlas), Science and Technology Park (STeP), and Faculty of Agroindustry, Chiang Mai University, for laboratory equipment and resources support and the Center of Excellence in Materials Science and Technology, Chiang Mai University. This research is supported by the Hub of Talents for Plasma Technology 2024 funded by the National Research Council of Thailand (NRCT), contract number N23E670083/2024.

Conflicts of Interest: The authors declare no conflicts of interest.

References

1. AlMaadeed, M.A.A.; Ponnamma, D.; El-Samak, A.A. Chapter 1—Polymers to Improve the World and Lifestyle: Physical, Mechanical, and Chemical Needs. In *Polymer Science and Innovative Applications*; AlMaadeed, M.A.A., Ponnamma, D., Carignano, M.A., Eds.; Elsevier: Amsterdam, The Netherlands, 2020; pp. 1–19.
2. Andrady, A.L.; Neal, M.A. Applications and Societal Benefits of Plastics. *Philos. Trans. R. Soc. Lond. B Biol. Sci.* **2009**, *364*, 1977–1984. [CrossRef] [PubMed]
3. Montemor, M.F. Corrosion Issues in Joining Lightweight Materials: A Review of the Latest Achievements. *Phys. Sci. Rev.* **2016**, *1*, 20150011. [CrossRef]
4. Gross, R.A.; Kalra, B. Biodegradable Polymers for the Environment. *Science* **2002**, *297*, 803–807. [CrossRef]
5. Formela, K.; Zedler, Ł.; Hejna, A.; Tercjak, A. Reactive Extrusion of Bio-Based Polymer Blends and Composites-Current Trends and Future Developments. *Express Polym. Lett.* **2018**, *12*, 24–57. [CrossRef]
6. Rai, P.; Mehrotra, S.; Priya, S.; Gnansounou, E.; Sharma, S.K. Recent Advances in the Sustainable Design and Applications of Biodegradable Polymers. *Bioresour. Technol.* **2021**, *325*, 124739. [CrossRef]
7. Xu, J.; Guo, B.H. Poly(butylene succinate) and Its Copolymers: Research, Development and Industrialization. *Biotechnol. J.* **2010**, *5*, 1149–1163. [CrossRef]
8. Rafiqah, S.A.; Khalina, A.; Harmaen, A.S.; Tawakkal, I.A.; Zaman, K.; Asim, M.; Nurrazi, M.N.; Lee, C.H. A Review on Properties and Application of Bio-Based Poly(Butylene Succinate). *Polymers* **2021**, *13*, 1436. [CrossRef] [PubMed]
9. de Matos Costa, A.R.; Crocitti, A.; Hecker de Carvalho, L.; Carroccio, S.C.; Cerruti, P.; Santagata, G. Properties of Biodegradable Films Based on Poly(butylene Succinate) (PBS) and Poly(butylene Adipate-*co*-Terephthalate) (PBAT) Blends. *Polymers* **2020**, *12*, 2317. [CrossRef]

10. Liminana, P.; Sanoguera, D.G.; Carrillo, L.Q.; Balart, R.; Montanes, N. Development and Characterization of Environmentally Friendly Composites from Poly(butylene succinate) (PBS) and Almond Shell Flour with Different Compatibilizers. *Compos. Part B Eng.* **2018**, *144*, 153–162. [CrossRef]
11. Deng, Y.; Yu, C.; Wongwiwattana, P.; Thomas, N.L. Optimising Ductility of Poly(lactic acid)/Poly(butylene adipate-co-terephthalate) Blends Through Co-Continuous Phase Morphology. *J. Polym. Environ.* **2018**, *26*, 3802–3816. [CrossRef]
12. Rodrigues, B.V.M.; Silva, A.S.; Melo, G.F.S.; Vasconscellos, L.M.R.; Marciano, F.R.; Lobo, A.O. Influence of Low Contents of Superhydrophilic MWCNT on The Properties and Cell Viability of Electrospun Poly(butylene adipate-co-terephthalate) Fibers. *Mater. Sci. Eng. C* **2016**, *59*, 782–791. [CrossRef] [PubMed]
13. Bumbudsanpharoke, N.; Wongphan, P.; Promhuad, K.; Leelaphiwat, P.; Harnkarnsujarit, N. Morphology and Permeability of Bio-Based Poly(butylene adipate-co-terephthalate)(PBAT), Poly(butylene succinate)(PBS) and Linear Low-Density Polyethylene (LLDPE) Blend Films Control Shelf-Life of Packaged Bread. *Food Control* **2022**, *132*, 108541. [CrossRef]
14. Phan, K.T.K.; Phan, H.T.; Brennan, C.S.; Regenstain, J.M.; Jantanasakulwong, K.; Boonyawan, D.; Phimolsiripol, Y. Gliding Arc Discharge Non-Thermal Plasma for Retardation of Mango Anthracnose. *LWT* **2019**, *105*, 142–148. [CrossRef]
15. Könczöl, L.; Döll, W.; Buchholz, U.; Mülhaupt, R. Ultimate Properties of Epoxy Resins Modified with a Polysiloxane–Polycaprolactone Block Copolymer. *J. Appl. Polym. Sci.* **1994**, *54*, 815–826. [CrossRef]
16. Chiou, K.C.; Chang, F.C. Reactive Compatibilization of Polyamide-6 (PA 6)/Polybutylene terephthalate (PBT) Blends by a Multifunctional Epoxy Resin. *J. Polym. Sci. Part B Polym. Phys.* **2000**, *38*, 23–33. [CrossRef]
17. Kodsangma, A.; Homsaard, N.; Nadon, S.; Rachtanapun, P.; Leksawasdi, N.; Phimolsiripol, Y.; Insomphun, C.; Seesuriyachan, P.; Chaiyaso, T.; Jantrawut, P.; et al. Effect of Sodium Benzoate and Chlorhexidine Gluconate on a Bio-Thermoplastic Elastomer Made from Thermoplastic Starch-Chitosan Blended with Epoxidized Natural Rubber. *Carbohydr. Polym.* **2020**, *242*, 116421. [CrossRef]
18. Thomas, R.; Yumei, D.; Yuelong, H.; Le, Y.; Moldenaers, P.; Weimin, Y.; Czigany, T.; Thomas, S. Miscibility, Morphology, Thermal, and Mechanical Properties of A DGEBA Based Epoxy Resin Toughened with A Liquid Rubber. *Polymer* **2008**, *49*, 278–294. [CrossRef]
19. Li, H.; Ning, N.; Zhang, L.; Wang, Y.; Liang, W.; Tian, M. Different Flame Retardancy Effects and Mechanisms of Aluminium Phosphinate in PPO, TPU and PP. *Polym. Degrad. Stab.* **2014**, *105*, 86–95. [CrossRef]
20. Battegazzore, D.; Frache, A.; Carosio, F. Layer-by-Layer Nanostructured Interphase Produces Mechanically Strong and Flame Retardant Bio-Composites. *Compos. Part B Eng.* **2020**, *200*, 108310. [CrossRef]
21. Feng, J.; Sun, Y.; Song, P.; Lei, W.; Wu, Q.; Liu, L.; Yu, Y.; Wang, H. Fire-Resistant, Strong, and Green Polymer Nanocomposites Based on Poly(lactic acid) and Core–Shell Nanofibrous Flame Retardants. *ACS Sustain. Chem. Eng.* **2017**, *5*, 7894–7904. [CrossRef]
22. Jiang, D.; Pan, M.; Cai, X.; Zhao, Y. Flame Retardancy of Rice Straw-Polyethylene Composites Affected by In Situ Polymerization of Ammonium Polyphosphate/Silica. *Compos. Part A Appl. Sci. Manuf.* **2018**, *109*, 1–9. [CrossRef]
23. Pradhan, S.P.; Shubhadarshinee, L.; Mohapatra, P.; Mohanty, P.; Jali, B.R.; Mohapatra, P.; Barick, A.K. Conducting Polymer Composites for Antistatic Application in Aerospace. In *Aerospace Polymeric Materials*; Inamuddin, Altalhi, T., Adnan, S.M., Eds.; John Wiley & Sons: Hoboken, NJ, USA, 2022; pp. 155–187.
24. Bharati, B.; Vijaykumar, B.T.; Ramabai, N.; Basavaraj, S. Synthesis, Characterizations, and Physical Properties of Magnesium Oxide with Poly Aniline Nanocomposites. *Mater. Today Proc.* **2023**, *92*, 1640–1645. [CrossRef]
25. Stöckel, S.; Ebert, S.; Böttcher, M.; Seifert, A.; Wamser, T.; Krenkel, W.; Schulze, S.; Hietschold, M.; Gnaegi, H.; Goedel, W.A. Coating of Alumina Fibres with Aluminium Phosphate by a Continuous Chemical Vapour Deposition Process. *Chem. Vap. Depos.* **2014**, *20*, 388–398. [CrossRef]
26. Turkoglu, S.H.; Alazzawi, M.; Kadim, A.A.N. Atmospheric Pressure Plasma Surface Treatment of Polymers and Influence on Cell Cultivation. *Molecules* **2021**, *26*, 1665. [CrossRef] [PubMed]
27. Can-Herrera, L.A.; Ávila-Ortega, A.; de la Rosa-García, S.; Oliva, A.I.; Cauich-Rodríguez, J.V.; Cervantes-Uc, J.M. Surface Modification of Electrospun Polycaprolactone Microfibers by Air Plasma Treatment: Effect of Plasma Power and Treatment Time. *Eur. Polym. J.* **2016**, *84*, 502–513. [CrossRef]
28. Dhanumalayan, E.; Trimukhe, A.M.; Deshmukh, R.; Joshi, G.M. Disparity in Hydrophobic to Hydrophilic Nature of Polymer Blend Modified by K2Ti6O13 as a Function of Air Plasma Treatment. *Prog. Org. Coat.* **2017**, *111*, 371–380. [CrossRef]
29. Al Ghufais, I.A.; Rahaman, M.; Aldalbahi, A. Evaluation of Physical Properties of Recycled Polyethylene Waste Films and Application of Its Carbon Filled Composites as Anti-Static Material in Electronic Packaging. Bachelor's Thesis, King Saud University, Riyadh, Saudi Arabia, 2018.
30. Silva, T.F.d.; Menezes, F.; Montagna, L.S.; Lemes, A.P.; Passador, F.R. Preparation and Characterization of Antistatic Packaging for Electronic Components Based on Poly(lactic acid)/Carbon Black Composites. *J. Appl. Polym. Sci.* **2019**, *136*, 47273. [CrossRef]
31. Zoubek, M.; Kudláček, J.; Kreibich, V.; Jirout, T.; Abramov, A. The Influence of Mixing Method and Mixing Parameters in Process of Preparation of Anti-static Coating Materials Containing Nanoparticles. In *Advances in Manufacturing II. Lecture Notes in Mechanical Engineering*; Gapiński, B., Szostak, M., Ivanov, V., Eds.; Springer: Berlin/Heidelberg, Germany, 2019; pp. 154–196.

32. Kumpika, T.; Kantarak, E.; Sroila, W.; Panthawan, A.; Jhuntama, N.; Sanmuangmoon, P.; Thongsuwan, W.; Singjai, P. Superhydrophilic/Superhydrophobic Surfaces Fabricated by Spark-coating. *Surf. Interface Anal.* **2018**, *50*, 827–834. [CrossRef]
33. Kumpika, T.; Thongsuwan, W.; Singjai, P. Optical and Electrical Properties of ZnO Nanoparticle Thin Films Deposited on Quartz by Sparking Process. *Thin Solid Film.* **2008**, *516*, 5640–5644. [CrossRef]
34. Jantanasakulwong, K.; Thanakkasaranee, S.; Seesuriyachan, P.; Singjai, P.; Saenjaiban, A.; Photphroet, S.; Pratinthong, K.; Phimolsiripol, Y.; Leksawasdi, N.; Chaiyaso, T.; et al. Sparking Nano-Metals on a Surface of Polyethylene Terephthalate and Its Application: Anti-Coronavirus and Anti-Fogging Properties. *Int. J. Mol. Sci.* **2022**, *23*, 10541. [CrossRef]
35. Underwriters Laboratories. *UL 94 Standard for Tests for Flammability of Plastic Materials for Parts in Devices and Appliances*, 7th ed.; Underwriters Laboratories: Northbrook, IL, USA, 2023; Available online: https://omnexus.specialchem.com/polymer-property/flammability-ul94 (accessed on 12 May 2025).
36. *JIS K 6251-7;* Rubber, Vulcanized or Thermoplastic-Determination of Tensile Stress-Strain Properties. Japanese Industrial Standards Committee: Tokyo, Japan, 2017.
37. González, M.G.; Cabanelas, J.C.; Baselga, J. Applications of FTIR on Epoxy Resins-Identification, Monitoring the Curing Process, Phase Separation and Water Uptake. *Infrared Spectrosc.-Mater. Sci. Eng. Technol.* **2012**, *2*, 261–284.
38. Cecen, V.; Seki, Y.; Sarikanat, M.; Tavman, I.H. FTIR and SEM Analysis of Polyester- and Epoxy-Based Composites Manufactured by VARTM Process. *J. Appl. Polym. Sci.* **2008**, *108*, 2163–2170. [CrossRef]
39. Yin, Q.; Chen, F.; Zhang, H.; Liu, C. Fabrication and Characterisation of Thermoplastic Starch/Poly(butylene succinate) Blends with Maleated Poly(butylene succinate) as Compatibiliser. *Plast. Rubber Compos.* **2015**, *44*, 362–367. [CrossRef]
40. Ostrowska, J.; Sadurski, W.; Paluch, M.; Tyński, P.; Bogusz, J. The Effect of Poly(butylene succinate) Content on the Structure and Thermal and Mechanical Properties of Its Blends with Polylactide. *Polym. Int.* **2019**, *68*, 1271–1279. [CrossRef]
41. Cai, Y.; Lv, J.; Feng, J. Spectral Characterization of Four Kinds of Biodegradable Plastics: Poly (Lactic Acid), Poly (Butylenes Adipate-Co-Terephthalate), Poly (Hydroxybutyrate-Co-Hydroxyvalerate) and Poly (Butylenes Succinate) with FTIR and Raman Spectroscopy. *J. Polym. Environ.* **2012**, *21*, 108–114. [CrossRef]
42. Boonprasertpoh, A.; Pentrakoon, D.; Junkasem, J. Effect of PBAT on Physical, Morphological, and Mechanical Properties of PBS/PBAT Foam. *Cell. Polym.* **2019**, *39*, 31–41. [CrossRef]
43. Liu, B.; Guan, T.; Wu, G.; Fu, Y.; Weng, Y. Biodegradation Behavior of Degradable Mulch with Poly (Butylene Adipate-co-Terephthalate) (PBAT) and Poly (Butylene Succinate) (PBS) in Simulation Marine Environment. *Polymers* **2022**, *14*, 1515. [CrossRef]
44. Hajibeygi, M.; Mousavi, M.; Shabanian, M.; Habibnejad, N.; Vahabi, H. Design and Preparation of New Polypropylene/Magnesium Oxide Micro Particles Composites Reinforced with Hydroxyapatite Nanoparticles: A Study of Thermal Stability, Flame Retardancy and Mechanical Properties. *Mater. Chem. Phys.* **2021**, *258*, 123917. [CrossRef]
45. Dong, Q.; Gao, C.; Ding, Y.; Wang, F.; Wen, B.; Zhang, S.; Wang, T.; Yang, M. A Polycarbonate/Magnesium Oxide Nanocomposite with High Flame Retardancy. *J. Appl. Polym. Sci.* **2012**, *123*, 1085–1093. [CrossRef]
46. Thajai, N.; Rachtanapun, P.; Thanakkasaranee, S.; Chaiyaso, T.; Phimolsiripol, Y.; Leksawasdi, N.; Sommano, S.R.; Sringarm, K.; Chaiwarit, T.; Ruksiriwanich, W.; et al. Antimicrobial Thermoplastic Starch Reactive Blend with Chlorhexidine Gluconate and Epoxy Resin. *Carbohydr. Polym.* **2023**, *301 Part B*, 120328. [CrossRef]
47. Zhou, X.; He, T.; Jiang, Y.; Chang, S.; Yu, Y.; Fang, X.; Zhang, Y. A Novel Network-Structured Compatibilizer for Improving the Interfacial Behavior of PBS/Lignin. *ACS Sustain. Chem. Eng.* **2021**, *9*, 8592–8602. [CrossRef]
48. Promhuad, K.; Phothisarattana, D.; Laorenza, Y.; Bumbudsanpharoke, N.; Harnkarnsujarit, N. Zinc Oxide Enhanced the Antibacterial Efficacy of Biodegradable PBAT/PBS Nanocomposite Films: Morphology and Food Packaging Properties. *Food Biosci.* **2023**, *55*, 103077. [CrossRef]
49. Kontou, E.; Christopoulos, A.; Koralli, P.; Mouzakis, D.E. The Effect of Silica Particle Size on the Mechanical Enhancement of Polymer Nanocomposites. *Nanomaterials* **2023**, *13*, 1095. [CrossRef]
50. Bajaj, P.; Paliwal, D.; Gupta, A. Influence of Metal Ions on Structure and Properties of Acrylic Fibers. *J. Appl. Polym. Sci.* **1998**, *67*, 1647–1659. [CrossRef]
51. Jia, J.; Yang, J.; Zhao, Y.; Liang, H.; Chen, M. The Crystallization Behaviors and Mechanical Properties of Poly(L-lactic acid)/Magnesium Oxide Nanoparticle Composites. *RSC Adv.* **2016**, *6*, 43855–43863. [CrossRef]
52. John, J.; Mani, R.; Bhattacharya, M. Evaluation of Compatibility and Properties of Biodegradable Polyester Blends. *J. Polym. Sci. Part A Polym. Chem.* **2002**, *40*, 2003–2014. [CrossRef]
53. Muthuraj, R.; Misra, M.; Mohanty, A.K. Biodegradable Poly (butylene succinate) and Poly (butylene adipate-co-terephthalate) Blends: Reactive Extrusion and Performance Evaluation. *J. Polym. Environ.* **2014**, *22*, 336–349. [CrossRef]
54. Thajai, N.; Rachtanapun, P.; Thanakkasaranee, S.; Punyodom, W.; Worajittiphon, P.; Phimolsiripol, Y.; Leksawasdi, N.; Ross, S.; Jantrawut, P.; Jantanasakulwong, K. Reactive Blending of Modified Thermoplastic Starch Chlorhexidine Gluconate and Poly(butylene succinate) Blending with Epoxy Compatibilizer. *Polymers* **2023**, *15*, 3487. [CrossRef]

55. Kiattipornpithak, K.; Thajai, N.; Kanthiya, T.; Rachtanapun, P.; Leksawasdi, N.; Phimolsiripol, Y.; Rohindra, D.; Ruksiriwanich, W.; Sommano, S.R.; Jantanasakulwong, K. Reaction Mechanism and Mechanical Property Improvement of Poly(Lactic Acid) Reactive Blending with Epoxy Resin. *Polymers* **2021**, *13*, 2429. [CrossRef]
56. Szlachetka, I.; Dobrev, J.W.; Baryła, A.; Dohojda, M. Low-Density Polyethylene (LDPE) Building Films—Tensile Properties and Surface Morphology. *J. Build. Eng.* **2021**, *44*, 103386. [CrossRef]
57. Yongcheng, Y. The Thermal Stability of Polyethylene Blends During Heat-Processing and Short-Term Photooxidation. *Polym. Degrad. Stab.* **1993**, *39*, 193–198. [CrossRef]
58. Sanchis, M.R.; Blanes, V.; Blanes, M.; Garcia, D.; Balart, R. Surface Modification of Low Density Polyethylene (LDPE) Film by Low Pressure O_2 Plasma Treatment. *Eur. Polym. J.* **2006**, *42*, 1558–1568. [CrossRef]
59. Ashraf, M.A.; Peng, W.; Zare, Y.; Rhee, K.Y. Effects of Size and Aggregation/Agglomeration of Nanoparticles on the Interfacial/Interphase Properties and Tensile Strength of Polymer Nanocomposites. *Nanoscale Res. Lett.* **2018**, *13*, 214. [CrossRef]
60. Paul, D.R.; Robeson, L.M. Polymer Nanotechnology: Nanocomposites. *Polymer* **2008**, *49*, 3187–3204. [CrossRef]
61. Qiu, Z.; Ikehara, T.; Nishi, T. Miscibility and Crystallization Behaviour of Biodegradable Blends of Two Aliphatic Polyesters. Poly(3-hydroxybutyrate-co-hydroxyvalerate) and Poly(butylene succinate) Blends. *Polymer* **2003**, *44*, 7519–7527. [CrossRef]
62. Kennouche, S.; Moigne, N.L.; Kaci, M.; Quantin, J.C.; Caro-Bretelle, A.S.; Delaite, C.; Lopez-Cuesta, J.M. Morphological Characterization and Thermal Properties of Compatibilized Poly(3-hydroxybutyrate-co-3-hydroxyvalerate) (PHBV)/Poly(butylene succinate) (PBS)/Halloysite Ternary Nanocomposites. *Eur. Polym. J.* **2016**, *75*, 142–162. [CrossRef]
63. Xiang, S.; Feng, L.; Bian, X.; Li, G.; Chen, X. Evaluation of PLA Content in PLA/PBAT Blends Using TGA. *Polym. Test.* **2020**, *81*, 106211. [CrossRef]
64. Jakić, M.; Vrandečić, N.S.; Klarić, I. Thermal Degradation of Poly(vinyl chloride)/Poly(ethylene oxide) Blends: Thermogravimetric Analysis. *Polym. Degrad. Stab.* **2013**, *98*, 1738–1743. [CrossRef]
65. Borhade, A.V.; Kanade, K.G.; Tope, D.R.; Patil, M.D. A Comparative Study on Synthesis, Characterization and Photocatalytic Activities of MgO and Fe/MgO Nanoparticles. *Res. Chem. Intermed.* **2012**, *38*, 1931–1946. [CrossRef]
66. Radhakrishnan, S.; Thorat, S.; Desale, A.; Desai, P.; Kulkarni, M. Structure and Properties of PBS/PBAT Blends and Nanocomposites. *IOP Conf. Ser. Mater. Sci. Eng.* **2022**, *1248*, 012013. [CrossRef]
67. Zeng, J.-B.; Jiao, L.; Li, Y.-D.; Srinivasan, M.; Li, T.; Wang, Y.-Z. Bio-Based Blends of Starch and Poly(butylene succinate) with Improved Miscibility, Mechanical Properties, and Reduced Water Absorption. *Carbohydr. Polym.* **2011**, *83*, 762–768. [CrossRef]
68. Kumar, S.; Gautam, C.; Chauhan, B.S.; Srikrishna, S.; Yadav, R.S.; Rai, S.B. Enhanced Mechanical Properties and Hydrophilic Behavior of Magnesium Oxide Added Hydroxyapatite Nanocomposite: A Bone Substitute Material for Load Bearing Applications. *Ceram. Int.* **2020**, *46 Part B*, 16235–16248. [CrossRef]
69. Meindrawan, B.; Suyatma, N.E.; Muchtadi, T.R.; Iriani, E.S. Preparation and Characterization of Bionanocomposite Films Made from Carrageenan, Beeswax and ZnO Nanoparticles. In *Materials Science Forum*; Trans Tech Publications Ltd.: Zürich, Switzerland, 2016; Volume 872, pp. 157–161.
70. Wen, W.; Luo, B.; Qin, X.; Li, C.; Liu, M.; Ding, S.; Zhou, C. Strengthening and Toughening of Poly(L-lactide) Composites by Surface Modified MgO Whiskers. *Appl. Surf. Sci.* **2015**, *332*, 215–223. [CrossRef]
71. Gravogl, G.; Knoll, C.; Welch, J.M.; Artner, W.; Freiberger, N.; Nilica, R.; Eitenberger, E.; Friedbacher, G.; Harasek, M.; Werner, A.; et al. Cycle Stability and Hydration Behavior of Magnesium Oxide and Its Dependence on the Precursor-Related Particle Morphology. *Nanomaterials* **2018**, *8*, 795. [CrossRef]
72. Morgan, A.B.; Bundy, M. Cone Calorimeter Analysis of UL-94 V-Rated Plastics. *Fire Mater. Int. J.* **2007**, *31*, 257–283. [CrossRef]
73. Najim, M.N.; Sabr, O.H.; Kadhim, B.J. The Effect of MgO Nanoparticle on PVA/PEG-Based Membranes for Potential Application in Wound Healing. *J. Biomater. Sci. Polym. Ed.* **2024**, *13*, 1963–1977. [CrossRef] [PubMed]
74. Peng, W.; Ren, S.; Zhang, Y.; Fan, R.; Zhou, Y.; Li, L.; Xu, X.; Xu, Y. MgO Nanoparticles-Incorporated PCL/Gelatin-Derived Coaxial Electrospinning Nanocellulose Membranes for Periodontal Tissue Regeneration. *Front. Bioeng. Biotechnol.* **2021**, *9*, 668428. [CrossRef]
75. Hosseini, S.M.; Bagheripour, E.; Ansari, M. Adapting the Performance and Physico-Chemical Properties of PES Nanofiltration Membrane by Using of Magnesium Oxide Nanoparticles. *Korean J. Chem. Eng.* **2017**, *34*, 1774–1780. [CrossRef]
76. Hong, S.; Yang, J.; Ahn, S.; Mun, Y.; Lee, G. Flame Retardant Performance of Various UL94 Classified Materials Exposed to External Ignition Sources. *Fire Mater.* **2004**, *28*, 25–31. [CrossRef]
77. Li, S.; Wang, C.; Wang, G.; Wang, Y.; Han, Z. Polycarbosilane/Divinylbenzene-Modified Magnesium Hydroxide to Enhance the Flame Retardancy of Ethylene–Vinyl Acetate Copolymer. *Polymers* **2023**, *15*, 4440. [CrossRef]
78. Mirmohseni, A.; Azizi, M.; Seyed Dorraji, M.S. A Promising Ternary Nanohybrid of Copper@ Zinc Oxide Intercalated with Polyaniline for Simultaneous Antistatic and Antibacterial Applications. *J. Coat. Technol. Res.* **2019**, *16*, 1411–1422. [CrossRef]
79. Galletti, A.M.R.; Antonetti, C.; Marracci, M.; Piccinelli, F.; Tellini, B. Novel Microwave-Synthesis of Cu Nanoparticles in the Absence of Any Stabilizing Agent and Their Antibacterial and Antistatic Applications. *Appl. Surf. Sci.* **2013**, *280*, 610–618. [CrossRef]

80. Zhang, Y.; Yuan, H.; Yan, W.; Chen, S.; Qiu, M.; Liao, B. Atomic-Oxygen-Durable and Antistatic α-Al x Ti y O/γ-NiCr Coating on Kapton for Aerospace Applications. *ACS Appl. Mater. Interfaces* **2021**, *13*, 58179–58192. [CrossRef] [PubMed]
81. Mazur, M.; Domaradzki, J.; Borna, A. Electrical and Antistatic Properties of Magnetron Sputtered Thin Films Based on TiO_2:(V, Ta). In Proceedings of the 2011 International Students and Young Scientists Workshop "Photonics and Microsystems", Cottbus, Germany, 8–10 July 2011; pp. 89–93.
82. Samanta, K.K.; Jassal, M.; Agrawal, A.K. Antistatic Effect of Atmospheric Pressure Glow Discharge Cold Plasma Treatment on Textile Substrates. *Fibers Polym.* **2010**, *11*, 431–437. [CrossRef]
83. Belkind, A.; Freilich, A.; Lopez, J.; Zhao, Z.; Zhu, W.; Becker, K. Characterization of Pulsed DC Magnetron Sputtering Plasmas. *New J. Phys.* **2005**, *7*, 90. [CrossRef]
84. Abdelrahman, M. Study of Plasma and Ion Beam Sputtering Processes. *J. Phys. Sci. Appl.* **2015**, *5*, 128–142.
85. Han, J.G. Recent Progress in Thin Film Processing by Magnetron Sputtering with Plasma Diagnostics. *J. Phys. D Appl. Phys.* **2009**, *42*, 043001. [CrossRef]

MDPI AG
Grosspeteranlage 5
4052 Basel
Switzerland
Tel.: +41 61 683 77 34

Polymers Editorial Office
E-mail: polymers@mdpi.com
www.mdpi.com/journal/polymers

www.ingramcontent.com/pod-product-compliance
Lightning Source LLC
Chambersburg PA
CBHW061515210926
51850CB00008B/1928
* 9 7 8 3 7 2 5 8 4 8 5 3 9 *